Flexible and Wearable Electronics for Smart Clothing

# Flexible and Wearable Electronics for Smart Clothing

*Edited by*
*Gang Wang*
*Chengyi Hou*
*Hongzhi Wang*

WILEY-VCH

**Editors**

***Dr Gang Wang***
Professor, Eastern Scholar, State Key Laboratory for Modification of Chemical Fibres and Polymer Materials
College of Materials Science and Engineering
Donghua University
2999 North Renmin Road
Songjiang
201620 Shanghai
People's Republic of China

***Prof. Chengyi Hou***
Associate Professor, State Key Laboratory for Modification of Chemical Fibres and Polymer Materials
College of Materials Science and Engineering
Donghua University
2999 North Renmin Road
Songjiang
201620 Shanghai
People's Republic of China

***Prof. Hongzhi Wang***
Professor, Eastern Scholar, Director of Research Management Department, State Key Laboratory for Modification of Chemical Fibres and Polymer Materials
College of Materials Science and Engineering
Donghua University
2999 North Renmin Road
Songjiang
201620 Shanghai
People's Republic of China



**Library of Congress Card No.:**
applied for

**British Library Cataloguing-in-Publication Data**
A catalogue record for this book is available from the British Library.

**Bibliographic information published by the Deutsche Nationalbibliothek**
The Deutsche Nationalbibliothek lists this publication in the Deutsche Nationalbibliografie; detailed bibliographic data are available on the Internet at <http://dnb.d-nb.de>.



**Print ISBN:** 978-3-527-34534-2
**ePDF ISBN:** 978-3-527-81858-7
**ePub ISBN:** 978-3-527-81856-3
**oBook ISBN:** 978-3-527-81855-6

**Cover Design** Formgeber, Mannheim, Germany
**Typesetting** SPi Global, Chennai, India
**Printing and Binding** Markono Print Media Pte Ltd, Singapore
Printed on acid-free paper

10 9 8 7 6 5 4 3 2 1

# Contents

# Preface

When we study the history of clothing, we notice that its development depends on the revolution of materials and is relative to the industrial revolution. Materials development brings new functions to clothes and new opportunities to human society. We cannot help imagine what clothes will be like in the future. Many people have successfully taken forward the technology revolution in the clothing industry. For example, the so-called smart and wearable devices have emerged and are integrated into clothes; these devices are able to monitor our health conditions or charge mobile phones or displays, which takes us one step closer to the smart clothing that we expect. In the future, most functions that digital devices offer will be integrated into smart clothes; with all these functions together, as our second skin, the new clothes will be an intelligent interface between humans and nature.

How likely is it that it will become a reality? Be aware of this example – a fantasy shoe that can deform was demonstrated in a science fiction movie "Back to the Future" in 1989, and 17 years later, the exact same shoe was released by Nike.

To this extent, we shall believe that smart clothing, delivered by wearable technologies, is coming very soon.

Actually, wearable and flexible electronics have achieved considerable developments within the past decade and have now come to a point where they can be utilized in substantial applications, including energy conversion, artificial skin, health monitoring, and so on. In view of their versatility, ease of use, and the vast market of practical applications, wearable electronics have rapidly "exploded" in the United States, China, Korea, and all over the world. The research fields of wearable and flexible electronics are now covered from applied physics, chemistry, mechanical engineering, to material science, biomedicine, and clothing technology. Researchers and the public now pay increasing attention to wearable and flexible electronics, due to their tremendous achievement and vast potential.

This book will give an overview of recent developments in wearable and flexible electronics and their potentialities in smart clothing. The book will consist of an introductory overview followed by four sections: from Sensing (Part I), Energy (Part II), Interacting (Part III) to Integrating and Connecting (Part IV).

Sensing is one of the most typical characters of smart clothes. Part I looks first at the emergence of wearable organic nano-sensors (Chapter 1), and then moves forward to the stimuli-responsive electronic skins (Chapter 2) and flexible thermoelectrics and thermoelectric textiles for various sensing applications (Chapter 3).

Then, the next key step is to enable the wearable electronics function continuously, e.g. what is the energy supply for the smart clothes applications? Part II then moves on to energy-related topics. The first chapter in this part focuses on self-powered triboelectric nanogenerators for energy harvesting in smart textiles (Chapter 4). Besides the mechanical energy-harvesting techniques, solar cells and supercapacitors are also introduced for the smart clothes applications (Chapter 5). Lithium-ion battery is the most widely and well-commercialized energy supply for the current smart clothes industry; the development history, materials and microstructure design, and future prospects are well included in Chapter 6.

Then, how smart clothes interact with the human body is discussed in Part III. The first chapter in this section looks at the thermal and humidity management for the next-generation smart clothes (Chapter 7). The second chapter looks at the functionalization of fiber materials for washable smart wearable textiles (Chapter 8). The two chapters (Chapters 7 and 8) bring together how to make smart clothes behave humanization design and particle for the real commercialization. Microfluidics in wearable electronics is discussed in the next chapter (Chapter 9) – one of the most emerging techniques for wearable electronics in health-care applications, enabling the interaction between smart clothes and body fluids.

Even though many emerging technologies have been adopted for the development of smart clothes, only very limited real products are truly ubiquitous in current markets, compared with the conventional textile industry. How can the time-induced ultimate demise of laboratory-based wearable techniques be avoided? It is important to look at what end-users really need, the promising strategy of moving from the lab to fab for industrialization, and the future challenges that should be conquered.

Part IV considers the following themes: "how to integrate?"– the authors focus on flexible bio-integrated electronics for the health-care targeted smart clothes (Chapter 10). Then, "how to fabricate?" – flexible and printed electronics are well discussed (Chapter 11). Chapter 12 and 13 complete the picture by looking at the fabrication of the actual products that are available in the markets or the cutting-edge materials and processing for next-generation e-textiles devices.

# Part I

# Sensing

# 1

# Wearable Organic Nano-sensors

*Wei Huang*[1], *Liangwen Feng*[2], *Gang Wang*[3,4], *and Elsa Reichmanis*[4]

[1] *University of Electronic Science and Technology of China (UESTC), School of Optoelectronic Science and Technology, No. 4, Sec. 2, North Jianshe Rd, Chengdu 610054, China*
[2] *Chinese Academy of Sciences, The State Key Laboratory of Organometallic Chemistry, Shanghai Institute of Organic Chemistry, 345 Lingling Lu, Shanghai 200032, China*
[3] *Donghua University, State Key Laboratory for Modification of Chemical Fibers and Polymer Materials, College of Material Science and Engineering, 2999 North Renmin Road, Shanghai 201620, People's Republic of China*
[4] *Georgia Institute of Technology, School of Chemical and Biomolecular Engineering, School of Chemistry and Biochemistry, and School of Materials Science and Engineering, Atlanta, GA 30332, USA*

## 1.1 Introduction

The development of unconventional electronics, or the Internet of Things (IoT), which enables active connections between various electronic devices and massive information flows from both the environment and humans, will likely lead to another technology revolution in the near future [1]. To realize effective connections between factors such as environmental parameters and human vital statistics and electronic devices, sensors serving as an inevitable bridge that converts environmental/human body signals to electronic signals and their characteristics will determine the future of IoT [2].

Nowadays, numerous sensors, including optical, chemical, electrical, gas, heat, and mechanical, already play an important role in our daily life [3]. For instance, a typical smartphone contains an accelerometer, gyroscope, magnetometer, proximity sensor, ambient light sensor, microphone, touchscreen sensor, etc. These sensors enable us to live better, learn quicker, and work faster [4]. Even though enormous achievements have been realized with these state-of-the-art sensors, they continue to provide us with emerging applications, including humidity sensors that can be integrated into a cellphone and optical sensors that are able to recognize a human face. These successfully commercialized sensors rely on traditional inorganic mechanically rigid materials and complex fabrication processes, which are usually not compatible with flexible/stretchable substrates [5].

In recent decades, another kind of electronic material consisting of organic components has emerged as an attractive alternative [6]. Compared to its inorganic counterparts, organic materials, in theory, possess unlimited species, since the properties of organic materials can be manipulated by simply regulating the chemical structures and/or compositions [7]. Organic semiconductors (OSCs)

*Flexible and Wearable Electronics for Smart Clothing*, First Edition.
Edited by Gang Wang, Chengyi Hou, and Hongzhi Wang.

have already accelerated the development of flat panel displays due to the incredibly high performance of organic light-emitting diodes (OLEDs). Organic thin-film transistor (OTFT) based back panel displays also give rise to the commercialization of the first ever bendable smartphone [8]. A plastic empire is on the rise, as a large fraction of the organic materials are polymers, and they are utilized in almost every aspect of daily life in the human society [8]. Polymers can be designed to be stronger than metals, as reliable as ceramics, and as soft as liquids. More interestingly, they can be constructed to be insulators, semiconductors, or even conductors. Unconventional properties, such as self-healing, and stretchable semiconductors/conductors can also be accomplished with organic materials [9].

Taking advantage of the wide range of properties of organic materials, multi-functional sensors with flexibility/stretchability are rapidly developing and as a result, the design of wearable sensors has become a major research target. The next-generation sensors should be low-cost, lightweight, low energy consuming devices, which together with capabilities such as being bendable or even stretchable, will ensure compatibility with wearable technologies. Wearable sensors can not only simplify the daily life routine but also provide a great tool for in situ monitoring of either external or internal parameters [10]. People have already partially benefited from wearable sensors; applications to monitor heart rate (or even electrocardiography) and the number of steps walked are already commonplace in smart watches. In the future, wearable sensors will enable real-time monitoring of the health of the human body, including blood pressure, oxygen levels or if someone has suddenly fallen, and will likely enable quick diagnosis of common diseases [11]. Furthermore, environmental conditions and hazardous atmospheric conditions can be monitored, which can act as information collection terminals in IoT, improving public safety [12].

## 1.2 Wearable Organic Sensors Based on Different Device Architectures

Owing to the boost in the area of organic electronic materials, different device architectures can be adopted to realize flexible and stretchable organic sensors for potentially wearable applications. Among various device architectures, resistors, transistors, electrochemical, and diode-based devices are intensively investigated, owing to their compatibility with traditional Si processing. While different device architectures have different advantages, they can all be designed to function as effective sensors. Sensing applications, including motion detection, hazardous gas monitoring, disease diagnosis, temperature recording, etc. will be summarized. Furthermore, the integration of wearable sensors with portable power sources and data processing hardware/software to enable real-time information collection and processing has been reported [13].

In the following sections, we will focus on different sensor device structures, illuminating the functional principles, popular organic materials, and practical applications. Representative examples for designing flexible/stretchable sensors and the strategies to enhance the sensing performance will be demonstrated.

## 1.2.1 Resistor-Based Sensors

### 1.2.1.1 Definitions and Important Parameters

One of the simplest electronic devices is the resistor, where resistance is the key parameter for any type of resistor. By utilizing resistors as sensors, the variation in resistance is recorded when the resistor is in contact with the analyte. Not only the value of the resistance is important for sensor applications, but several other parameters are also equally critical. These include response time and recovery time, which indicate how long the sensor will generate effective processable signals or recover to its original state, and sensitivity, which indicates the detection limit of a target analyte. As resistors are easily fabricated and characterized, they represent the most widely studied device structure for sensing applications, including sensors with stretchable and wearable properties.

### 1.2.1.2 Materials and Applications

Materials are the most important factor influencing sensor performance, contributing to the direct interaction with the analyte and electronic signal transformation. Among all the materials that can be used as active layers in a resistor-based sensor, graphene has exhibited emerging advantages due to its high surface to volume ratio, controllable conductivity by adjusting the composition/morphology/geometry, and excellent thermal conductivity. Wang and coworkers successfully demonstrated a highly conductive, flexible, and compressible all-graphene thin-film sensor [14]. This sensor can sense heat and cold, measure the dimensions of the heated/cooled area, discern human touch from other pressures, and enable human touch location and measurement of pressure level under zero working voltage. Graphene was also coupled with some organic materials to improve the sensing performance. The passive all-graphene flexible thin-film sensor (FTS) technology may pave new pathways for the development of electronic skin.

In another example, Zhu and coworkers fabricated highly sensitive graphene woven fabrics (GWFs) by using a crisscross copper mesh substrate with chemical vapor deposition (CVD) of graphene [15]. A flexible and wearable strain sensor was assembled by adhering the GWFs on polydimethylsiloxane (PDMS) and medical composite tape. The ultralight sensor exhibited features including relatively good sensitivity, high reversibility, superior physical robustness, easy fabrication, ease to follow human skin deformation without irritation, and so on. As a consequence of the piezoresistive effect of GWFs, the sensors were used as electronic skin covering the human body to detect body motions (Figure 1.1a,b). The signals of GWF resistance change depend on the deformation strain that is formed by the motions. The stronger the motion, the larger the strain, and the motion signals can be recorded more easily. Some weak human motions were chosen to test the notable resistance change, including hand clenching, phonation, expression change, blink, breath, and pulse. Because of the distinctive features of high sensitivity and recyclability, the GWFs–PDMS–tape based piezoresistive sensors exhibited wide potential applications in displays, robotics, fatigue detection, body monitoring, in vitro diagnostics, and advanced therapies.

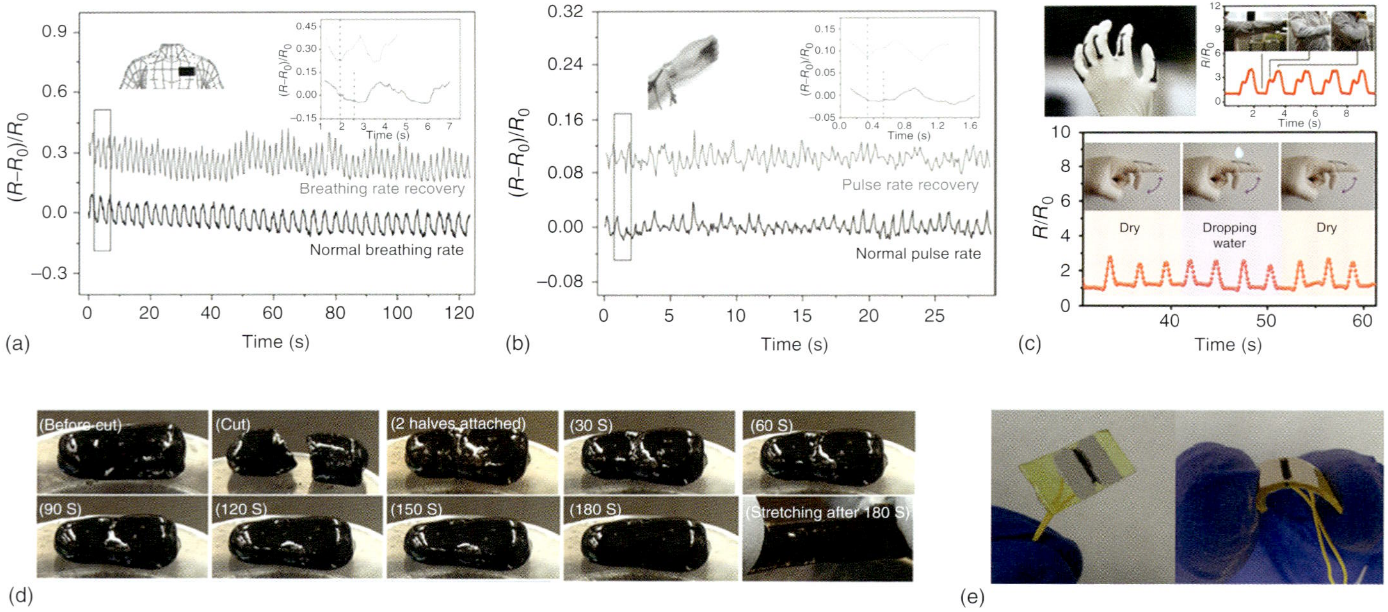


**Figure 1.1** Relative change in resistance of respiration and pulse, in (a) and (b), respectively, at still state and exercise state. (c) Real-time human motion detection using superhydrophobic MWCNT/TPE-film sensors, including an optical photograph of a latex glove with five film sensors coated on each finger; normalized relative resistance as a function of time; and real-time variation of the normalized relative resistance. (d) Self-healing property of H2010h1 hydrogel. The hydrogel was cut into two halves and then brought together again; every 30 seconds, the hydrogel is shown enabling us to track the vestige of cut. (e) Photograph of the self-healing chemiresistor consisting of a transparent (yellowish) self-healing substrate, jelly-like self-healing electrode, and pliable induced self-healing AuNP film. Source: (a, b) Wang et al. 2014 [15]. Reproduced with permission of John Wiley & Sons; (c) Li et al. 2017 [19]. Reproduced with permission of John Wiley & Sons; (d) Darabi et al. 2017 [20]. Reproduced with permission of John Wiley & Sons; (e) Huynh and Haick 2016 [21]. Reproduced with permission of John Wiley & Sons.

Further, Lee and coworkers reported a flexible and transparent chemical sensor comprising reduced graphene oxide (rGO) coupled with organic dye molecules (bromophenol blue) [16]. This device possesses promising properties such as high mechanical flexibility (>5000 bending cycles with a bending radius of 0.95 cm) and optical transparency (>60% in the visible region). Stacking the water-trapping dye layer on rGO enabled a higher response in a large relative humidity range (up to 80%), and dual-mode detection capabilities of colorimetric and electrical sensing for $NH_3$ gas (5–40 ppm). These advantages were attributed to the flexible and transparent rGO sensor coupled with organic dye molecules, providing great potential for real-time monitoring of toxic gas/vapor in future practical chemical sensing at room condition in wearable electronics.

Carbon nanotubes (CNTs) are another attractive material [17]. As one of the most promising semiconducting materials that may replace traditional Si, CNTs, which can act as core active materials in next-generation electronics, have also been adopted in sensing. Karimov et al. designed a CNT-based Al/CNT/Al pressure sensor [18]. This sensor was fabricated by depositing CNTs on an adhesive elastic polymer tape and placing it into an elastic casing. The diameter of multiwalled nanotubes varied between 10 and 30 nm. The nominal thickness of the CNT layers in the sensors was in the range ~300–430 μm. The interelectrode distance (length) and the width of the surface-type sensors were in the ranges 4–6 and 3–4 mm, respectively. The resistance of the sensors decreased by three to fourfold as the pressure was increased up to 17 $kN/m^2$. Similar to graphene, researchers also mixed CNTs with other materials to enable enhanced performance. Zhang and coworkers fabricated a highly flexible multifunctional smart coating by spray-coating multiwalled CNTs dispersed in a thermoplastic elastomer solution, followed by treatment with ethanol [19]. The coatings not only endowed various substrate materials with superhydrophobic surfaces but also responded to stretching, bending, and torsion – properties useful for flexible sensor applications (Figure 1.1c). The coatings exhibited superior sensitivity (gauge factor of 5.4–80), high resolution (1° of bending), fast response time (<8 ms), stable response over 5000 stretching–relaxing cycles, and wide sensing ranges (stretching: over 76%, bending: 0°–140°, torsion: 0–350 rad/m). Moreover, multifunctional coatings with thicknesses of only 1 μm can be directly applied to clothing for full-range and real-time detection of human motions. These sensors also showed extreme repellency to water, acid, and alkali, improving the work stability under wet and corrosive conditions.

Polymers are another widely used material for resistor-based sensing, since most polymers are intrinsically flexible/stretchable and can be engineered to be either insulators or (semi)conductors. Most of the research concerning polymer sensors is realized by utilizing/incorporating active polymer materials. In recent years, self-healing polymers represent a highlight in wearable sensors, since the self-healing characteristics provide durability upon bending and stretching. Among all potential candidates, hydrogels with excellent biocompatibility and mechanical features close to human tissues constitute a promising avenue for realizing health-care-oriented electronic functionalities. Xing and coworkers reported the development of a mechanically and electrically self-healing hydrogel based on physically and chemically cross-linked networks (Figure 1.1d) [20].

Autonomous intrinsic self-healing of the hydrogel was attained through dynamic ionic interactions between carboxylic groups of poly(acrylic acid) and ferric ions. Covalent cross-linking was used to support the mechanical structure of the hydrogel. Establishing a fair balance between the chemical and physical cross-linking networks together with the conductive nanostructure of polypyrrole networks led to a double network hydrogel with bulk conductivity, mechanical and electrical self-healing properties (100% mechanical recovery in two minutes), ultrastretchability (1500%), and pressure sensitivity. The practical potential of hydrogels is further revealed by their application in human motion detection and their 3D-printing performance. Mixing self-healing polymers with other active materials was also adopted in sensors. Huynh and Haick synthesized a self-healing polymer and composite (with Au nanoparticles, AuNPs) and assembled a bendable and stretchable self-healing chemiresistor for pressure/strain, temperature, and volatile organic compounds (VOCs) sensing (Figure 1.1e) [21]. Pressure/strain and temperature sensitivity was highly comparable to available flexible sensors. The limit of detection for VOCs in the parts per billion range makes the device useful for sensing VOCs. Healing efficiency of this chemiresistor is high so that the sensor survived after cutting several times at random positions. Moreover, the sensor was environmentally stable, i.e. the sensitivity slightly decreased (<10%) after six months. Three different chemiresistive AuNPs were used for this chemical sensor, proving its versatile combination of self-healing polymers with different sensing materials. The reported self-healing sensor raises expectations that flexible devices might one day become self-administered, thus increasing their reliability in various applications, such as durable-transparent touch-screens, self-healing e-skins, and implantable health-care electronics.

On the other hand, biocompatible polymers are also of interest for wearable sensors. To date, most reported skin-like pressure sensors are based on nanomaterials and microstructured PDMS films, limiting their wide practical applications due to unknown biotoxicity and the redundant fabrication procedure. A cost-effective, large-area-capable, and biocompatible approach for fabrication of high-performance skin-like pressure sensors is highly desired. Silk fibroin (SF) is a natural protein that has recently drawn attention due to its application as a substrate for flexible electronics.

As shown in Figure 1.2, Zhang and coworkers demonstrated the fabrication of skin-like pressure sensors using SF-derived active materials [22]. Flexible and conformal pressure sensors were fabricated using transparent carbonized silk nanofiber membranes (CSilkNM) and unstructured PDMS films through a cost-effective and large-scale capable approach (Figure 1.2a–e). Owing to the N-doped carbon nanofiber network structure of CSilkNM, the obtained pressure sensor shows superior performance, including ultrahigh sensitivity ($34.47\,kPa^{-1}$) for a broad pressure range, an ultralow detection limit (0.8 Pa), rapid response time (<16.7 ms), and high durability (>10 000 cycles). Based on its superior performance, the applications in monitoring human physiological signals, sensing subtle touch, and detecting spatial distribution of pressure were also demonstrated. Not only can the chemical structure be modified to realize high-performance sensors, but the dimensional structural design can also lead

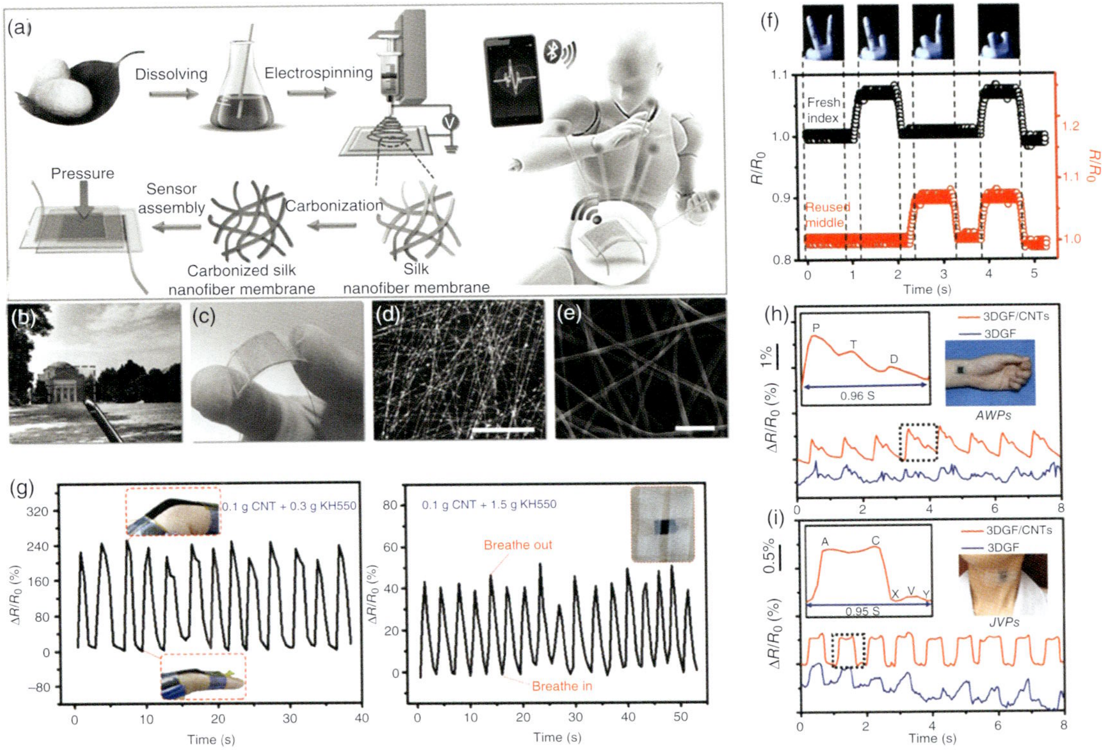


**Figure 1.2** Fabrication process and structure of the CSilkNM pressure sensor. (a) Schematic illustration showing the fabrication process of CSilkNM pressure sensors. Photographs showing the (b) transparency and (c) flexibility of the obtained sensor. (d) Optical image of a silk nanofiber membrane. Scale bar: 100 μm. (e) SEM image of CSilkNM. Scale bar: 1 μm.(f) The dual sensing result from a fresh hydrogel sensor (index finger, black) and a reusable (after 12 drying–soaking cycles) hydrogel sensor (middle finger, red) reveals that the as-fabricated hydrogel electronics is able to simultaneously detect motions of multiple objects without cross-impact. (g) The relative resistance change of the sensor (0.1 g CNT/0.3 g KH550) in hand motion from stretch to curve, and the relative resistance change of the sensor (0.1 g CNT/1.5 g KH550) in response to breathing. (h) Real-time and in situ AWP measurement with the 3DGF/CNT and 3DGF networked skin-attachable strain sensor attached on the wrist. The inset shows the photograph and zoomed waveform. (i) Measurement of JVPs with strain sensors attached on the neck. The inset is the photograph and zoomed waveform. Source: (a–e) Wang et al. 2017 [22]. Reproduced with permission of John Wiley & Sons; (f) Liu et al. 2018 [23]. Reproduced with permission of John Wiley & Sons; (g) Zhou et al. 2018 [25]. Reproduced with permission of American Chemical Society; (h, i) Cai et al. 2017 [26]. Reproduced with permission of John Wiley & Sons.

to superior performance. Xu and coworkers proposed a simple paradigm to prototype stretchable electronics with an embedded three-dimensional (3D) helical conductive layout based on biocompatible and stretchable hydrogels [23]. Thanks to the 3D helical structure, hydrogel electronics presents satisfactory mechanical and electrical robustness under stretch. In addition, reusability of stretchable electronics is realized with the proposed scenario benefiting from the swelling property of hydrogel. Although losing water would induce structural shrinkage of the hydrogel network and further undermine the function of the hydrogel in various applications, the worn-out hydrogel electronics can be reused by simply casting it in water. Through such a rehydration procedure, the dehydrated hydrogel can absorb water from the surroundings and then the hydrogel electronics can achieve resilience in mechanical stretchability and electronic functionality. Also, the ability to reflect pressure and strain changes has revealed hydrogel electronics to be promising for advanced wearable sensing applications (Figure 1.2f).

One classical way using multiple active materials within a single device is widely adopted in resistor-based sensors. A mixture of graphene and stretchable polymers was investigated by Ren and coworkers, who demonstrated a mechanical sensor fabricated using a graphene porous network (GPN) combined with PDMS [24]. Using nickel foam as a template and a chemical etching method, the GPN can be created in the PDMS-nickel foam coated with graphene. The resultant material achieved both pressure and strain sensing properties. Because of the pores in the GPN, the composite sensor exhibited a wide pressure sensing range and highest sensitivity among graphene foam-based sensors. In addition, it showed potential for use in applications such as monitoring or even recognizing walking states, the degree of finger bending, and wrist blood pressure. Fu and coworkers reported a new method to realize control on the local conductive networks of strain sensors, and thus their sensing behavior [25]. They spray-coated a mixture of CNTs and 3-aminopropyltriethoxysilane (KH550) with various ratios onto PDMS to prepare multifunctional crack-based sensors. The conductive CNT/KH550 layer exhibited brittle mechanical behavior, which triggered the formation of cracks upon stretching. This is thought to be responsible for the observed electromechanical behavior. These sensors exhibited adjustable gauge factors of 5–1000, stretchability ($\varepsilon$) of 2–250%, linearity (nonlinearity–linearity), and high durability over 1000 stretching–releasing cycles for mechanical deformation. Washable, wearable, and water-repellent sensors were prepared through such a method to successfully detect human physiological activities (Figure 1.2g). Moreover, the variation in temperature or the presence of solvent could also be detected due to the thermal expansion and swelling of the PDMS layer. It is expected that such a concept could be used to fabricate sensors for multiple applications, thanks to its multifunctionality, adjustable and robust performance, and simple and low-cost fabrication strategy.

Separately, Dong and coworkers demonstrated epidermal sensors based on an all-carbon collaborative percolation network, which consists of 3D graphene foam and CNTs obtained by two-step CVD processes [26]. The nanoscaled CNT networks largely enhance the stretchability and signal-to-noise ratio (SNR) of the 3D microarchitectural graphene foams, endowing the strain sensor with a gauge factor as high as 35, a wide reliable sensing range up to 85%, and excellent cycling stability (>5000 cycles). The flexible and reversible strain sensor can be easily mounted on human skin as a wearable electronic device for real-time and high-accuracy detection of electrophysiological stimuli and even for acoustic vibration recognition (Figure 1.2h,i). The rationally designed all-carbon nanoarchitectures are scalable, low cost, and promising in practical applications requiring extraordinary stretchability and ultrahigh SNRs. Lubineau and coworkers proposed transformation of an electrically conductive material from a sensor to a conductor using electrical welding (e-welding) [27]. This method is demonstrated in the case of a thermoplastic polymer sponge decorated with silver nanowires. The sensor-like behavior of the sponge was programmed by e-welding into conductor-like behavior, i.e. suppressing the gauge factor by 86%, without varying the density of the silver nanowires. An application of e-welding in the fabrication of a sensor-conductor hybrid material that may be applied as soft artificial skin in robotics was demonstrated.

### 1.2.2 Organic Field-Effect Transistor Based Sensors

#### 1.2.2.1 Definitions and Important Parameters

The transistor, the base of modern electronics, is usually configured as a three-terminal device (including a gate electrode, and source and drain electrodes), where the gate electrode accompanied with a dielectric can effectively modify the carrier concentration of the semiconductor, leading to controllable drain current. Key parameters associated with transistor operation are mobility, threshold voltage, on/off ratio, and subthreshold swing. However, in a transistor-based sensor, the sensitivities and the response/recovery time are the dominant parameters for practical applications [28]. Compared with resistors, even though transistor-based sensors are more complicated in device structure, they have the advantage of multiparameter sensing capabilities [29]. All the key parameters, which include but are not limited to mobility, threshold voltage, turn-on voltage, off-current, and on-current, can be used to detect the influence of analytes on the device [30]. Moreover, since transistors can be operated under various gate biases and may be designed to amplify the current, such sensors have the potential to realize ultrahigh sensitivity [31]. Furthermore, since the active components in a transistor are richer than in a resistor, one can enhance the sensing performance by not only changing the semiconducting materials, but also the dielectric and electrode materials.

#### 1.2.2.2 Strategy and Applications

Transistors based on OSCs are widely explored in the area of sensing [32]. Modifying the organic semiconducting layer is the most common way to improve transistor sensing parameters, where strategies including using thinner OSC films, organic nanowires, coupling with other analyte receptors are generally used. Perhaps the most common way to optimize sensor performance is to utilize an OSC that is intrinsically sensitive. As shown in Figure 1.3a, Noh and coworkers reported a highly sensitive printed ammonia ($NH_3$) gas sensor based on OTFTs fabricated with the fluorinated difluorobenzothiadiazole-dithienosilole polymer (PDFDT) [33]. The sensor detected $NH_3$ down to 1 ppm with high sensitivity using bar-coated ultrathin (<4 nm) PDFDT layers in the absence of any receptor additives (Figure 1.3b,c). The sensing mechanism was confirmed by cyclic voltammetry, hydrogen/fluorine nuclear magnetic resonance, and UV/visible absorption spectroscopy. PDFDT–$NH_3$ interactions comprise hydrogen bonds and electrostatic interactions between the PDFDT polymer backbone and $NH_3$ gas molecules, thus lowering the highest occupied molecular orbital levels, leading to hole trapping in the OTFT active layer. Additionally, density functional theory calculations demonstrated that gaseous $NH_3$ molecules were captured via cooperation of fluorine atoms and dithienosilole units in PDFDT. They verified that incorporation of functional groups that interact with a specific gas molecule in a conjugated polymer presents a promising strategy for producing high-performance printed OTFT gas sensors. Sensitive organic semiconducting materials were also applied in photo-sensors.

Qiu and coworkers prepared flexible and low-voltage near-infrared organic phototransistors (NIR OPTs) with a low-bandgap donor–acceptor conjugated polymer as the semiconductor layer and *n*-octadecyl phosphonic acid modified anodic alumina ($AlO_x$/ODPA) as the insulating layer (Figure 1.3d) [34]. The phototransistors exhibited typical n-channel transistor characteristics at a voltage below 5 V. The photosensitivity can be enhanced by regulating the packing densities of the ODPA self-assembled monolayers and forming different trap states. The enhanced organic phototransistors (OPTs) exhibited good photosensitivity to 808–980 nm near-infrared (NIR) with the photocurrent/dark current ratio and photoresponsivity as high as $5 \times 10^3$ and 20 mA/W, respectively, benefiting from the charge-trapping effect at the $AlO_x$/ODPA interface, as shown in Figure 1.3e. The OPTs also presented a fast-optical switching speed of 20/30 ms and excellent mechanical flexibility. The outstanding performance of the NIR OPTs indicates that the development of wearable electronics is, indeed, possible.

Research on dielectrics has also been applied to the development of transistor-based sensors as a means to enhance performance. Huang and coworkers demonstrated a highly thermally stable organic transistor by applying a three-arm stereocomplex polylactide (tascPLA) as the dielectric and substrate material [35]. The resulting flexible transistors were stable up to 200 °C, while devices based on traditional polylactide (PLA) were damaged at 100 °C. Furthermore, the charge-trapping effect induced by polar groups of the dielectric was also utilized to significantly enhance the temperature sensitivity of the electronic devices. A skin-like temperature sensor array was successfully demonstrated based on such transistors, which also exhibited good biocompatibility in cytotoxicity measurements. By presenting the combined advantages of transparency, flexibility, thermal stability, temperature sensitivity, degradability, and biocompatibility, these organic transistors possess broad applicability in applications such as environmentally friendly electronics, implantable medical devices, and artificial skin. Good dielectric performance can enhance the sensitivity and lower the driving voltage at the same time. Zheng and coworkers developed controllable polyelectrolyte composites based on poly(ethylene glycol) (PEG) and polyacrylic acid (PAA) as a type of high capacitance dielectric for flexible OTFTs and ultrasensitive pressure sensors with sub-1 V operation (Figure 1.3f) [36].

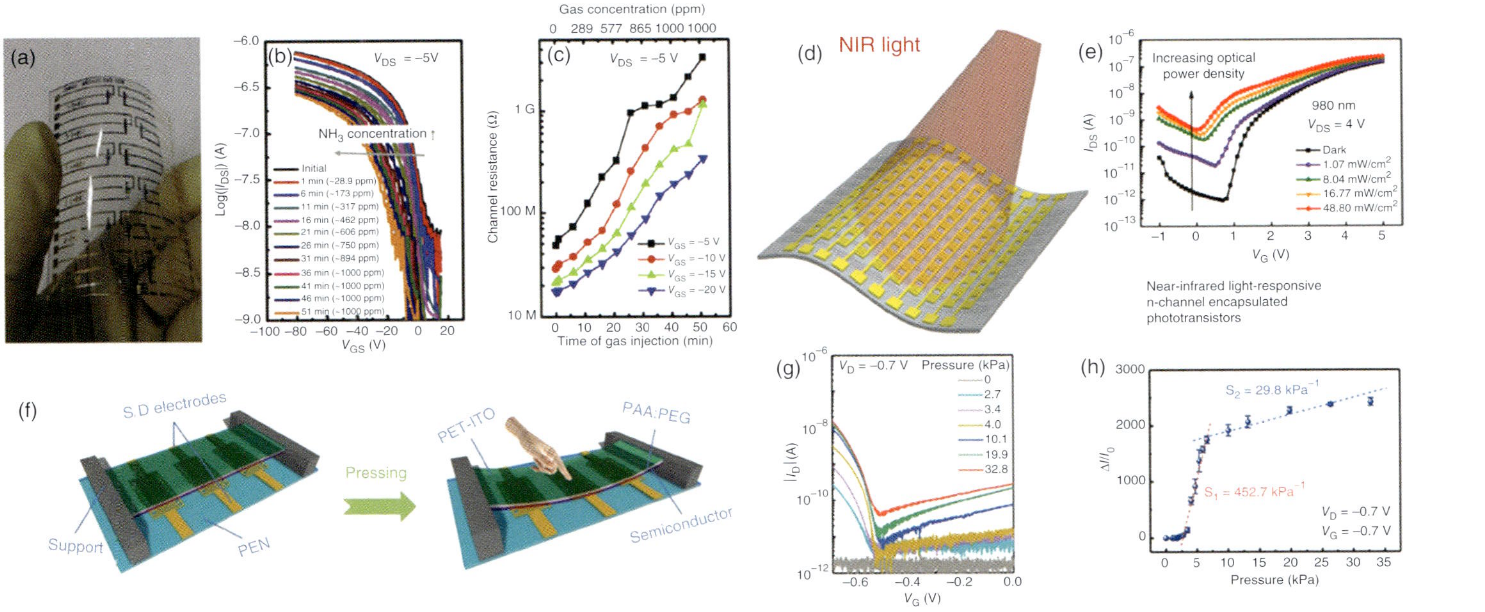


**Figure 1.3** Various OTFT-based flexible sensors. (a) Image of P-29-DPP-SVS OFET showing the BG/TC geometry for the gas sensor, (b) transfer curves, and (c) channel resistance of P-29-DPP-SVS OFET. (d) Near-infrared organic phototransistors array. (e) Increased optical power density when exposed to different density of light. (f) Schematic illustration of the OTFT-based pressure sensor (gap: ≈310 μm) in the initial state (left) and pressed state (right). (g) Transfer characteristics of the OTFT sensor under different pressures. (h) Relative change of $I_D$ in response to external pressure for the OTFT sensor at both constant $V_D$ and $V_G$ of −0.7 V Source: (a–c) Nketia-Yawson et al. 2017 [33]. Reproduced with permission of American Chemical Society; (d, e) Wang et al. 2018 [34]. Reproduced with permission of American Chemical Society; (f–h) Liu et al. 2018 [36]. Reproduced with permission of John Wiley & Sons.

Flexible OTFTs using the PAA:PEG dielectrics showed good universality and greatly enhanced electrical performance under a much smaller operating voltage of −0.7 V than those with a pristine PAA dielectric. The low-voltage OTFTs also exhibited excellent flexibility and bending stability under various bending radii and long cycles. Flexible OTFT-based pressure sensors with low-voltage operation and superhigh sensitivity were demonstrated by using a suspended semiconductor/dielectric/gate structure in combination with the PAA:PEG dielectric, as shown in Figure 1.3g,h. The sensors delivered a record high sensitivity of 452.7 $kPa^{-1}$ under a low voltage of −0.7 V, and excellent operating stability over 5000 cycles. The OTFT sensors were built into a wearable sensor array for spatial pressure mapping, which demonstrates the bright potential of flexible electronics for wearable devices and smart skins.

Apart from polymer and small molecule based transistors, CNTs have also been explored in transistor applications. Han and coworkers demonstrated a large-area high-performance flexible pressure sensor built on an active matrix of 16 × 16 carbon nanotube thin-film transistors (CNT TFTs) [37]. The active matrix exhibited superior flexible thin-film transistor (TFT) performance with high mobility and large current density, along with a high device yield of nearly 99% over a 4-in. sample area. The fully integrated flexible pressure sensor operated within a small voltage range of 3 V and exhibited superb performance featuring a high spatial resolution of 4 mm, faster response than human skin (<30 ms), and excellent accuracy in sensing complex objects on both flat and curved surfaces (Figure 1.4a,b). This work paved the road for future integration of high-performance electronic skin in smart robotics and prosthetic solutions.

Because of its narrow band gap, graphene generally does not exhibit semiconducting behavior. However, in the area of sensing, graphene-based transistors can be successfully fabricated as only the sensing functionality is needed. Also, graphene can be easily functionalized. Lee and coworkers demonstrated a flexible strain sensor based on a reduced graphene oxide field-effect transistor (rGO FET) with ultrasensitivity, stability, and repeatability for the detection of tensile and compressive strains [38]. The novelty of the rGO FET strain sensor is the incorporation of an rGO channel as a sensing layer in which the electrical resistance can be greatly modified upon application of an extremely low level of strain resulting in an intrinsically amplified sensing signal. The rGO FET device was ultrasensitive to extremely low strain levels, as low as 0.02%. Owing to weak coupling between adjacent nanosheets, therefore, upon applying small levels of strain into the rGO thin film, a modulation of the inter-nanosheet resistance is expected, inducing a large change in the transconductance of the rGO FET. Using a simple printing and self-assembly process, the facile fabrication of an rGO FET array over a large area was also demonstrated. In addition, the device was shown to be able to detect small and rapid physical movements of the human body.

A mixture of the organic semiconducting layer or dielectric layers is also a general strategy for improving sensing performance. Moreover, this approach can also impart mechanical stretchability leading to stretchable electronics. Yu and coworkers introduced an all solution-processed type of electronics and sensors that are rubbery and intrinsically stretchable as an outcome

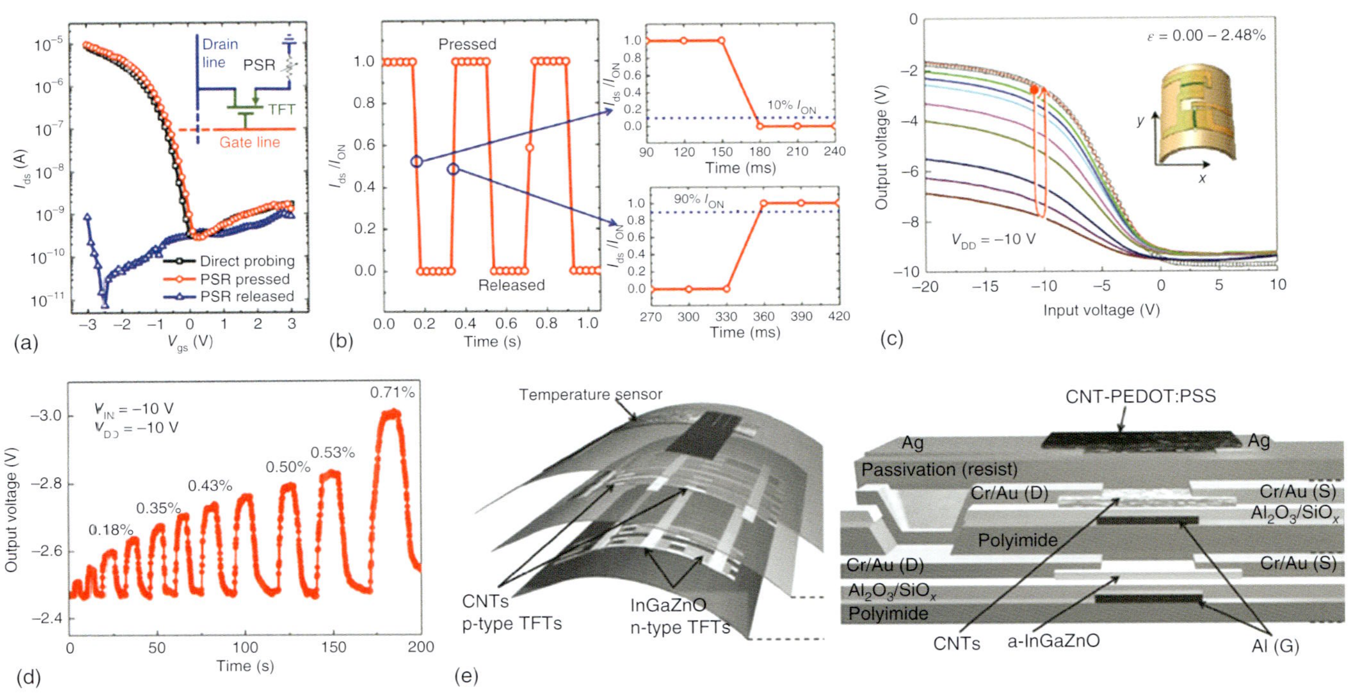


**Figure 1.4** (a) $I_{cs} - V_{gs}$ transfer curves at $V_{ds} = -1$ V for an individual pixel when directly probing through the source pad (black), when the PSR is pressed (red), and when the PSR is released (blue), respectively. The inset shows the circuit diagram for a single pixel in the pressure sensor. (b) Time-resolved measurement of the sensor response between pressed and released states. The time resolution of the measurement equipment is 30 ms. The zoomed insets are used to extract the fall time when the weight is lifted (top panel) and the rise time when the weight is dropped (bottom panel). (c) Voltage transfer characteristics of the inverter-type heptazole-based strain gauge, as obtained by a sweeping input voltage from $V_{IN}$ −20 to 10 V: vertical strain bending on driver TFT but horizontal bending on load as shown in the inset, which leads to (d) the output voltage dynamics. (e) Schematic of a vertically integrated 3D flexible CMOS and sensor device. Source: (a, b) Nela et al. 2018 [37]. Reproduced with permission of American Chemical Society; (c, d) Nam et al. 2014 [40]. Reproduced with permission of John Wiley & Sons; (e) Honda et al. 2015 [41]. Reproduced with permission of John Wiley & Sons.

from all the elastomeric materials in percolated composite formats with poly(3-hexylthiophene-2,5-diyl) nanofibrils (P3HT-NFs) and Au nanoparticles with conformally coated silver nanowires (AuNP–AgNW) in PDMS [39]. The fabricated TFTs retained more than 55% of their electrical performances upon 50% stretching and exhibited one of the highest P3HT-based field-effect mobilities of $1.4\,cm^2/(V\,s)$, owing to improved crystallinity. Rubbery sensors, which include strain, pressure, and temperature sensors, exhibit reliable sensing capabilities and can be exploited as smart skins that enable translation of gestures for sign language alphabet and haptic sensing for robotics, illustrating one application of these sensors.

The advantages of transistors also include their ability to be integrated into facial systems for signal amplification. Im and coworkers reported a nonclassical organic strain gauge as a voltage signal sensor using an inverter-type TFT circuit, which sensitively measured a large quantity of elastic strain (up to ≈2.48%) [40]. Heptazole-based organic TFTs were chosen to be incorporated in this gauge circuit, because of the small domain size of the organic solid heptazole. While large crystal domain-pentacene TFTs seldom show sufficient current variation upon mechanical bending for tensile strain, these heptazole TFTs demonstrated a significant variation for the same strain condition as applied to the pentacene alternatives, as shown in Figure 1.4c,d. In addition, the pentacene channel does not recover to its original electric state after bending. In contrast, heptazole channels are very elastic and reversible, even after serious bending. More interesting is the observation that the heptazole TFTs show only minimum variation of signal current under horizontal direction strain, while they make a significant amount of current decrease under vertical direction strain. Utilizing the anisotropic response to the tensile bending strain, an ultrasensitive voltage output strain gauge composed of a horizontally and vertically oriented TFT couple was demonstrated. A complementary inverter using n-type InGaZnO- and p-type CNT-TFTs was also adopted in the area of sensing. Takei and coworkers reported this vertically integrated 3D complementary metal-oxide semiconductor (CMOS) with a temperature sensor device on a flexible substrate; the structure is shown in Figure 1.4e [41]. Relative to conventional lateral integration, this vertical integration process for a flexible device realized high-density integration by adding only ≈4 μm thickness of the flexible passivation layers. The mechanical reliability and temperature dependence of the 3D CMOS inverters were characterized for the first time. These inverters exhibited good mechanical flexibility and relatively stable temperature dependence without a change in performance.

More importantly, vertical integration of both TFTs and sensors facilitated the fabrication of highly integrated and flexible devices for practical low-cost electronics. The introduction of sensor arrays is also a general strategy to improve sensing capability and selectivity. Park and coworkers successfully demonstrated a general approach to fabricate e-skin that can (i) extract effects from the target sensing signals, such as P or T, while the flexible sensor is under multimode stimulus; and (ii) enable real-time bimodal sensing using a single field-effect transistor (FET) device by extracting parameters associated with mechanical deformation [42]. The advantages in integrating FET arrays with multimodal sensing elements in flexible e-skin greatly reduced the complexity in structural integration, eliminated or minimized the signal interference coupled by strain, significantly reduced the power consumption, and decreased the failure rate in production due to facile integration of FET devices into the circuits. Furthermore, it has potential for reducing fabrication costs of large-area flexible e-skins. This approach may be extended to realize multimodality in large-area flexible e-skin with heterogeneous input stimuli of physical, chemical, or biological nature, and also to solve problems associated with strains induced during operative service of flexible electronic systems.

### 1.2.3 Electrochemical Sensors

#### 1.2.3.1 Definitions and Important Parameters

Similar to OFETs, electrochemical sensors usually consist of three electrodes, including the reference, the counter, and the working electrodes [43]. Among all the geometries of electrochemical-based sensors, organic electrochemical transistors (OECTs) have received much more attention due to the structure similar to that of conventional transistors, which facilitates fabrication and characterization [44]. Generally, OECTs are comprised of a (semi)conductive polymer, and an electrolyte acting as the dielectric. The gate electrode is immersed in the electrolyte to apply gate bias. Owing to the ultra-large capacitance, OECTs can be effectively driven at voltages lower than 1 V. Moreover, as a kind of transistor, they have all the advantages of transistors. In addition, a fascinating feature of OECTs is that the electrolyte acts as a dielectric and can be used as a sensing medium. Owing to the double layer capacitor characteristics, the geometry of an OECT can be easily adjusted [45]. In the following section, we will focus more on OECT-based sensors.

#### 1.2.3.2 Strategy and Applications

Owing to their unique device properties, OECTs are more attractive when applied in the area of stretchable and wearable sensors, as their dielectrics are either liquid or ultra-stretchable gel materials. Furthermore, OECTs are extremely compatible with biosensor applications, as their behavior is very similar to that of the synapse. Lee et al. reported an organic optoelectronic sensorimotor synapse that uses an organic optoelectronic synapse and a neuromuscular system based on a stretchable organic nanowire synaptic transistor (s-ONWST) [46]. The voltage pulses of a self-powered photodetector triggered by optical signals drive the s-ONWST, and the resultant informative synaptic outputs were used not only for optical wireless communication of human–machine interfaces but also for light-interactive actuation of an artificial muscle actuator in the same way that a biological muscle fiber contracts. This organic optoelectronic sensorimotor synapse suggests a promising strategy toward developing bioinspired soft electronics, neurologically inspired robotics, and electronic prostheses.

Other than organic small molecules and polymers, CNTs were also applied in electrochemical sensors. Huang and coworkers reported an attractive stretchable electrochemical sensor that takes advantage of a hierarchical percolation network of carbon nanotubes and gold nanotubes (CNTsAu NTs) [47]. This hybrid nanostructure provided the sensor with excellent time-reproducible mechanical and electrochemical performance while granting very good cellular compatibility, making it perfectly apt to induce and monitor simultaneously transient biochemical signals. This is validated by monitoring the stretch-induced transient release of small signaling molecules by both endothelial and epithelial cells cultured on this sensor and submitted to stretching strains of different intensities, as shown in Figure 1.5a–d. This work demonstrates that the hybrid CNTs-Au NTs platform offers a versatile and highly sensitive approach to characterize and quantify short-time mechano-transduction responses.

Natural organic materials have also been widely utilized as active sensing materials of electrochemical-based sensors. Wang and coworkers developed a flexible glove-based electrochemical biosensor with a highly stretchable printed electrode system as a wearable point-of-use screening tool for defense and food security applications by adopting an enzyme [48]. This disposable, mechanically robust "lab-on-a-glove" integrated a stretchable printable enzyme-based biosensing system and active surface for swipe sampling on different fingers and was coupled with a compact electronic interface for electrochemical detection and real-time wireless data transmission to a smartphone device (Figure 1.5e–j). Stress enduring inks were used to print the electrode system and the long serpentine connections to the wireless electronic interface. Dynamic mechanical deformation, bending, and stretching studies illustrated the resilience and compliance of the printed traces against extreme mechanical deformations

**Figure 1.5** (a) Fluorescence microscopic image of mechanically strained HUVECs (30% stretch) after staining with Alexa Fluor 488 phalloidin (green: actin fibers) and Hoechst 33342 (blue: nuclei). (b) Schematic diagram of cellular mechanotransduction mechanism stimulated in stretched cells. (c) Current responses detected from HUVECs submitted to different stretch moduli (I–III) (see text). (d) Current responses detected from HBECs submitted to different stretch moduli. The vertical gray dashed lines in (c) and (d) indicate the beginning points of mechanical stretches; it took five seconds to achieve the desired strains. The electrode was submitted to a cycle of stretch and release process before applying another stretch episode to the cells. (e) Image of the serpentine stencil design employed for printing the glove-based stretchable device. (f) Schematic of (left) the biosensing scan finger (index finger) containing smiling face shaped carbon-based counter (CE), working (WE) electrodes and Ag/AgCl-based reference electrode (RE), and (right) collecting thumb finger with its printed carbon pad; scale bar 10 mm. (g) Photographs of the biosensing index finger under 0% (left) and 50% (right) linear stretch; scale bar, 10 mm. (h) On-glove swiping protocol for sampling chemical threat residues from tomato and stainless steel surfaces. (i) On-glove sensing procedure by joining the index (scan) and thumb (collector) fingers to complete the electrochemical cell. (j) Photographs of the wearable glove biosensor, consisting of a sensing finger, containing the immobilized OPH enzyme layer, and the collector/sampling finger. (k) Schematic illustration of the current change mechanism for detecting target PSA molecules attached to the sensor surface. (l, m) Kinetic monitoring of PSA binding to rGO@SFP-based biosensor without and with immobilized anti-PSA antibodies. (n) The response time of rGO@SFP-based biosensor detection against $1 \times 10^{-12}$ M target PSA. Source: (a–d) Liu et al. 2017 [47]. Reproduced with permission of John Wiley & Sons; (e–j) Mishra et al. 2017 [48]. Reproduced with permission of Royal Society of Chemistry; (k–n) Wang et al. 2016 [49]. Reproduced with permission of John Wiley & Sons.

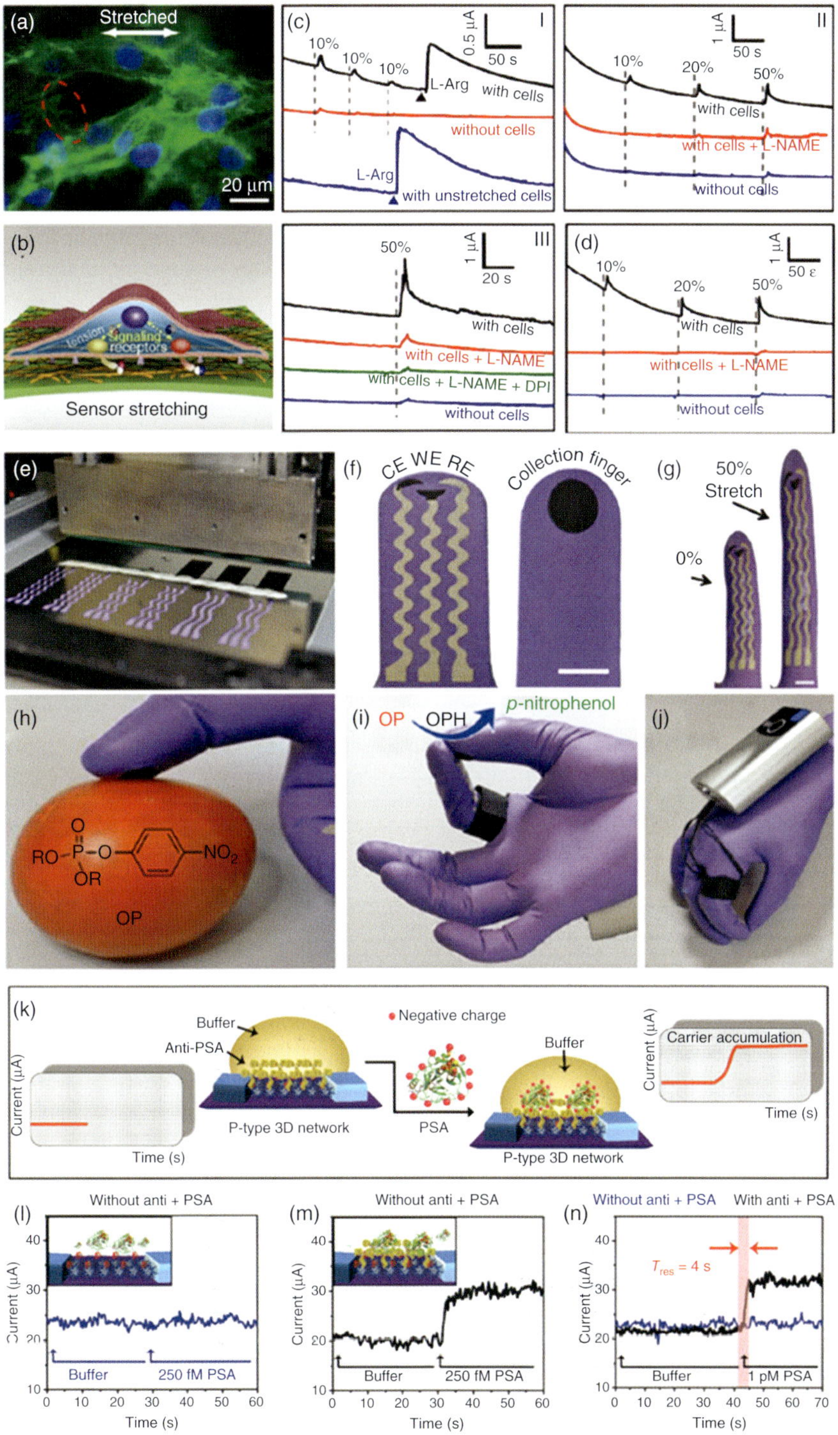
(a)
Stretched
20 μm
(b)
Tension
Signaling receptors
Sensor stretching
(c)
I
0.5 μA
50 s
10%
10%
10%
L-Arg
with cells
without cells
L-Arg
with unstretched cells
II
1 μA
50 s
10%
20%
50%
with cells
with cells + L-NAME
without cells
III
1 μA
20 s
50%
with cells
with cells + L-NAME
with cells + L-NAME + DPI
without cells
(d)
1 μA
50 ε
10%
20%
50%
with cells
with cells + L-NAME
without cells
(e)
(f)
CE WE RE
Collection finger
(g)
50% Stretch
0%
(h)
RO-P-O
OR
NO2
OP
(i) OP
OPH
p-nitrophenol
(j)
(k)
Current (μA)
Time (s)
Buffer
Anti-PSA
P-type 3D network
Negative charge
PSA
Buffer
P-type 3D network
Carrier accumulation
Current (μA)
Time (s)
(l)
Without anti + PSA
Current (μA)
Buffer
250 fM PSA
Time (s)
(m)
Without anti + PSA
Current (μA)
Buffer
250 fM PSA
Time (s)
(n)
Without anti + PSA
With anti + PSA
T_res = 4 s
Current (μA)
Buffer
1 pM PSA
Time (s)

expected for such on-glove sampling/sensing operation. An organophosphorus hydrolase (OPH)-based biosensor system on the index finger enabled rapid on-site detection of organophosphate (OP) nerve agent compounds on suspicious surfaces and agricultural products following their swipe collection on the thumb finger. The new wireless glove-based biosensor system offers considerable promise for field screening of OP nerve agents and pesticides in defense and food safety applications, with significant speed and cost advantages. Such "lab-on-a-glove" demonstration opened the area of flexible wearable sensors to future on-the-hand multiplexed chemical detection in diverse fields. Other than pure natural organic materials for active layer, they were also coupled with graphene to enhance sensing performance. Cho and coworkers described an ultrasensitive, flexible sensor as a 3D hierarchical biocomposite comprising of hollow, natural pollen microcapsules that were coated with a conductive graphene layer [49]. Modular assembly of the graphene-coated microcapsules onto an ultrathin polyethylene terephthalate layer enables a highly flexible sensor configuration with tunable selectivity afforded by subsequent covalent immobilization of antibodies against target antigens. In a proof-of-concept example, the biosensor demonstrated ultrahigh sensitivity detection of prostate-specific antigen (PSA) down to $1.7 \times 10^{-15}$ m with real-time feedback and superior performance over conventional two-dimensional graphene-coated sensors (Figure 1.5k–n). Importantly, the device performance was consistently high across various bending conditions. Taken together, the results demonstrated in this work highlight the merits of employing lightweight biocomposites as modular building blocks for the design of flexible biosensors with highly responsive and sensitive molecular detection capabilities.

### 1.2.4 Diode-Based Sensors

#### 1.2.4.1 Definitions and Important Parameters

A diode is a two-terminal device that contains a p-type and an n-type material to form a p–n junction and has the property of only conducting current in one direction. In the sensing area, diodes have been widely used in the form of photodetectors, in which the current of the diode is extremely sensitive to the light illumination [50]. Key parameters include photosensitivity, response/recovery time. A single photodetector can be utilized to realize environmental light sensing. By coupling with a light-emitting diode, it can be used to detect human health information such as heart rate, blood pressure, and oxygen and glucose levels in the blood. This kind of device has already been commercialized for decades. In recent years, with the emergence of organic materials, further development with flexible, even wearable photosensors has occurred, targeting next-generation portable health monitoring.

#### 1.2.4.2 Strategy and Applications

For the acquisition of health information, organic materials-based light-emitting diodes have been designed for emission of purer light, leading to improved accuracy of the photodetectors. Arias and coworkers presented a method to print two materials of different functionality during the same printing step [51]. In printed electronics, devices are built layer by layer and conventionally only one type of material is deposited in one pass. Here, the challenges involving the printing of two emissive materials to form polymer light-emitting diodes (PLEDs) that emit light of different wavelengths without any significant changes in the device characteristics were described. The surface-energy-patterning technique was utilized to print materials in regions of interest. This technique proved beneficial in reducing the amount of ink used during blade coating and improving the reproducibility of printed films. A variety of colors (green, red, and near-infrared) were demonstrated and characterized. This is the first known attempt to print multiple materials by blade coating. These devices were further used in conjunction with a commercially available photodiode to perform blood oxygenation measurements on the wrist, where common accessories are worn. Prior to the actual application, the threshold conditions for each color were determined in order to acquire a stable and reproducible photoplethysmogram (PPG) signal. Finally, based on the conditions, PPG and oxygenation measurements were successfully performed on the wrist with green and red PLEDs. Yoo and coworkers exploited the design freedom offered by organic technologies to realize a reflective patch-type pulse oximetry sensor with ultralow power consumption [52]. On the basis of flexible OLEDs and organic photodiodes designed via an optical simulation of color-sensitive light propagation within the human skin, the proposed monolithically integrated organic pulse oximetry sensor heads exhibit successful operation at electrical power as low as 24 mW on average.

### 1.2.5 Other Devices and System Integration

Other than the abovementioned four types of devices, other devices can also be fabricated to function as sensors [53]. Dickey and coworkers described the fabrication and characterization of soft and stretchable capacitive sensors of torsion, strain, and touch using hollow elastomeric fibers [54]. Twisting or elongating an intertwined bundle of two fibers increased the contact area between the fibers and therefore the capacitance. Additionally, bundles of fibers filled with LM can serve as capacitive touch sensors along the length of the bundle by using fibers filled with metal to various lengths. These sensors were able to detect touch via self-capacitance with the finger, rather than by physical deformation. Because these fiber sensors are extremely soft and stretchable, as well as small ($\approx$200–800 μm diameter), they could be used with artificial muscles, soft robotics, and stretchable devices. They have higher detection limits than state-of-the-art sensors; however, they can measure large ranges of torsion and strain and have an advantageous fiber shape that can conform to a variety of complex surfaces. As the modes for sensing torsion, strain, and touch all rely on capacitance, it is currently not possible to decouple them, although future work could measure other electrical properties, such as resistance, to help decouple their effects. For these fibers to be utilized in fabrics, further work must be done to understand their cyclic behavior and wash resistance; however, these fibers offer a promising opportunity to create woven and wearable sensors in stretchable textiles for a variety of sensing functions.

To reveal the potential of various devices and materials, a systems integration strategy has been extensively utilized [55]. With systems integration, ultrahigh selectivity and in-situ multiparameter monitoring can be realized, even though these are not accessible within a single device [56]. System integration is applied in force sensors at the very beginning. Kyung and coworkers developed a fast responding force sensor array that detects multiple points; it is thin, highly transparent, and highly flexible; and it is based on polymer waveguides that can be applied to curvilinear interfaces [57]. The force sensor detects contact forces by monitoring light intensity transmitted through sensing areas that allow a touch layer to be in contact with the bare core of the waveguide patterns. The force sensor is capable of working without any electronic components on the sensing areas. The response characteristics including sensitivity, response time, and hysteresis depend on the mechanical properties of the touch layers. The use of the touch layer allowed the force sensor to provide a fast and sensitive response without significant hysteresis to the programmed sinusoidal input force (c. 0–3 N) with frequency in the range of 1–16 Hz. The response was also highly durable over 3600 continuous loading and unloading cycles and resistant to bending. The thin-film architecture with the combination of the waveguide sensing mechanism and soft optical materials allowed the demonstration of detection of simultaneous and multiple forces on a curvilinear surface such as the human arm. When the sensor is practically applied to electronic devices as a flexible input interface, it may additionally be applicable to the analysis of environmental effects such as temperature and humidity.

We also need to investigate the optimal spatial pattern for power consumption and materials for rapid response with low hysteresis. Wang and coworkers reported a self-powered triboelectric sensor based on flexible thin-film materials. It relies on contact electrification to generate a voltage signal in response to a physical contact without using an external power supply [58]. Enabled by the sensing mechanism and surface modification by polymer nanowires, the triboelectric sensor exhibited an exceptional pressure sensitivity of 44 mV/Pa (0.09% $Pa^{-1}$) and a maximum touch sensitivity of 1.1 V/Pa (2.3% $Pa^{-1}$) in the extremely low-pressure region (<0.15 kPa). Through the integration of the sensor with a signal-processing circuit, a complete tactile sensing system was further developed. Diverse applications were demonstrated, explicitly indicating a variety of immediate uses in the human–electronics interface, automatic control, surveillance, remote operation, and security systems, as shown in Figure 1.6a–f.

Chan and coworkers had demonstrated a temperature-sensor device based on a large-area flexible OFET array [59]. By utilizing an ultrathin PEN substrate, the device can be conformally attached to various objects or the human body. The electrochemically grown alumina was adopted as the dielectric, which allowed the device to operate below 4 V and maintain a leakage current approximately in the range of tens of pA. The fabrication technique is not limited to temperature sensors, but can also be utilized as a driving/switching circuit for other devices in array form, such as light-emitting diode (LED) back panels. By controlling the nominal thickness of the Ag NPs, both the conductivity and sensitivity of thermistors can be tuned. The integrated temperature sensor maintained more than

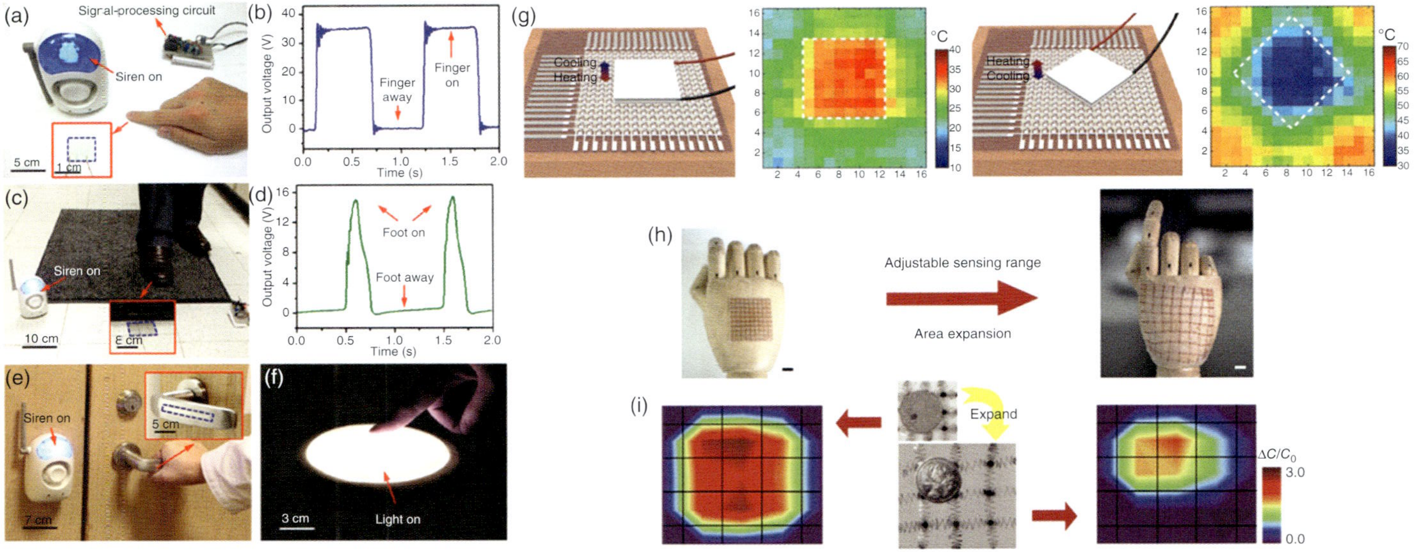


**Figure 1.6** (a) Triggering a wireless alarm system by gentle finger tapping on a TES (1 cm in side length). Inset: enlarged view of the TES on a piece of paper. (b) Output voltage of the TES generated by finger tapping. (c) Triggering a wireless alarm system by stepping on top of a TES (10 cm in side length) that is embedded underneath a carpet. Inset: enlarged view of the TES on the floor. (d) Output voltage generated by stepping on the TES covered by the carpet. (e) Triggering a wireless alarm system by grabbing the TES (5 cm by 3 mm) that is applied on a door handle. Inset: enlarged view of the TES on the door handle. (f) Switching a panel light by finger tapping on the TES (1 cm in side length) that is applied on the surface of the light. (g) Schematic diagram of temperature measurement setup; a Peltier heater is placed on top of the sensor array, and the base of the Peltier heater warms up/cools down the region of the sensor array, with temperature distribution measured from the sensor array; the dashed lines indicate where the Peltier heater is located. (h) Schematic illustration of an SCMN as an artificial electronic skin on a hand, showing sensing adjustability and expandability (scale bar: 1 cm). (i) Pressure mapping before and after the 300% expansion of an SCMN; the position of the pressure load is also identified after expansion. Source: (a–f) Zhu et al. 2014 [58]. Reproduced with permission of American Chemical Society; (g) Xia et al. 2016 [59]. Reproduced with permission of Royal Society of Chemistry; (h, i) Ye et al. 2018 [60]. Reproduced with permission of Springer Nature.

20 times the output current change, while the operating temperature changed from 20 to 100 °C. The actively addressed 16 × 16 OFET array attained a 100% yield rate and provided 2D temperature information of the contacted objects, including those with irregular shapes. As shown in Figure 1.6g, the current device enabled defect prediction of electronic devices, remote sensing of harsh environments, food supply chain management, and e-skin applications. Attaching the temperature-sensor array onto the outside of the human body or its organs during surgery may provide valuable thermal distribution information that could be useful for diagnostic and therapeutic purposes.

A multifunctional sensor array has also been realized with systems integration. Wang and coworkers presented a skin-inspired highly stretchable and conformable matrix network (SCMN) that successfully expanded the e-skin sensing functionality including but not limited to temperature, in-plane strain, humidity, light, magnetic field, pressure, and proximity [60]. The actualized specific expandable sensor units integrated on a structured polyimide network, potentially in a three-dimensional (3D) integration scheme, can also fulfill simultaneous multi-stimulus sensing and achieve an adjustable sensing range and large-area expandability (Figure 1.6h,i). A personalized intelligent prosthesis was further constructed and demonstrated its use in real-time spatial pressure mapping and temperature estimation. Looking forward, this SCMN has broader applications in humanoid robotics, new prosthetics, human–machine interfaces, and health-monitoring technologies. Takei and coworkers presented a flexible, multifunctional printed health-care sensor equipped with a three-axis acceleration sensor to monitor physical movement and motion [61]. Because the device was designed to be attached directly onto the skin, it had a modular design with two detachable components: one device component was nondisposable, whereas the other one was disposable and designed to be worn in contact with the skin. The design of this disposable sensing sheet takes into account hygiene concerns and low-cost materials and fabrication methods as well as features including integrated, printed sensors to monitor for temperature, acceleration, electrocardiograms, and a kirigami structure, which allows for stretching on the skin. The reusable component contains more expensive device components, featuring an ultraviolet light sensor that is controlled by CNT TFTs and has a mechanically flexible and stable liquid metal contact for connection to the disposable sensing sheet. After characterizing the electrical properties of the transistors and flexible sensors, a proof-of-concept device was demonstrated that is capable of health-care monitoring combined with detection of physical activity, showing that this device provides an excellent platform for the development of commercially viable, wearable health-care monitors.

## 1.3 Summary and Perspective

The development of organic nano-sensors is well on its way to realizing commercialization. Owing to the flexible, stretchable properties of organic functional materials, they are among the most promising materials that can be used in wearable sensors. Currently, there is tremendous demand for better monitoring of human health and recording of environmental parameters. As the most important interconnection part of this process, sensors play a key role in timely, accurate information collection. Accompanied with other wearable active constituent parts, including a wearable power source, stretchable electrical interconnection, and lightweight information processor, the practical commercialization of smart clothing can be expected in the near future.

Until now, several major challenges continue to impede the further development of wearable sensors [62]. Stability is one of the most stubborn problems for organic materials. Stability specifically refers to the long-lived activities under working conditions, e.g. the bias stability of transistors under light exposure or analyte chemical exposure. Another important stability parameter relates to whether the sensor performance is robust when functioning under situations of bending, twisting, or stretching. A well-recognized standard to quantify the stability properties is still missing, but it is believed that such standard will soon be established, along with the massive production of smart clothing. Biocompatibility is another important issue for organic materials-based sensors. For accurate measurement of human health information, most of the organic nano-sensors are directly applied to human skin, and even implanted into tissues and organs. Thus, the toxicity of organic materials must be carefully examined, and any massive commercialization and experimental trials should take safety as the first priority. Multidisciplinary collaboration will serve as the most effective strategy to promote the further development of wearable organic nano-sensors, with the participation of organic chemists for new materials development, the materials, mechanical and electrical engineers for materials evaluation, optimization, and device fabrication, as well as the clinicians for applications in human health monitoring.

## References


1 Han, S.T., Peng, H., Sun, Q. et al. (2017). *Adv. Mater.* 29: 1700375.
2 (a) Zhao, S., Li, J., Cao, D. et al. (2017). *ACS Appl. Mater. Interfaces* 9: 12147. (b) Zang, Y.P., Zhang, F.J., Di, C.A., and Zhu, D.B. (2015). *Mater. Horiz.* 2: 140.
3 (a) Xu, M., Obodo, D., and Yadavalli, V.K. (2019). *Biosens. Bioelectron.* 124–125: 96. (b) Xie, M., Hisano, K., Zhu, M. et al. (2019). *Adv. Mater. Technol.* 4: 1800626.
4 Kim, S.J., Choi, S.J., Jang, J.S. et al. (2017). *Acc. Chem. Res.* 50: 1587.
5 (a) Wang, X., Liu, Z., and Zhang, T. (2017). *Small* 13: 1602790. (b) Ramasamy, M., Kumar, P.S., and Varadan, V.K. (2017). *Nanosens Biosens Info Tech Sens 3D Syst* 10167: 1016703.
6 (a) Matzeu, G., Florea, L., and Diamond, D. (2015). *Sens. Actuators, B* 211: 403. (b) Lou, Z., Wang, L., and Shen, G. (2018). *Adv. Mater. Technol.* 3: 1800444.
7 (a) Lee, Y.H., Kweon, O.Y., Kim, H. et al. (2018). *J. Mater. Chem. C* 6: 8569. (b) Kim, J., Jeerapan, I., Sempionatto, J.R. et al. (2018). *Acc. Chem. Res.* 51: 2820.

8 Kleinschmidt, A.T. and Lipomi, D.J. (2018). *Acc. Chem. Res.* 51: 3134.
9 Andrew, T.L., Zhang, L., Cheng, N. et al. (2018). *Acc. Chem. Res.* 51: 850.
10 (a) Tian, L., Li, Y., Webb, R.C. et al. (2017). *Adv. Funct. Mater.* 27: 1701282. (b) Tessarolo, M., Gualandi, I., and Fraboni, B. (2018). *Adv. Mater. Technol.* 3: 1700310.
11 Neethirajan, S. (2017). *Sens. Biosens. Res.* 12: 15.
12 Nag, A., Mukhopadhyay, S.C., and Kosel, J. (2017). *IEEE Sens. J.* 17: 3949.
13 Rim, Y.S., Bae, S.H., Chen, H. et al. (2016). *Adv. Mater.* 28: 4415.
14 Hou, C., Wang, H., Zhang, Q. et al. (2014). *Adv. Mater.* 26: 5018.
15 Wang, Y., Wang, L., Yang, T.T. et al. (2014). *Adv. Funct. Mater.* 24: 4666.
16 Duy, L.T., Trung, T.Q., Dang, V.Q. et al. (2016). *Adv. Funct. Mater.* 26: 4329.
17 Lipomi, D.J., Vosgueritchian, M., Tee, B.C. et al. (2011). *Nat. Nanotechnol.* 6: 788.
18 Karimov, K.S., Saleem, M., Karieva, Z.M. et al. (2011). *Phys. Scr.* 83: 065703.
19 Li, L., Bai, Y., Li, L. et al. (2017). *Adv. Mater.* 29: 1702517.
20 Darabi, M.A., Khosrozadeh, A., Mbeleck, R. et al. (2017). *Adv. Mater.* 29: 1700533.
21 Huynh, T.P. and Haick, H. (2016). *Adv. Mater.* 28: 138.
22 Wang, Q., Jian, M., Wang, C., and Zhang, Y. (2017). *Adv. Funct. Mater.* 27: 1605657.
23 Liu, H., Li, M., Ouyang, C. et al. (2018). *Small* 14: e1801711.
24 Pang, Y., Tian, H., Tao, L. et al. (2016). *ACS Appl. Mater. Interfaces* 8: 26458.
25 Zhou, X., Zhu, L., Fan, L. et al. (2018). *ACS Appl. Mater. Interfaces* 10: 31655.
26 Cai, Y., Shen, J., Dai, Z. et al. (2017). *Adv. Mater.* 29: 1606411.
27 Chellattoan, R., Lube, V., and Lubineau, G. (2018). *Adv. Electron. Mater.* 5: 1800273.
28 Yu, X.G., Zhou, N.J., Han, S.J. et al. (2013). *J. Mater. Chem. C* 1: 6532.
29 Zang, Y., Huang, D., Di, C.A., and Zhu, D. (2016). *Adv. Mater.* 28: 4549.
30 Lee, Y.H., Jang, M., Lee, M.Y. et al. (2017). *Chem* 3: 724.
31 Lee, M.Y., Lee, H.R., Park, C.H. et al. (2018). *Acc. Chem. Res.* 51: 2829.
32 Ryu, G.S., Park, K.H., Park, W.T. et al. (2015). *Org. Electron.* 23: 76.
33 Nketia-Yawson, B., Jung, A.R., Noh, Y. et al. (2017). *ACS Appl. Mater. Interfaces* 9: 7322.
34 Wang, G., Huang, K., Liu, Z. et al. (2018). *ACS Appl. Mater. Interfaces* 10: 36177.
35 Wu, X.H., Ma, Y., Zhang, G.Q. et al. (2015). *Adv. Funct. Mater.* 25: 2138.
36 Liu, Z., Yin, Z., Wang, J., and Zheng, Q. (2018). *Adv. Funct. Mater.*: 1806092.
37 Nela, L., Tang, J., Cao, Q. et al. (2018). *Nano Lett.* 18: 2054.
38 Trung, T.Q., Tien, N.T., Kim, D. et al. (2014). *Adv. Funct. Mater.* 24: 117.
39 Kim, H.J., Sim, K., Thukral, A., and Yu, C. (2017). *Sci. Adv.* 3: e1701114.
40 Nam, S.H., Jeon, P.J., Min, S.W. et al. (2014). *Adv. Funct. Mater.* 24: 4413.
41 Honda, W., Harada, S., Ishida, S. et al. (2015). *Adv. Mater.* 27: 4674.
42 Tien, N.T., Jeon, S., Kim, D.I. et al. (2014). *Adv. Mater.* 26: 796.
43 Yang, M., Jeong, S.W., Chang, S.J. et al. (2016). *ACS Appl. Mater. Interfaces* 8: 34978.
44 Yang, A., Li, Y., Yang, C. et al. (2018). *Adv. Mater.* 30: e1800051.
45 Shiwaku, R., Matsui, H., Nagamine, K. et al. (2018). *Sci. Rep.* 8: 6368.

46 Lee, Y., Oh, J.Y., Xu, W. et al. (2018). *Sci. Adv.* 4: eaat7387.
47 Liu, Y.L., Qin, Y., Jin, Z.H. et al. (2017). *Angew. Chem. Int. Ed. Engl.* 56: 9454.
48 Mishra, R.K., Hubble, L.J., Martín, A. et al. (2017). *ACS Sens.* 2: 553.
49 Wang, L.L., Jackman, J.A., Ng, W.B., and Cho, N.J. (2016). *Adv. Funct. Mater.* 26: 8623.
50 Wu, Z., Zhai, Y., Kim, H. et al. (2018). *Acc. Chem. Res.* 51: 3144.
51 Han, D., Khan, Y., Ting, J. et al. (2017). *Adv. Mater.* 29: 1606206.
52 Lee, H., Kim, E., Lee, Y. et al. (2018). *Sci. Adv.* 4: eaas9530.
53 Nakata, S., Shiomi, M., Fujita, Y. et al. (2018). *Nat. Electron.* 1: 596.
54 Cooper, C.B., Arutselvan, K., Liu, Y. et al. (2017). *Adv. Funct. Mater.* 27: 1605630.
55 (a) Xu, S., Zhang, Y., Jia, L. et al. (2014). *Science* 344: 70. (b) Yang, Y., Zhang, H.L., Lin, Z.H. et al. (2013). *ACS Nano* 7: 9213.
56 Sarwar, M.S., Dobashi, Y., Preston, C. et al. (2017). *Sci. Adv.* 3: e1602200.
57 Yun, S., Park, S., Park, B. et al. (2014). *Adv. Mater.* 26: 4474.
58 Zhu, G., Yang, W.Q., Zhang, T. et al. (2014). *Nano Lett.* 14: 3208.
59 Ren, X., Pei, K., Peng, B. et al. (2016). A low-operating-power and flexible active-matrix organic-transistor temperature-sensor array. *Adv. Mat.* 28 (24): 4832–4838.
60 Ye, L., Hu, H., Ghasemi, M. et al. (2018). *Nat. Mater.* 17: 253.
61 Yamamoto, Y., Harada, S., Yamamoto, D. et al. (2016). *Sci. Adv.* 2: e1601473.
62 Bandodkar, A.J., Jeerapan, I., and Wang, J. (2016). *ACS Sens.* 1: 464.

# 2

# Stimuli-Responsive Electronic Skins

*Zhouyue Lei and Peiyi Wu*

*Donghua University, State Key Laboratory for Modification of Chemical Fibers and Polymer Materials, College of Chemistry, Chemical Engineering and Biotechnology, Center for Advanced Low-Dimension Materials, Shanghai 201620, China*

## 2.1 Introduction

Human skin, as the largest sensory organ in our bodies, is adaptable to dynamic nonplanar surfaces, responsive to a wide range of stimuli, and can self-regulate upon environmental changes. Inspired from human skin and motivated by increasing demands for artificially intelligent technologies, electronic skins have gained significant scientific interest during the last two decades on account of their potential applications in smart clothing, wearable devices, soft robotics, and Internet of Things (IoTs).

In order to mimic the human skin's adaptability, sensitivity, and self-regulation, people have devoted much effort into material innovation on the electronic skins, for example, by developing conductive materials that can change their chemical or physical properties and functionalities upon stimuli, and by designing circuit structures that can convert environmental stimuli into real-time electrical signals. In this chapter, we will discuss the fundamental materials of electronic skins, the sensing principles of external stimuli, the macroscopic stimuli-responsive behaviors, and understanding the microscopic mechanism. Finally, the potential applications and future research directions of stimuli-responsive electronic skins are presented.

## 2.2 Materials for Electronic Skins

In the development of electronic skins, one of the most intense research interests is to create conductive materials that are intrinsically flexible, stretchable, and even self-healable, which are absent in conventional rigid electronics. Ranging from liquids to gels and to elastomers, many novel types of fundamental materials for electronic skins have been produced in the past decade, by combining conductive components and traditional soft materials with a wide and tunable spectrum of Young's moduli. Additionally, they also show some attractive features

*Flexible and Wearable Electronics for Smart Clothing*, First Edition.
Edited by Gang Wang, Chengyi Hou, and Hongzhi Wang.

such as shapeability, toughness, fatigue resistance, and stimuli responsivity. Here we will discuss the materials including liquid metals, hydrogels, ionogels, conductive elastomers, conductive polymers, and inorganic materials (Figure 2.1).

### 2.2.1 Liquid Metals

Liquid metals are the best examples for combining deformability and conductivity owing to their fluidic and conductive nature [1–3]. They are metal materials while in liquid phase at room temperature. For example, Hg as the most well-known liquid metal, has been extensively applied for electronic devices [4, 5]. It has a high electrical conductivity up to $10^6$ S/m and the melting point is as low as −39 °C. However, Hg is toxic and the high surface tension (487 mN/m) makes it incompatible with other materials and difficult to process. In recent years, Ga and its alloys, e.g. EGaIn (75% gallium and 25% indium) and Galinstan (68.5% gallium, 21.5% indium, and 10% tin), have been widely studied. They also have high electrical conductivities up to $10^6$ S/m [6]. Fortunately, they are less toxic and have negligible vapor pressure [7], and thus they are much safer than Hg. Moreover, upon exposure to air, an atomically thin oxide layer can form on the surfaces of Ga-based alloys [8, 9], which provides mechanical stability and makes it easy for patterning [6, 10].

Since the liquid metals have extremely low moduli and fluidic nature, they are usually printed as conductive channels and embedded in elastomer matrixes such as poly(dimethylsiloxane) (PDMS), Ecoflex, and polyacrylates to form intrinsically stretchable conductors, as shown in Figure 2.2a [3, 11–13]. The mechanical properties and stimuli-responsive behaviors of the whole devices are also determined by the elastomer matrixes while the liquid metals show negligible influence. One of the most important advantages of the electronic skins using liquid metals as conductive channels is the high and constant conductivity during deformation.

### 2.2.2 Hydrogels

Hydrogels that are composed of aqueous polymeric networks have structural similarity to biological tissues, and they have found great potential as skin-like

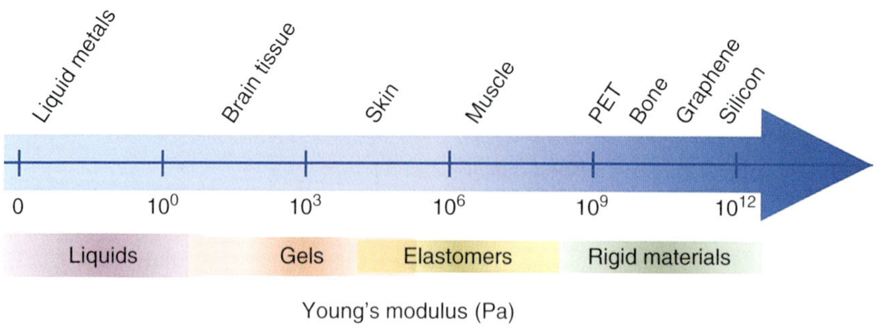


**Figure 2.1** Comparison of Young's modulus of different materials.

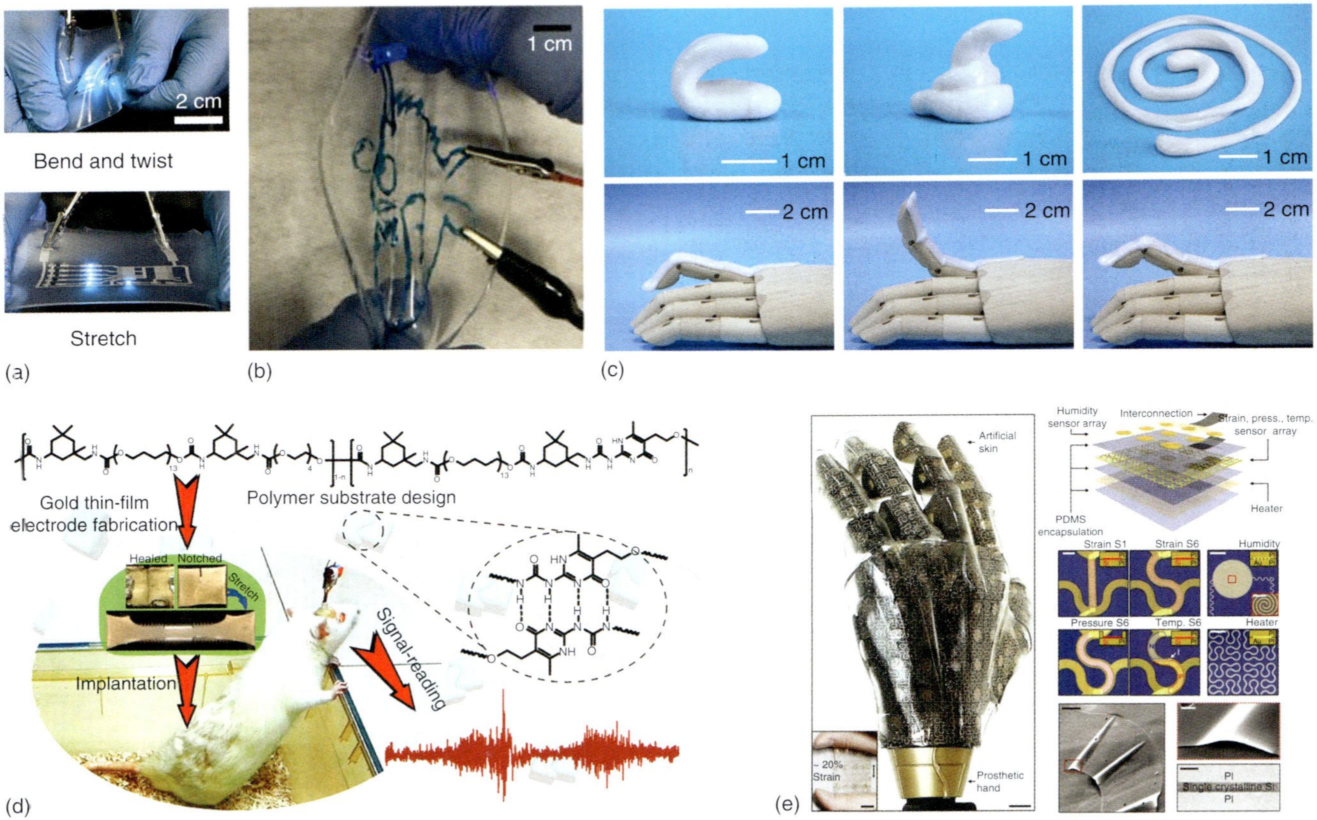


**Figure 2.2** Different materials for electronic skins. (a) Liquid metals, (b) hydrogel and elastomer hybrids, (c) physically cross-linked hydrogels. (d) Self-healing elastomers coated with Au. (e) Silicon nanoribbons encapsulated in PDMS. Source: (a) Tang et al. 2018 [3]. Reproduced with permission of Elsevier; (b) Liu et al. 2018 [29]. https://www.nature.com/articles/s41467-018-03269-x#rightslink. https://creativecommons.org/licenses/by/4.0/. Licensed under CCBY 4.0; (c) Lei et al. 2017 [23]. Copyright 2017, Reproduced with permission of John Wiley & Sons; (d) Yan et al. 2018 [45]. Copyright 2018, Reproduced with permission of American Chemical Society; (e) Kim et al. 2014 [51]. Copyright 2014, Reproduced with permission of Springer Nature.

iontronics to address several challenges encountered in traditional electronic skins [14–17]. In 2013, Researchers at Harvard University demonstrated that polyacrylamide hydrogels using inorganic ions as carriers can report electrical signals and respond to the applied electric fields [14]. After that, the hydrogels with intrinsic transparency and stretchability have gradually attracted research interests in the fields of iontronics and soft bioelectronics (Figure 2.2b). Electronic skins based on hydrogels are also referred to as ionic skins [15]. Owing to their excellent physicochemical properties such as biocompatibility, constant conductivity during deformation, and water-rich nature [14, 15, 18], they have shown potential applications as electroluminescent skins [19], ionic touch panels [20], triboelectric nanogenerators [21], liquid-crystal devices [22], etc.

Moreover, the hydrogels' physicochemical properties are widely tunable and there is great room for improvement. In addition to traditional chemically cross-linked polyacrylamide hydrogels with salts doping, some physically cross-linked hydrogels with better mechanical adaptability, self-healability, and stimuli-responsivity have been utilized in the hydrogel-based electronic skins (or called ionic skins) and exhibited more functionalities, as shown in Figure 2.2c. In our recent works, we improved the self-healability, mechanical adaptability, and sensitivity of ionic skins by developing a type of bioinspired mineral hydrogels [23], proposed a general strategy to transduce the volume phase transition behaviors of stimuli-responsive hydrogels into reliable electrical signals [24], and further optimized the mechanical properties based on a supramolecular polyelectrolyte hydrogel and extended the sensory capabilities using a bimodal design [25].

Some other functional hydrogels are also developed. For example, conductive hydrogels prepared from functionalized single-wall carbon nanotubes, biocompatible polyvinyl alcohol, and polydopamine have exhibited impressive self-healing ability and robust adhesiveness [26]. They can further assemble as healable, adhesive, and soft human-motion sensors. A double network hydrogel combining poly(acrylic acid) networks and conductive nanostructure of polypyrrole shows 3D printability, bulk conductivity, mechanical and electrical self-healing properties, ultrastretchability (1500%), and pressure sensitivity [27]. A composite hydrogel incorporating MXene ($Ti_3C_2T_x$) further achieves asymmetrical strain sensitivity coupled with viscous deformation and adds new dimensions to the sensing capability of hydrogel-based electronic skins.

Although many advantages and significant improvements of hydrogel-based electronic skins have been witnessed in the past five years, there remain some challenges. First of all, hydrogels containing free water are usually instable in ambient condition owing to the evaporation of free water [25, 28] and thus extra protection layers are required [29, 30]. However, this increases the complexity for the device fabrication process and may sacrifice the intrinsic properties of the hydrogels [25]. Therefore, it is important to prepare hydrogel-based electronic skins with much better stability in ambient condition. Secondly, compared with electronic skins using electronic conductors, the skins using ionic conductors show better mechanical adaptability but limited sensory capabilities, i.e. so far only strain, stress, and temperature sensations have been reported in hydrogel-based electronic skins. It is challenging to extend their sensory capabilities. Last but not least, many interesting properties and functionalities of hydrogels should be combined in their skin-like devices, such as multiple stimuli-responsive behaviors [31], shape memory, toughness [32], fatigue resistance [33], photochromic [34], antimicrobial and nonfouling properties, and freeze tolerance [35, 36]. They will further empower the hydrogels for a myriad of potential applications for soft bioelectronics.

### 2.2.3 Ionogels

In order to solve the problem of dehydration of hydrogels in open air, Suo and Chen proposed to use ionogels as nonvolatile, stretchable, transparent conductors [37]. They exhibit high transparency, low elastic modulus, high stretchability, as well as negligible hysteresis and degradation over cyclic deformation. Pan et al. used an ionogel with electrical double layer (EDL) capacitance design and improved the sensitivity and stability of an epidermal electronic. The device realizes extremely high sensitivity (up to 3.1 nF/kPa), a rapid mechanical response, excellent repeatability (>30 000 testing cycles), and thermal stability under temperatures ranging from 25 to 50 °C. Researchers at Chinese Academy of Sciences prepared a double-network ionogel with both good mechanical strength and high conductivity [38]. It can be used for a flexible electronic skin even under harsh conditions.

Ionogels with nonvolatile nature show their advantages when applied as electronic skins, but there are also some concerns. The ionic liquids contained in the ionogels are at risk of migration and are unstable for a long period and toxic toward human beings [39]. For the next generation of ionic conductors, it would be a rational design to discard dissimilar components (e.g. free water or small-molecule ionic liquids), use flexible macromolecular backbones to achieve mechanical adaptability, and transmit information relaying on free carriers.

### 2.2.4 Elastomers

Elastomers generally refer to elastic polymers with higher modulus than gels. Insulative elastomers with electronically conductive fillers embedded in the 3D elastic polymer networks have been extensively studied for electronic skins. Typically, silver nanowires [40], carbon nanotubes [41, 42], graphene [43], and metal nanoparticles [44] have been blended with elastomers. Alternatively, conductive layers can also be coated on the flexible elastomers [45]. The PDMS-based thin-film metal electrodes demonstrate their practical utility via both in vitro and in vivo measurements of electromyography signals (Figure 2.2d).

However, the electronically conductive filler particles usually have limited mobilities, and there is a trade-off between the stretchability and conductivity of the elastomers. Generally, the amount of the conductive fillers has to be high enough, since the electrical signals can only transmit within the conductive percolating networks; unfortunately, too much of fillers may sacrifice the stretchability of the composites. Therefore, the most significant drawbacks of the electronically conductive elastomers are the degradation of their percolating networks and the instability of the conductivities during deformation [46, 47].

### 2.2.5 Conductive Polymers

Conductive polymers can transmit electrical signals through conjugated structures. However, the conjugated backbones increase the rigidity of the macromolecular chains and thus make it difficult to process and usually sacrifice the transparency and stretchability [48]. In order to make such electronic materials mechanically compliant, conductive polymers are usually modified with additives such as plasticizers or by adjusting the chemical structures. Ionic liquids have been reported for acting as both plasticizers and conductivity-enhancing dopants in the case of poly(3,4-ethylenedioxythiophene):poly(styrenesulfonate) (PEDOT:PSS). The resulting skin-like electronics are highly durable under cyclic loading, with the conductivity maintained at 3600 S/cm even after 1000 cycles to 100% strain. The stretchability and conductivity are superior to even the best carbon nanotube- or silver nanowire-based conductive elastomers. The combination of the excellent mechanical and electrical properties allows it to serve as electronic skin or as interconnect for field-effect transistor arrays with high device density. Besides being modified by physical blending, chemical modification can also address the trade-off between electrical properties and mechanical compliance. In 2017, Bao and coworkers reported a type of highly stretchable polymeric semiconductor films through the nanoconfinement effect [49]. The nanoconfinement of polymers significantly reduces the moduli of the conjugated polymer and improves the stretchability of the polymeric semiconductors, without affecting their charge transport mobility. As a result, the semiconducting film can be stretched up to 100% strain and retain charge transport mobility comparable to that of amorphous silicon. Furthermore, the intrinsically stretchable electronic polymers can be integrated into a transistor array with unprecedented high device density [50]. The charge-carrier mobility of the transistor array varies only slightly (less than one order of magnitude) when subjected to 100% strain for 1000 cycles, and there is negligible current–voltage hysteresis. The transistor array enables the fabrication of a new generation of intrinsically stretchable electronic skins.

There are several demonstrations for the electronic skins based on conductive polymers, but the trade-off among stretchability, conductivity, processability, and transparency is still a persisting challenge. For future electronic skins and their large-scale applications, facile preparation and more sophisticated functionalities are highly recommended.

### 2.2.6 Inorganic Materials

Inorganic materials, such as silver, gold, platinum, graphene, and carbon nanotubes are intrinsically conductive, but their atomic packing structures also exhibit high moduli that are usually incompatible to biological tissues [51, 52]. The inorganic materials are not intrinsically stretchable. Instead, they need hybridization with soft platforms and require structural design for stretchable electronics, including wrinkle, island–bridge, origami, kirigami, textile, crack, etc. (Figure 2.2e) [53].

The wrinkle design developed for inorganic-material-based electronic skins is inspired from the fold or ridge commonly seen on fabrics or skin. In this design, inorganic thin films are bonded on the surface of elastomeric substrates with pre-expansion, and the wrinkle patterns of the inorganic thin films are formed when the pre-strain is released [4]. Both single and multiple wrinkle patterns can be produced, corresponding to different pre-strain modes. The island–bridge structure is probably the most widely studied for electronic skins. High stretchability can be achieved by accommodating functional components at the islands that are connected by the conductive bridges [52]. Upon strain, the conductive bridges deform both in plane and out of plane, whereas the islands remain still, and thus it enables the mechanical isolation of the functional components in the islands [54]. Origami and kirigami designs mainly focus on the innovation of the structures of the elastomeric substrates in the fabrication process [55]. Origami is an ancient art of paper folding with ori- meaning fold and -gami meaning paper, which transforms a planar sheet into complex 3D structures based on the defined hinge crease patterns. Therefore, it allows the 3D structure to be folded and unfolded simultaneously without inducing large strain in the functional areas of the electronics located at the facets [56, 57]. Similar to origami, kirigami is an art combining paper cutting and folding to create artistic models, which emerges as an effective method toward skin-like electronics [58–60]. Textiles are naturally indispensable and stretchable structures, which also enable structural optimization and function integration for the traditional inorganic materials. In addition, cracks with controlled distribution provide percolated conductive meshes in electronic skins [61–63].

Overall, inorganic materials with rigid nature can provide excellent electronic properties, and their stretchability can be achieved through various structural designs and hybridization with flexible substrates.

## 2.3 Stimuli-Responsive Behaviors

The current electronic skins can sense strain, stress, tactility, temperature, lights, humidity, and biomolecules. The sensing principles are based on different signal transduction to convert external stimuli into electrical signals, including resistance, current, capacitance, and voltage [47, 64]. The details are presented in this section 2.3. Besides, external stimuli can also induce changes in physicochemical properties, which render the electronic skins adaptable to different environments and mimic the dynamics of biological tissues. So far, the physicochemical property changes of these electronic skins have mainly focused on the self-healing, tunable optical appearances, actuations, and processability.

### 2.3.1 Electrical Signals in Response to Environmental Stimuli

The sensing mechanisms of the strain and stress sensors include piezoelectricity, capacitance, and piezoresistivity. The piezoelectric sensors mainly rely on piezoelectric materials, e.g. ZnO, which can convert mechanical energy into electrical energy. Compared with other types of sensors, piezoelectric sensors exhibit high sensitivity, ultrafast response, and low power consumption [65]. The capacitive sensor usually uses a sandwiched parallel-plate configuration containing two conductive layers and an intermediate dielectric layer. The capacitive strain sensors have simple and robust structures, feature high resolution and sensitivity, and allow drift-free sensing even when humidity and temperature change [15, 20]. The piezoresistive effect has also been widely used for strain and stress sensors. In general, the conductors including electronically and ionically conductive materials that can generate different resistance during deformation are suitable for piezoresistive sensors [66]. There are too many reports on such sensors or electronic skins, and we will not go into the details here.

Tactile sensors, on the one hand, can share the same mechanism with the stress sensors that convert the pressure-induced deformation into the piezoelective, capacitive, or piezoresistive signals. On the other hand, even without large deformation, they can also detect the gentle touch based on triboelectricity [67, 68], disturbance of a fringing electric field [69, 70], or electrical-double-layer capacitance [71]. Triboelectricity can be generated by charge transfer between two surfaces with the opposite charges. When two different charged surfaces are brought close to each other, current flows from the negative to the positive side, and the cyclic contact and separation between the two surfaces drive the output current flow back and forth. By incorporating triboelectric nanogenerators and field-effect transistor units into a polyimide substrate, Wang and coworkers developed a flexible tribotronic transistor array as an active tactile sensing system [68]. The drain–source current of each transistor can be individually modulated with the corresponding external contact, which induces a local electrostatic potential to serve as the conventional gate voltage. However, the most popular example based on tactile sensing is capacitive touch screens, which can detect touch based on the change of fringing capacitance [69, 70]. Different from the strain and pressure sensing capabilities relaying on the dimension changes, when a finger, or to be more general, any of a grounded conducting medium, approaches the tactile receptor, the fringing electric field (through the medium) is partially intercepted and shunted to ground by the finger and leads to a decrease in capacitance. The bare finger produces a decrease in the capacitance upon finger touching (no force applied) owing to the conductivity of the human body. Moreover, in many cases, a gloved finger can also be detected with less capacitance decrease. In addition, as a unique characteristic of the ionic-electronic contact, the interfacial capacitance lying on the EDL can report the tactile or very small pressure through ultrahigh EDL capacitance increase [71].

Besides detecting mechanical stimuli, electronic skins can also monitor physiological signals such as body temperature, blood glucose, and sweat levels. Temperature changes can be precisely measured by resistors based on the temperature coefficient of the resistance. Rogers and coworkers demonstrated a type of epidermal electronic skins for body temperature monitoring relaying on PIN diodes [72]. Recently, our group also reported a thermal sensor based on a 3D printed thermoresponsive hydrogel [24]. The skin-like capacitive sensor integrates two grid-structured hydrogel layers with a dielectric layer. Therefore, the parallel-plate capacitance depends on the conductive area in contact with the dielectric layer. When the temperature increases, the volume of the hydrogels gradually expands, which results in the increment of the capacitive area, and the capacitance is enhanced. This strategy employing capacitance–temperature relationship not only develops the first example of a thermoresponsive ionic skin, but also provides a facile method to realize electrical signal transduction on stimuli-responsive materials, i.e. the parallel-plate capacitance can report environmental changes based on the volume phase transition behaviors upon stimuli. Furthermore, on the basis of electrochemical analysis, there are many reports on the invasive or noninvasive methods for real-time blood glucose and sweat monitoring [73–75]. For example, people can detect the glucose level from body fluids using immobilized glucose oxidase layers [76]. Reacting the glucose in body fluids with oxygen generates gluconolactone and hydrogen peroxide. The $H_2O_2$ electrochemically reacts at the working electrode and generates $2H^+$, $O_2$, and $2e^-$; thus the glucose level can be extracted by reading the electron ($e^-$) concentration.

### 2.3.2 Stimuli-Responsive Self-healing

Self-healing is the ability to recover damage, which is essential for biological tissues. Incorporating the stimuli-responsive self-healing into electronic skins can enhance their durability and mimic the natural skins. Ideally, self-healing electronic skins need to "sense" a wound to initiate repair, similar to the active biological systems with a smart response. Some self-healing electronic skins can repair themselves autonomously, but here we mainly discuss the categories that require to be activated by external stimuli, including heat, light, redox, and electrical stimulus. The stimuli-induced self-healing electronic skins are mechanically stable in normal conditions. After cracking, the original polymer networks can revert to the monomeric, oligomeric, or un-cross-linked states upon external stimuli. The monomeric, oligomeric, or un-cross-linked units can reorganize to form new networks after the removal of stimuli and thus achieve self-healing (Figure 2.3) [77].

One important mechanism of heat-triggered self-healing is based on thermally activated Diels–Alder cycloaddition and retro-Diels–Alder reaction [78, 79]. A conjugated diene and a dienophile can spontaneously undergo

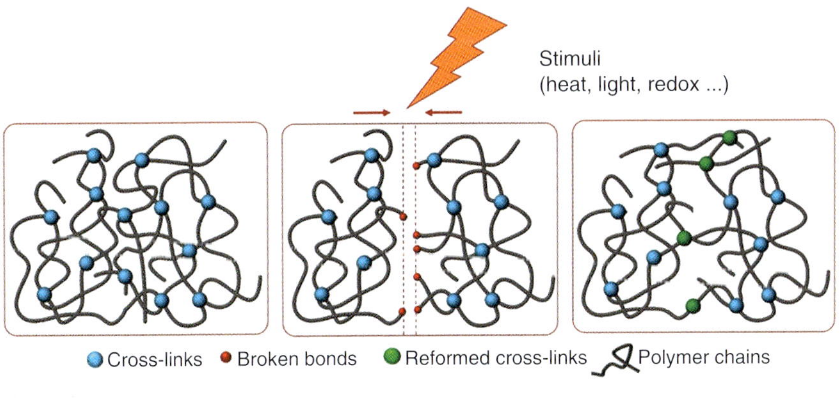


**Figure 2.3** Schematic illustration of the self-healing process upon stimuli.

[4+2] cycloaddition and form unsaturated six-membered rings with more energetically stable σ-bonds. Heating the damaged sample not only results in disconnection between diene and dienophile owing to the retro-Diels–Alder reaction, but also improves the mobility of disconnected groups. Subsequently, the disconnected groups reform adducts to heal the crack during the cooling process. For example, Zhu and coworkers prepared a flexible self-healing electronic skin by incorporating surface-modified $CaCu_3Ti_4O_{12}$ nanoparticles in self-healing polymer matrix based on dynamic Diels–Alder adducts [80]. In addition, a novel concept of vitrimers, which indicates permanent organic networks with glass-like fluidity, also draws great attention as heat-triggered self-healing materials and is successfully applied in electronic skins [81]. Some self-healing electronic skins show even more rapid healing rate and higher efficiency at a gentle heating temperature owing to the higher mobility of the polymer chains in physically cross-linked networks [42].

Light stimulation is another powerful and convenient method. The stimulation is easy to handle and can repair the targeted areas without the influence of the uncracked parts. Generally, light stimulation occurs at room temperature. The first example of photochemical cycloaddition for crack healing is based on a photo-cross-linkable monomer with three optically active cinnamoyl groups [82]. The crack healing can be achieved via re-photocycloaddition of the cinnamoyl groups upon exposing to UV lights. Besides, when some photothermal-responsive materials are exposed to lights, the heat induced by irradiation assists the self-healing and thus this process can be controlled by lights. For example, due to the photothermal effect and the quickly generated heat by near-infrared irradiation, the heat-induced dynamic disulfide exchange reactions result in a zipper-like self-healing process in epoxy networks [83].

Among the dynamic interactions responsible for self-healing, the host–guest interactions can be altered by redox potential. Harada and coworkers have reported a type of redox-responsive self-healing supramolecules that achieves a reversible sol–gel phase transition induced by the reversible host–guest interaction between side chains of poly(acrylic acid) modified with cyclodextrins and ferrocene [84]. As we have discussed above, such polyelectrolyte hydrogels can be applied for redox-responsive self-healing ionic skins.

Conductive polymer composites have an alternative self-healing route. In the conductive composites, it decreases the number of pathways for carrier transportation upon material damage, and thus increases the resistance at the site of damage. Local heating can be generated by applying an electrical field to the site of damage, which increases chain mobility and initiates the self-healing induced by heat [85].

### 2.3.3 Stimuli-Responsive Optical Appearances

Stimuli-responsive optical appearances are demonstrated in the phase transition systems. Suo and coworkers reported a skin-like display screen based on Fredericks transition behaviors of liquid crystals and ionically conductive hydrogels with high transmittance [22]. They encapsulated liquid crystals in a dielectric elastomer cell, which was then sandwiched between two hydrogel layers. This skin-like soft device can respond to applied electric fields. When voltage is off, the area covered by the cholesteric liquid crystals is opaque, and the area covered by the hydrogel and elastomer is transparent. When voltage is on, the cholesteric parts become transparent. It can maintain their electrooptical performance even under biaxial stretching. Besides, hydrogels with upper critical solution temperature (UCST) type or lower critical solution temperature (LCST) type behaviors also enable the ionic skins with tunable visual effect at different temperatures (Figure 2.4) [24].

On the basis of electroluminescent or electrochromic materials, the electronic skins exhibit significant color change upon electrical stimuli. Suo and coworkers reported a type of stretchable electroluminescent skins by sandwiching phosphor particles between two dielectric layers and two ionically conductive hydrogel layers [86]. Both of the dielectric and the ionic layers are transparent and stretchable, allowing the luminescence produced from the phosphor particles to come out, as well as the overall structure to deform. The ionically conductive layers connect to a power source through electronic conductors. This electroluminescent skin can display luminescence in several colors by using widely available phosphor particles and exhibit different brightness under the control of applied voltages, while Shepherd and coworkers reported a type of stretchable electroluminescent skin using a different five-layer structure [19]. It can change their shapes and colors under the control of applied voltages, and also provide sensory feedbacks for touch using capacitive signals. This artificial device is quite like octopuses, which combine a stretchable skin and color tuning organs to control both posture and color for visual communication and disguise. In addition, many types of skin-like electrochromic devices have emerged in the past decade [87]. Wang and coworkers reported a type of transparent stretchable electronic conductor prepared from the nanocomposites of sliver nanowires (AgNWs) and PDMS [88]. With the integration of electrothermal chromatic ink, this device displays good color-changing and actuating properties under a small voltage stimulus. Some other ion-based electrochromic skin-like devices also show great potential for writing boards and IoTs applications [87, 89].

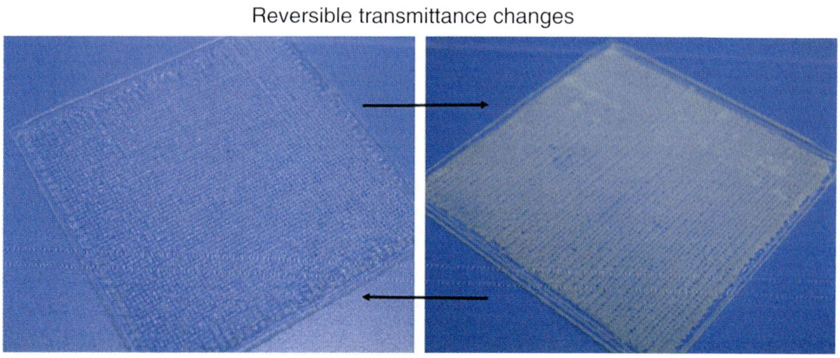


**Figure 2.4** Representative photographs of the stimuli-responsive optical appearances.

### 2.3.4 Stimuli-Responsive Actuations

Since abundant materials have been integrated into the electronic skins, actuations can be achieved through the stimuli of pressure, temperature, applied electric field, or magnetic field. For example, pressurized gas or liquid can easily drive the motions of fluidic materials (Figure 2.5) [90]; temperature changes generate the volume expansion or contraction of thermoresponsive components [91]; electric field can induce the electrostatic Maxwell stress, which pressurizes and displaces the dielectric elastomers [92]; and the magnetic field is also able to control the responsive behaviors of artificial skins with the addition of ferromagnetic domains [93]. Research on electronic skins combined with stimuli actuations is still in its infant stage, but it is quite important for the development of next-generation soft robotic devices with self-sensing capability.

### 2.3.5 Improved Processability Based on Stimuli-Responsive Behaviors

Traditional electronics based on rigid inorganic materials, e.g. silica or gold, usually require strict conditions to be processed into complex shapes, and thus hinder the low-cost and large-scale production of electronic skins. In the past two decades, the emergence of new materials has brought more possibilities for processing. Their mechanical properties change upon stimuli such as pressure, shear, and extrusion. In general, dilute dispersions with extremely low viscosity allow for inkjet printing; thick pastes can be processed by blade coating, screen printing, and 3D printing; freestanding gels with shear-thinning behavior are suitable for direct ink writing; and playdough-like materials can be shaped with molds or even by hand [94]. Besides, some stimuli-responsive materials can respond to thermal or photo changes. For example, polycaprolactone macromonomers with conductive fillers can be printed by commercial stereolithography (SLA) technology to

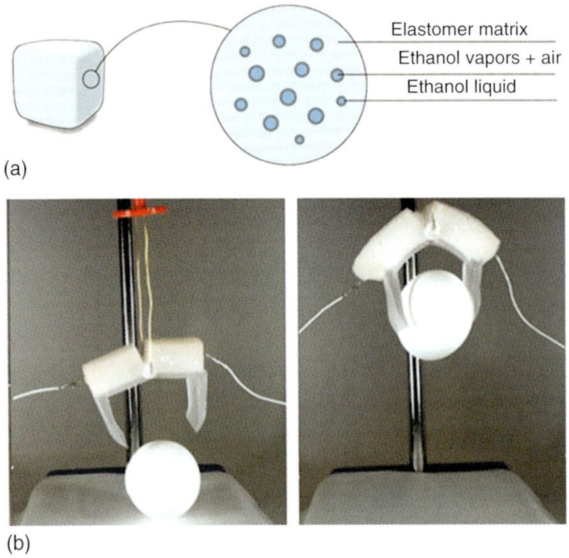


Figure 2.5 (a) Stimuli-responsive actuations driven by pressurized gas and liquid. Ethanol liquid is expanded to ethanol vapors upon heating. (b) The stimuli-responsive skin is attached to a gripper. (c) The gripper is actuated by the stimuli-responsive skin. Source: Meriyev et al. 2017 [90]. Copyright 2017, Reproduced with permission of Springer Nature.

Figure 2.6 A photograph of the 3D printing process by FDM.

form flexible devices; UV curable conductive elastomers can be used for digital light processing (DLP); and conductive inks with shear-thinning behavior can be processed by fused deposition modeling (FDM) as shown in Figure 2.6.

The core challenge for improving the processability is to change the materials' viscosity under different stimuli, in which low viscosity is suitable for processing into various patterns while high viscosity can retain the patterns after processing. From this perspective, hydrogels and elastomers exhibit different rheological behaviors upon different stimuli, and are promising candidates [95]. Walsh and Lewis used thermal curable elastomers to print a capacitive soft strain sensor with a multicore–shell fiber structure [95]; Suo and Vlassak reported a conductive heterogeneous hydrogel–elastomer system, which is fluid-like at high shear rates and solid-like after UV curing [96]; Chen and coworkers prepared a type of dual-responsive conductive elastomers that can be printed by FDM, and demonstrated that its shape memory behaviors can be triggered by sunshine [97]; we also developed a thermos-sensitive hydrogel showing UCST behaviors, which can be printed at higher temperature and solidify at room temperature [24].

## 2.4 Understanding the Mechanism of Stimuli-Responsive Materials Applied for Electronic Skins

For the next generation of electronic skins, it is important to implant multifunctions found in natural skins. For example, as a result of over millions of years of evolution, natural skins combine a variety of unique properties, including tunable visual effect, multiple sensations, and excellent mechanical properties [98, 99]. One of the reasons for such a wide scope of advanced properties is the intricate interplay, especially dynamic interactions and stimuli-responsive behaviors acting at multiple scales, which programs the biofunctions through some yet-to-be-understood mechanisms. Although people have devoted much effort into preparing electronic skins to imitate the mechanical properties and sensory capabilities, these artificial skins still struggle to meet the demands for more sophisticated stimuli-responsive behaviors [47, 98, 100].

To understand the mechanisms underlying the stimuli-responsive behaviors, two-dimensional correlation spectroscopy (2Dcos) provides detailed information on molecular interactions and can characterize nearly any kind of stimuli-responsive systems through spectral insights, e.g. time-dependent properties, thermo/photoresponsive behaviors, dynamic mechanical performance, and even catalytic mechanisms [101–105]. Compared with conventional 1D spectra, 2D spectra introduce Fourier transforms to resolve the spectral features with an additional dimension. Integration of 2D spectra of individual Fourier components results in the final overall synchronous and asynchronous spectra [101, 106]. 2Dcos not only significantly enhances spectral resolution and thus can distinguish complex responsive behaviors at different molecular levels but also provides comprehensive information on different dynamic interactions via the sequential order of spectral changes. In the 2Dcos analysis, synchronous spectra reflect the relative degree of in-phase response between two given wavenumbers, while asynchronous ones illustrate out-phase response similarly, that is, the synchronous spectra reflect simultaneous changes between two wavenumbers and the asynchronous spectra mainly enhance the spectral resolution. According to Noda's rule, when the cross-peaks of $\nu_1$ and $\nu_2$ ($\nu_1 > \nu_2$) in synchronous and asynchronous spectra have the same symbol (either positive or negative), then the change at peak $\nu_1$ has an earlier response than that at peak $\nu_2$, and vice versa [107].

During the past decades, 2Dcos analysis has been proved to be a powerful and robust analytical tool in studying diverse dynamic systems. From our perspective, it may also play an important role in the understanding of stimuli-responsive mechanisms and the design of electronic skin with more interesting stimuli-responsive behaviors. Here we show an example of the design and analysis of a stimuli-responsive electronic skin based on zwitterionic hydrogels, which astonishingly combine distinct stimuli-responsive behaviors and tunable mechanical properties.

In the past, it was widely known that one type of materials can be responsive to multiple stimuli, but it is difficult to combine distinct stimuli-responsive behaviors. For example, they can be either UCST type or LCST type, while a wide spectrum of customizable stimuli-responsive functionalities are rarely combined [108]. Here, we design a type of zwitterionic hydrogels that are composed of methacrylic acid (MAA) and 3-dimethyl (methacryloyloxyethyl) ammonium propane sulfonate (DMAPS) units and prepared by a facile one-step radical copolymerization. The zwitterionic copolymers' side groups provide potential hydrogen bonds, ionic interactions, and hydrophobic associations. The zwitterionic hydrogels achieve unusual distinct thermoresponsivity that can be switched from UCST to LCST types, and make the electronic skins exhibit tunable visual effect. To understand the mechanisms lying behind the distinct stimuli-responsive behaviors, the hydrogels with the mass ratios of 2 : 5 and 5 : 2 are taken as examples (Figure 2.7a,b), and 2Dcos analysis (Figure 2.7c,d) is performed [104, 105].

According to 2Dcos analysis, the specific locations of C=O groups in UCST- (MAA: DMAPS = 2 : 5) and LCST- (MAA: DMAPS = 5 : 2) type hydrogels are determined at 1736 and 1720 $cm^{-1}$, indicating that the C=O groups in UCST- and

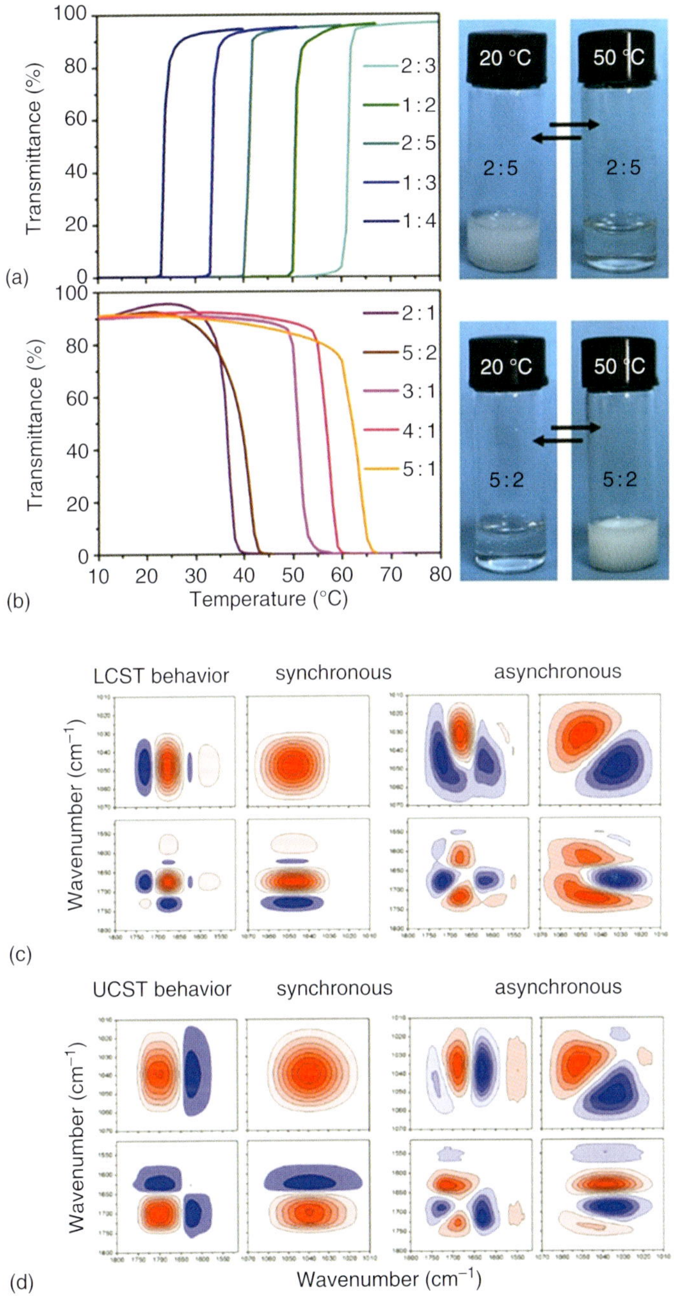


**Figure 2.7** (a) Transmittance vs. temperature for the UCST-type hydrogels with different monomer mass ratios. (b) Transmittance vs. temperature for the LCST-type hydrogels with different monomer mass ratios. (c) 2D synchronous and asynchronous spectra of a UCST-type hydrogel (MAA: DMAPS = 2 : 5) during heating from 25 to 55 °C. (d) 2D synchronous and asynchronous spectra of an LCST-type hydrogel (MAA: DMAPS = 5 : 2) during heating from 25 to 55 °C. Warm colors (red) are defined as positive intensities, and cool colors (blue) as negative ones. Source: Lei and Wu 2018 [109], Copyright 2018, Reproduced with permission of American Chemical Society.

LCST-type hydrogels are dominated by free (disassociated) carbonyl groups and associated carbonyl groups, respectively. On the basis of Noda's judging rule [107], in the UCST-type hydrogel, associated $SO_3^-$ groups (1049 $cm^{-1}$) respond prior to both disassociated $SO_3^-$ (1037 $cm^{-1}$) and disassociated C=O (1736 $cm^{-1}$) groups, while the disassociated $SO_3^-$ and C=O groups have no sequence order upon the temperature increasing (Figure 2.2a). Thus, the UCST phase transition is probably driven by the disassociation of $SO_3^-$ groups, that is, hydrogen bonds breaking between $SO_3^-$ groups and water as well as ionic interactions decreasing between $SO_3^-$ and quaternary ammonium groups. In contrast, the LCST-type hydrogel with more cross-links from the associated carbonyl groups shows a sequence order of the stretching modes during phase transition as: $\nu$(associated C=O) (1720 $cm^{-1}$) → $\nu_s$(associated $SO_3^-$) (1045 $cm^{-1}$) → $\nu_s$ (disassociated $SO_3^-$) (1031 $cm^{-1}$) (Figure 2.2b). It suggests that the associated C=O groups firstly respond to the temperature perturbation and lead to the disassociation of the $SO_3^-$ groups. Therefore, the LCST phase behavior is driven by the hydrogen bonds breaking and ionic disassociation of C=O groups, and involves stepwise response of the $SO_3^-$ groups accompanied with decreasing hydrogen bonds and ionic interactions. This conclusion is also in line with the relatively broad transition region observed in turbidity measurements of LCST-type hydrogels.

On the basis of 2Dcos analysis and turbidity results, the reasons for the distinct phase behaviors and tunable mechanical properties are ascribed to mutual influences of hydrogen bonds, ionic interactions, and hydrophobic associations: (i) since PDMAPS is reported to be a UCST-type polyzwitterion [110], with the addition of a small amount of MAA, hydrophobic associations imposed by $CH_3$ groups of MAA improve the moduli of hydrogels and tune their UCST phase behaviors [111]; (ii) with the increase of MAA, the hydrophobic associations result in microphase separation, and the hydrogel (MAA:DMAPS = 1 : 1) is opaque over a wide temperature range; (iii) when the MAA amount further increases, the associated C=O groups provide more dynamic cross-links for higher moduli and improve the solubility of the copolymers at low temperature, but as temperature increases, the disassociation of C=O groups leads to microphase separation and the LCST phase behavior is observed.

The comprehensive understanding of such complex and subtle dynamic interactions in hydrogels and other stimuli-responsive conductors is difficult to be achieved via other methods, and here we reveal the mechanism through their spectral features according to 2Dcos analysis. In future work, we will try to reveal the influence of dynamic interactions on mechanical performance by hyphenated rheometry-FTIR characterizations and the corresponding 2Dcos analysis.

## 2.5 Conclusion

This chapter aims to provide a glance at the stimuli-responsive behaviors of electronic skins. As we have shown, the emergence of new materials including liquid metals, nanocomposites, and ionic conductors enables the multifunctions and diversity of electronic skins. So far, electronic skins have achieved sensations toward many types of stimuli such as strain, stress, tactility, temperature, humidity, light, and biomolecules. Besides, the stimuli-responsive behaviors in electronic skins also include the changes in physicochemical properties, and some of the electronic skins that are adaptable to different environments and can mimic the dynamics of biological tissues are also desirable. For example, stimuli-responsive electronic skins are reported for their self-healing upon stimuli, tunable optical appearances, actuations, and improved processability. In addition, 2Dcos may be an effective tool to provide understanding of the microscopic mechanism behind the macroscopic stimuli-responsive behaviors designed for electronic skins. For the next-generation stimuli-responsive electronic skins, there are enormous challenges and opportunities. (i) For instance, distinguishing between a variety of different stimuli. It may be an effective approach to introduce multimodal strategy, but it is important to make sure that the different signals are independent of each other. (ii) Precisely tuning the sensitivity and guaranteeing the reliability of each sensation, especially when the mechanical adaptability of the electronic skins is enhanced, which is often at the cost of accuracy. (iii) Making the stimuli-responsive devices not only to show real-time electrical signal feedback but also to respond to environmental changes with human-like driving behaviors. These are huge challenges and also issues that deserve further exploration. At last, we believe that, with further understanding about the microscopic mechanism and optimization of bionic design, artificial devices with much more sophisticated functionalities can be achieved in the future.

## References


1 Dickey, M.D. (2017). *Adv. Mater.* 29: 1606425.
2 Wang, X. and Liu, J. (2016). *Micromachines* 7: 206.
3 Tang, L., Cheng, S., Zhang, L. et al. (2018). *iScience* 4: 302.
4 Holmlin, R.E., Haag, R., Chabinyc, M.L. et al. (2001). *J. Am. Chem. Soc.* 123: 5075.
5 Kasirga, T.S., Ertas, Y.N., and Bayindir, M. (2009). *Appl. Phys. Lett.* 95: 214102.
6 Khoshmanesh, K., Tang, S.-Y., Zhu, J.Y. et al. (2017). *Lab Chip* 17: 974.
7 Lu, Y., Hu, Q., Lin, Y. et al. (2015). *Nat. Commun.* 6: 10066.
8 Chiechi, R.C., Weiss, E.A., Dickey, M.D., and Whitesides, G.M. (2007). *Angew. Chem. Int. Ed.* 47: 142.
9 Kazem, N., Hellebrekers, T., and Majidi, C. (2017). *Adv. Mater.* 29: 1605985.
10 Joshipura, I.D., Ayers, H.R., Majidi, C., and Dickey, M.D. (2015). *J. Mater. Chem. C* 3: 3834.
11 Zhu, S., So, J.-H., Mays, R. et al. (2013). *Adv. Funct. Mater.* 23: 2308.
12 Kramer, R.K., Majidi, C., and Wood, R.J. (2013). *Adv. Funct. Mater.* 23: 5292.
13 Wanliang, S., Tong, L., and Carmel, M. (2013). *Smart Mater. Struct.* 22: 085005.
14 Keplinger, C., Sun, J.-Y., Foo, C.C. et al. (2013). *Science* 341: 984.

15 Sun, J.-Y., Keplinger, C., Whitesides, G.M., and Suo, Z. (2014). *Adv. Mater.* 26: 7608.
16 Bisri, S.Z., Shimizu, S., Nakano, M., and Iwasa, Y. (2017). *Adv. Mater.* 29: 1607054.
17 Yang, C. and Suo, Z. (2018). *Nat. Rev. Mater.* 3: 125.
18 Sun, T.L., Kurokawa, T., Kuroda, S. et al. (2013). *Nat. Mater.* 12: 932.
19 Larson, C., Peele, B., Li, S. et al. (2016). *Science* 351: 1071.
20 Kim, C.-C., Lee, H.-H., Oh, K.H., and Sun, J.-Y. (2016). *Science* 353: 682.
21 Pu, X., Liu, M., Chen, X. et al. (2017). *Sci. Adv.* 3: e1700015.
22 Yang, C.H., Zhou, S., Shian, S. et al. (2017). *Mater. Horiz.* 4: 1102.
23 Lei, Z., Wang, Q., Sun, S. et al. (2017). *Adv. Mater.* 29 https://doi.org/10.1002/adma.201700321 *Adv. Mater.* 29, 1700321.
24 Lei, Z., Wang, Q., and Wu, P. (2017). *Mater. Horiz.* 4: 694.
25 Lei, Z. and Wu, P. (2018). *Nat. Commun.* 9: 1134.
26 Liao, M., Wan, P., Wen, J. et al. (2017). *Adv. Funct. Mater.* 27: 1703852.
27 Darabi, M.A., Khosrozadeh, A., Mbeleck, R. et al. (2017). *Adv. Mater.* 29: 1700533.
28 Le Floch, P., Yao, X., Liu, Q. et al. (2017). *ACS Appl. Mater. Interfaces* 9: 25542.
29 Liu, Q., Nian, G., Yang, C. et al. (2018). *Nat. Commun.* 9: 846.
30 Yuk, H., Zhang, T., Parada, G.A. et al. (2016). *Nat. Commun.* 7: 12028.
31 Döring, A., Birnbaum, W., and Kuckling, D. (2013). *Chem. Soc. Rev.* 42: 7391.
32 Sun, J.-Y., Zhao, X., Illeperuma, W.R.K. et al. (2012). *Nature* 489: 133.
33 Zhang, E., Bai, R., Morelle, X.P., and Suo, Z. (2018). *Soft Matter* 14: 3563.
34 Liu, G., Zhang, Y.-M., Xu, X. et al. (2017). *Adv. Opt. Mater.* 5: 1700149.
35 Gao, H., Zhao, Z., Cai, Y. et al. (2017). *Nat. Commun.* 8: 15911.
36 Morelle, X.P., Illeperuma, W.R., Tian, K. et al. (2018). *Adv. Mater.* 30: 1801541.
37 Chen, B., Lu, J.J., Yang, C.H. et al. (2014). *ACS Appl. Mater. Interfaces* 6: 7840.
38 Ding, Y., Zhang, J., Chang, L. et al. (2017). *Adv. Mater.* 29: 1704253.
39 Zhao, D., Liao, Y., and Zhang, Z. (2007). *CLEAN – Soil, Air, Water* 35: 42.
40 Yun, S., Niu, X., Yu, Z. et al. (2012). *Adv. Mater.* 24: 1321.
41 Yang, H., Qi, D., Liu, Z. et al. (2016). *Adv. Mater.* 28: 9175.
42 Tee, B.C.K., Wang, C., Allen, R., and Bao, Z. (2012). *Nat. Nanotechnol.* 7: 825.
43 Boland, C.S., Khan, U., Ryan, G. et al. (2016). *Science* 354: 1257.
44 Jeon, J., Lee, H.-B.-R., and Bao, Z. (2013). *Adv. Mater.* 25: 850.
45 Yan, X., Liu, Z., Zhang, Q. et al. (2018). *J. Am. Chem. Soc.* 140: 5280.
46 Sekitani, T., Noguchi, Y., Hata, K. et al. (2008). *Science* 321: 1468.
47 Wang, J., Lin, M.-F., Park, S., and Lee, P.S. (2018). *Mater. Today* 21: 508.
48 Swager, T.M. (2017). *Macromolecules* 50: 4867.
49 Xu, J., Wang, S., Wang, G.-J.N. et al. (2017). *Science* 355: 59.
50 Wang, S., Xu, J., Wang, W. et al. (2018). *Nature* 555: 83.
51 Kim, J., Lee, M., Shim, H.J. et al. (2014). *Nat. Commun.* 5: 5747.
52 Kim, D.-H., Lu, N., Ma, R. et al. (2011). *Science* 333: 838.
53 Wang, C., Wang, C., Huang, Z., and Xu, S. (2018). *Adv. Mater.* 30: 1801368.

54 Li, T., Suo, Z., Lacour, S.P., and Wagner, S. (2005). *J. Mater. Res.* 20: 3274.
55 Tang, R., Huang, H., Tu, H. et al. (2014). *Appl. Phys. Lett.* 104: 083501.
56 Mahadevan, L. and Rica, S. (2005). *Science* 307: 1740.
57 Bertoldi, K., Vitelli, V., Christensen, J., and van Hecke, M. (2017). *Nat. Rev. Mater.* 2: 17066.
58 Zhang, Y., Yan, Z., Nan, K. et al. (2015). *Proc. Natl. Acad. Sci. U.S.A.* 112: 11757.
59 Tang, Y., Lin, G., Yang, S. et al. (2017). *Adv. Mater.* 29: 1604262.
60 Humood, M., Shi, Y., Han, M. et al. (2018). *Small* 14: 1703852.
61 Graz, I.M., Cotton, D.P.J., and Lacour, S.P. (2009). *Appl. Phys. Lett.* 94: 071902.
62 Lu, N., Wang, X., Suo, Z., and Vlassak, J. (2007). *Appl. Phys. Lett.* 91: 221909.
63 Lacour, S.P., Chan, D., Wagner, S. et al. (2006). *Appl. Phys. Lett.* 88: 204103.
64 Hua, Q., Sun, J., Liu, H. et al. (2018). *Nat. Commun.* 9: 244.
65 Gullapalli, H., Vemuru, V.S.M., Kumar, A. et al. (2010). *Small* 6: 1641.
66 Yao, H.-B., Ge, J., Wang, C.-F. et al. (2013). *Adv. Mater.* 25: 6692.
67 Ha, M., Park, J., Lee, Y., and Ko, H. (2015). *ACS Nano* 9: 3421.
68 Yang, Z.W., Pang, Y., Zhang, L. et al. (2016). *ACS Nano* 10: 10912.
69 Cotton, D.P.J., Graz, I.M., and Lacour, S.P. (2009). *IEEE Sens. J.* 9: 2008.
70 Yao, S. and Zhu, Y. (2014). *Nanoscale* 6: 2345.
71 Nie, B., Li, R., Cao, J. et al. (2015). *Adv. Mater.* 27: 6055.
72 Webb, R.C., Bonifas, A.P., Behnaz, A. et al. (2013). *Nat. Mater.* 12: 938.
73 Gao, Y., Zhang, Y., Wang, X. et al. (2017). *Sci. Adv.*: 3.
74 Chen, Y., Lu, S., Zhang, S. et al. (2017). *Sci. Adv.*: 3.
75 Choi, J., Ghaffari, R., Baker, L.B., and Rogers, J.A. (2018). *Sci. Adv.*: 4.
76 Miyashita, M., Ito, N., Ikeda, S. et al. (2009). *Biosens. Bioelectron.* 24: 1336.
77 Bergman, S.D. and Wudl, F. (2008). *J. Mater. Chem.* 18: 41.
78 Chen, X., Dam, M.A., Ono, K. et al. (2002). *Science* 295: 1698.
79 Chen, X., Wudl, F., Mal, A.K. et al. (2003). *Macromolecules* 36: 1802.
80 Yang, Y., Zhu, B., Yin, D. et al. (2015). *Nano Energy* 17: 1.
81 Deng, J., Kuang, X., Liu, R. et al. (2018). *Adv. Mater.* 30: 1705918.
82 Chung, C.-M., Roh, Y.-S., Cho, S.-Y., and Kim, J.-G. (2004). *Chem. Mater.* 16: 3982.
83 Guan, Q., Dai, Y., Yang, Y. et al. (2018). *Nano Energy* 51: 333.
84 Nakahata, M., Takashima, Y., Yamaguchi, H., and Harada, A. (2011). *Nat. Commun.* 2: 511.
85 Murphy, E.B. and Wudl, F. (2010). *Prog. Polym. Sci.* 35: 223.
86 Yang, C.H., Chen, B., Zhou, J. et al. (2016). *Adv. Mater.* 28: 4480.
87 Fang, H., Zheng, P., Ma, R. et al. (2018). *Mater. Horiz.* 5: 1000.
88 Fan, H., Li, K., Li, Q. et al. (2017). *J. Mater. Chem. C* 5: 9778.
89 Li, R., Li, K., Wang, G. et al. (2018). *ACS Nano* 12: 3759.
90 Miriyev, A., Stack, K., and Lipson, H. (2017). *Nat. Commun.* 8: 596.
91 Mirvakili, S.M. and Hunter, I.W. (2016). *Adv. Mater.* 29: 1604734.
92 Acome, E., Mitchell, S.K., Morrissey, T.G. et al. (2018). *Science* 359: 61.
93 Kim, Y., Yuk, H., Zhao, R. et al. (2018). *Nature* 558: 274.

94 Chiou, K., Byun, S., Kim, J., and Huang, J. (2018). *Proc. Natl. Acad. Sci. U.S.A.* 115: 5703.
95 Frutiger, A., Muth, J.T., Vogt, D.M. et al. (2015). *Adv. Mater.* 27: 2440.
96 Tian, K., Bae, J., Bakarich, S.E. et al. (2017). *Adv. Mater.* 29: 1604827.
97 Yang, H., Leow, W.R., Wang, T. et al. (2017). *Adv. Mater.* 29: 1701627.
98 Someya, T., Bao, Z., and Malliaras, G.G. (2016). *Nature* 540: 379.
99 Liu, Y., He, K., Chen, G. et al. (2017). *Chem. Rev.* 117: 12893.
100 Chortos, A., Liu, J., and Bao, Z. (2016). *Nat. Mater.* 15: 937.
101 Noda, I. (1993). *Appl. Spectrosc.* 47: 1329.
102 Noda, I. (2008). *J. Mol. Struct.* 883: 2.
103 Rosenfeld, D.E., Gengeliczki, Z., Smith, B.J. et al. (2011). *Science* 334: 634.
104 Noda, I. (2016). *J. Mol. Struct.* 1124: 3.
105 Sun, S.-t. and Wu, P.-y. (2017). *Chin. J. Polym. Sci.* 35: 700.
106 Noda, I. (2012). *Vib. Spectrosc.* 60: 146.
107 Noda, I. (2000). *Appl. Spectrosc.* 54: 994.
108 Zhang, Y.S. and Khademhosseini, A. (2017). *Science* 356: eaaf3627.
109 Lei, Z. and Wu, P. (2018). *ACS Nano* 12 12860.
110 Seuring, J. and Agarwal, S. (2012). *Macromol. Rapid Commun.* 33: 1898.
111 Xiaobo, H., Mohammad, V.-V., Jing, Z. et al. (2015). *Adv. Mater.* 27: 6899.

# 3

# Flexible Thermoelectrics and Thermoelectric Textiles

*Fei Jiao*

*Tianjin University, Department of Chemistry, School of Science, Weijin 92, Tianjin 300072, China*

## 3.1 Introduction

In order to perform long lifetime built-in electronic functions of smart clothing, power supply is extremely important. State-of-the-art power supply is mainly from batteries that need to be recharged or replaced after a period of time. As an alternative way, energy harvesting from the surrounding environment is attracting intense attention to provide sustainable and maintenance-free power supply.

To this end, several different energy sources have been considered for the development of wearable power generators, such as thermal, vibration, radio frequency, and solar energy. With the development of different harvesters and the decreasing power consumption of electronic devices, it is becoming possible to harvest energy from the environment and power the load efficiently. Among different energy-harvesting technologies, thermoelectricity is especially suitable for smart clothing due to the existence of temperature difference between the human body and the environment. Besides, thermoelectric generators (TEGs) also have the merits of long operating lifetime, no noise, no moving parts, and easy maintenance.

In this chapter, we will give an overview of the basics of thermoelectric materials and devices. Recent advances in the development of thermoelectric materials and devices that are suitable for smart clothing as well as the prospects and challenges are presented.

## 3.2 Thermoelectricity and Thermoelectric Materials

Thermoelectric phenomenon is a set of old physical effects about the coupling between heat transport and charge carrier transport of materials, which could be dated back to as early as the nineteenth century [1]. In 1821, Thomas Seebeck (1770–1781) discovered that for a closed loop formed by two different metals with two joints, a compass needle would be deflected if one of the joints was heated. The so-called Seebeck effect describes the voltage generated when a

*Flexible and Wearable Electronics for Smart Clothing*, First Edition.
Edited by Gang Wang, Chengyi Hou, and Hongzhi Wang.

temperature gradient is applied to a thermoelectric material (heat to electricity). The induced voltage difference (electromotive force, EMF) is proportional to the temperature difference; thus the Seebeck coefficient (also known as thermopower) is defined by

$$S = \Delta V / \Delta T \tag{3.1}$$

and its unit is often given in microvolts per kelvin (μV/K). To give a better description of thermoelectric phenomenon, two other thermoelectric effects should also be included. In contrast to Seebeck effect, the Peltier effect describes the presence of heating or cooling when a current is passed through the material (electricity to heat), while Thomson effect describes the release or absorption of heat of a homogeneous current-carrying conductor under a temperature gradient. Essentially, the three effects are the same and are related by the Kelvin relationship.

For the evaluation of a given thermoelectric material, a dimensionless figure of merit $zT$ is defined as

$$zT = \frac{\sigma S^2 T}{\kappa} \tag{3.2}$$

where $\sigma$, $S$, $\kappa$, and $T$ are the electrical conductivity, the Seebeck coefficient, the thermal conductivity, and the absolute temperature, respectively. The higher the $zT$ the better the thermoelectric conversion and practical applications [2, 3].

Although all materials show thermoelectric effects, ideal candidates need to meet the requirements of high Seebeck coefficient, high electrical conductivity, and low thermal conductivity at the same time. The three parameters are not independent and are affected by band structure and carrier concentration. Usually, thermal conductivity and electrical conductivity are directly related, whereas the Seebeck coefficient and electrical conductivity are inversely related. As shown in Figure 3.1, metals and plastics are not ideal thermoelectric materials, since the electrical conductivity and thermal conductivity for metals are both high, while they are both low for plastics. For thermoelectrics, the choice falls into the semiconductor region [4]. P-type semiconductors with excess holes exhibit a positive Seebeck coefficient, while n-type semiconductors with excess electrons show a negative Seebeck coefficient.

Because of the interrelated relationship of $\sigma$, $S$, and $\kappa$, it is difficult to develop thermoelectric materials with high $zT$ values. However, great achievements have been made with the efforts of decades, and many different material systems and strategies to enhance $zT$ values have been developed for thermoelectrics. Inorganic semiconductor thermoelectric materials are well known and established, such as SnSe, oxides, chalcogenides, clathrates, silicides, Zintl phases, half-Heusler, MgAgSb, and BiCuSeO compounds [1, 5, 6]. For inorganic thermoelectrics, $zT$ values of larger than 2 have been reported, and the commercialized ones have a figure of merit, $zT$, above unity [7–18]. Although the performance of organic thermoelectrics is inferior, they have noteworthy advantages of low cost, solution processability, lightweight, and mechanical flexibility. Organic semiconductors as well as different organic–inorganic composites have spurred a growing interest in thermoelectric community and the results are encouraging. Table 3.1 gives an exemplary summary of the state-of-the-art high-performance

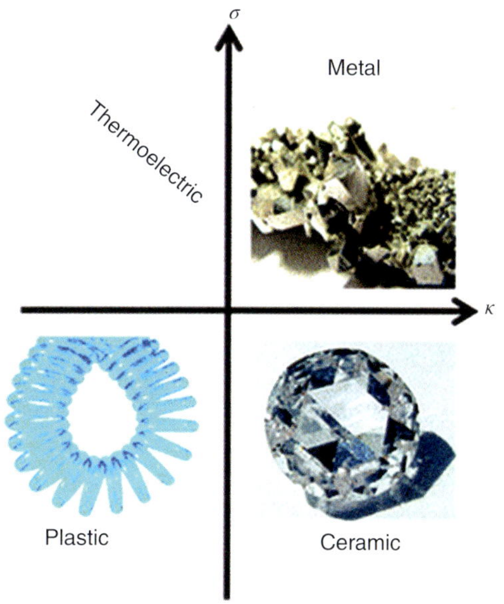


**Figure 3.1** Different materials classified with respect to their thermal $\kappa$ and electrical $\sigma$ conductivities. Source: Bubnova and Crispin 2012 [4]. Copyright 2012, Reprinted with permission of the Royal Society of Chemistry.

thermoelectric materials. For further information on the development of thermoelectric materials, the readers are referred to some outstanding reviews and the references therein [4, 31–38].

## 3.3 Thermoelectric Generators

TEGs are solid-state devices that can utilize the Seebeck effect to directly convert thermal energy to electricity. A TEG module is typically composed of a number of p and n legs that are connected electrically in series but thermally in parallel and sandwiched between two insulating ceramic layers, as shown in Figure 3.2 [22]. A direct electric current will flow in the circuit when there is a temperature difference between the two ends of the module. The conversion efficiency $\eta$ is one of the parameters used to gauge the performance of a TEG, and the maximum efficiency it can reach is defined by $zT$ and the working temperature:

$$\eta = \frac{\Delta T}{T_\mathrm{h}} \frac{\sqrt{1 + zT_\mathrm{m}} - 1}{\sqrt{1 + zT_\mathrm{m}} + \frac{T_\mathrm{c}}{T_\mathrm{H}}} \tag{3.3}$$

where $T_\mathrm{H}$, $T_\mathrm{C}$, $T_\mathrm{m}$, $\Delta T$, and $zT_\mathrm{m}$ are the temperature of the hot side, the temperature of the cold side, the average temperature of the device, the temperature difference, and the average $zT$ value, respectively.

For practical applications, the power output $P$ is more important than conversion efficiency. Assuming a TEG with m pairs of thermocouples, the power output of the generator is

$$P = \frac{(mS\Delta T)^2 R_\mathrm{L}}{(R_\mathrm{in} - R_\mathrm{L})^2 + 4R_\mathrm{L}} \tag{3.4}$$

**Table 3.1** Exemplary high-performance thermoelectric materials.

| Materials | Carrier type | $S$ (μV/K) | $\sigma$ (S/cm) | $zT$ | $T$ (K) | References |
|---|---|---|---|---|---|---|
| *Inorganic* | | | | | | |
| SnSe single crystal | P | 340 | 87 | 2.6 | 923 | [12] |
| PbTe–SrTe | P | 284 | 300 | 2.2 | 915 | [18] |
| $PbTe_{0.85}Se_{0.15}$–2% Na–4% SrTe | P | 250 | 360 | 2.3 | 923 | [8] |
| Na-doped $(PbTe)_{0.65}(PbS)_{0.25}(PbSe)_{0.1}$ | P | 280 | 140 | 2 | 850 | [9] |
| PbTe–PbS pseudobinary | P | 240 | 350 | 2.3 | 923 | [10] |
| K-doped $PbTe_{0.7}S_{0.3}$ | P | 300 | 160 | 2.2 | 923 | [13] |
| 2% Na-doped $(PbTe)_{0.86}(PbSe)_{0.07}(PbS)_{0.07}$ | P | 260 | 400 | 2 | 825 | [15] |
| $Cu_2Se$ | P | 170 | 300 | 2.3 | 400 | [17] |
| $Cu2S_{0.5}2Te_{0.48}$ | P | 200 | 200 | 2.1 | 1000 | [11] |
| $Ba_{0.3}In_{0.3}Co_4Sb_{12}$ with $BaFe_{12}O_{19}$ nanoparticles | N | 215 | 950 | 1.64 | 850 | [7] |
| $(Sr_{0.25}Ba_{0.25}Yb_{0.5})_{0.5}Co_4Sb_{12.5}$ | N | 121 | 791 | 1.9 | 835 | [14] |
| *Organic* | | | | | | |
| DMSO-mixed PEDOT:PSS (EG treatment) | P | 73 | 880 | 0.42 | 300 | [19] |
| Ar-plasma treated CNT paper | P | 350 | 9.9 | 0.4 | 670 | [20] |
| Anisotropic P3HT film | P | | | 0.1 | 365 | [21] |
| poly[$K_x$(Ni-ett)] pettet | N | 151.7 | 64 | 0.20 | 440 | [22] |
| Electrochemical prepared poly[$K_x$(Ni-ett)] film | N | 150 | 310 | 0.32 | 400 | [23] |
| *Composites* | | | | | | |
| PEDOT:PSS/Au NPs | P | 26.5 | 730 | 0.1 | 418 | [24] |
| PANi-CSA/SWNT hybrid | P | 65 | 769 | 0.12 | 300 | [25] |
| *n*-PETT/CNT/PVC hybrid film (MeOH treated) | P | 30.5 | 629.9 | 0.31 | 340 | [26] |
| Ni MW/PVDF nanocomposite film | N | 27 | 3000 | 0.15 | 380 | [27] |
| Cobaltocene-encapsulated SWNTs | N | 41.8 | 432 | 0.157 | 320 | [28] |
| CNT/$Ag_2Te$ buckypaper | N | 228 | 111 | 0.44 | 525 | [29] |
| PEDOT/CNT hybrid (TDAE treated) | N | 1200 | 7.3 | 0.5 | 300 | [30] |

where $S$, $R_L$, $R_{in}$, and $\Delta T$ are the Seebeck coefficient of the thermocouple, the load resistance, the inner resistance, and the temperature difference, respectively. From the equation it is obvious that the power output of the generator will reach a maximum when the load resistance and the inner resistance are equal.

As can be seen from Figure 3.3, the efficiency of TEGs is still much lower than other energy-conversion technologies because of the low $zT$ value [39]. However, TEGs have the merits of all solid state, no moving parts, no noise, no emissions, etc. Since the discovery of thermoelectricity, TEGs have found many applications

(a) (b)

**Figure 3.2** (a) Module structure and (b) photograph of a thermoelectric module consisting of 35 thermocouples. Source: Sun et al. 2012 [22]. Copyright 2012, Reprinted with permission of John Wiley & Sons.

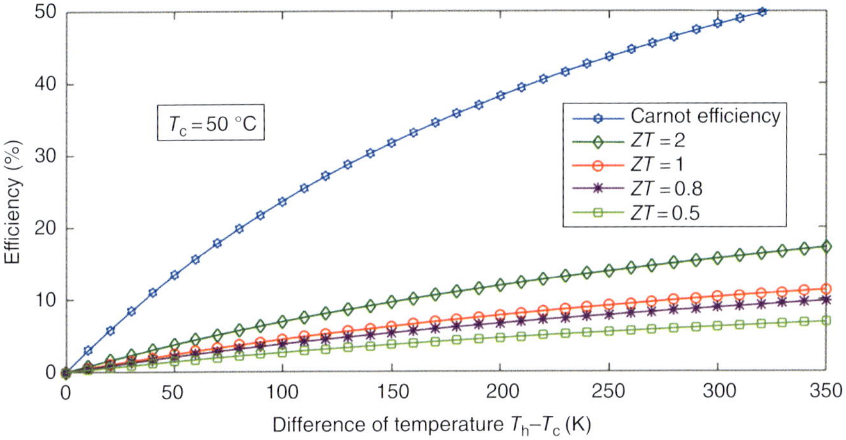


**Figure 3.3** Typical values of thermoelectric (TE) efficiency for different values of $zT$. Source: Champier 2017 [39]. Copyright 2017, Reprinted with permission of Elsevier.

where availability and reliability are more important than cost and efficiency, for example, industrial waste heat recovering, power supply for space missions, and waste-heat harvesting for automobiles. Broader applications of TEGs require the development of higher performance materials. At the same time, module configuration, and thermal and electrical contact also need to be optimized to achieve better efficiency and power output of TEGs.

## 3.4 Wearable Thermoelectric Generators for Smart Clothing

A living human body is a continuous heat source; take a male adult as an example; approximately 100–525 W of energy is released as heat depending on body condition and ambient temperature. Taking the skin area of about 2 m$^2$

**Table 3.2** Skin temperatures under ambient temperature (27 °C) and possible capabilities of thermoelectric harvesting power from different points of the human body.

| Skin location | Skin temperature (°C) | Power generation (mW) |
|---|---|---|
| Forehead | 35.2 | 2.3–27.6 |
| Chest | 34.4 | 3.1–36.6 |
| Arm | 33.2 | 1.7–20.2 |
| Forearm | 34.0 | 1.3–16.1 |
| Abdomen | 34.8 | 3.1–36.6 |
| Thigh | 33.7 | 2.4–28.8 |
| Foot | 30.4 | 2.1–25.2 |

into account, the heat flow varies from 50 to 262.5 $W/m^2$. Wearable TEGs are extremely suitable to harvest the thermal energy and can be used to power wireless and portable electronic devices and sensors. Considering a heat flow of 20 $mW/cm^2$ and a temperature difference of 12 °C, a wearable TEG can theoretically generate a maximum power of 180 $\mu W/cm^2$. However, the skin temperature of different points of the human body is not the same, as is the possible harvesting power by TEGs, as shown in Table 3.2. Although the power density of wearable TEGs is not very high, it is sufficient to be used as power supply for wearable electronics, with technological advances to enable ultralow power electronic devices and sensors [38, 40–43].

For the fabrication of wearable TEGs, vertical device structure is often used to maintain a high temperature difference between the human body and the environment. The devices can be built on thin plastic substrates or cotton/silk fibers and fabrics, which can provide them with enough flexibility to conform to the skin surface, or to be incorporated into garments. Great efforts have been made in recent years to develop efficient wearable TEGs; inorganic, organic, carbon-based and corresponding composites have been used as thermoelectric materials, and flexible TEGs can be deposited or printed as films on other flexible substrates. There are also some commercial products, such as wristwatch of Citizen Eco-Drive and Seiko Thermic [44]. Later in this chapter, we will give a brief summary of the advances in wearable thermoelectrics.

### 3.4.1 Flexible Thermoelectrics

#### 3.4.1.1 Inorganic Thermoelectric Materials Related

Since inorganic thermoelectric materials are well established and have higher $zT$ values, pioneer studies on wearable TEGs are based on inorganic materials. In 2004, the first wearable TEG was fabricated and equipped on the wrist for a body-powered wireless sensor [45]. However, the device was rigid and the thickness was more than 1 cm. For inorganic thermoelectric materials they are usually rigid, but they can be composited with other materials or be deposited onto flexible substrates for the fabrication of flexible TEGs.

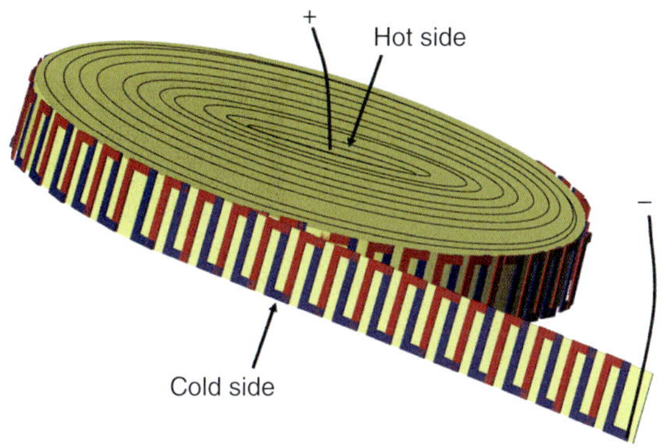


**Figure 3.4** Schematic of the coiled-up thermoelectric power generator. Source: Weber et al. 2006 [46]. Copyright 2006, Reprinted with permission of Elsevier.

In 2006, Weber and coworkers presented a coiled-up TEG using sputtered antimony and bismuth films on a thin polyimide foil, as shown in Figure 3.4 [46]. Using the human body as heat source, the coiled-up TEG could generate a high voltage and a large power output that was sufficient for a wristwatch. Furthermore, they developed antimony and $Bi_{0.85}Sb_{0.15}$-alloy pastes that could be fabricated by a low-cost screen-printing method.

Later in 2009, Carmo et al. developed a technique for the fabrication of a planar, wearable, and integrated circuit (IC)-compatible TEG with high $zT$ value [47]. The n-type bismuth telluride ($Bi_2Te_3$) and p-type antimony telluride ($Sb_2Te_3$) thin-film legs were deposited by the thermal co-evaporation of bismuth/antimony (Bi/Sb) and telluride (Te) on Kapton substrate. The measurement results showed that the $zT$ values were as high as 0.97 and 0.56 for the n and p legs at room temperature, respectively. Then, Francioso and coworkers experimentally validated a prototype of flexible TEG fabricated with $Bi_2Te_3$ and $Sb_2Te_3$ thin-film legs to be used for biometric sensors [48]. The prototype TEG, which integrated an array of 100 thermocouples, generated an open circuit voltage of 430 mV and an electrical power output of 32 nW at 40 K temperature difference (as shown in Figure 3.5), while in a real operation condition with a much lower temperature difference of only 15 K, the open circuit voltage and power output of the prototype TEG could still reach a relatively high value of 160 mV and 4.18 nW, respectively.

Cho and coworkers, for the first time, demonstrated a flexible TEG that was fabricated on a glass-fabric using a screen-printing technique [49]. Inorganic $Bi_2Te_3$ and $Sb_2Te_3$ pastes with high thermoelectric performances were screen-printed with subsequent annealing, resulting in a flexible and self-sustaining device without top and bottom substrates, as shown in Figure 3.6. At the same time, it was thin (~500 μm) and lightweight (0.13 g/cm$^2$). At a temperature difference of 50 K, the resulting TEG exhibited a relatively high output density of 3.8 mW/cm$^2$.

In their series of work, Dun et al. investigated the thermoelectric performances of flexible polyvinylidene fluoride (PVDF) composites with different inorganic thermoelectric materials [50–52]. The as-obtained p-type tellurium/PVDF fabric showed a high room temperature power factor (PF) of 45.8 W/(m K$^2$), while for the n-type $Se_2Bi_3$ nanoplate/PVDF and $Cu_{0.1}Bi_2Se_3$ nanoplate/PVDF composite fabrics, high power factors of 30 and 103 W/(m K$^2$) were reached at room

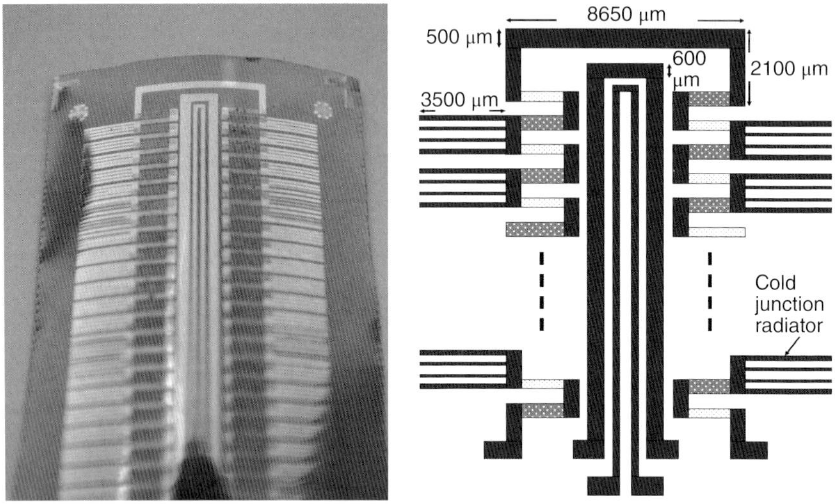


Figure 3.5 (a) Photograph of fabricated flexible TEG on Kapton and (b) its schematic. Source: Francioso et al. 2010 [48]. Copyright 2010, Reprinted with permission of IEEE.

temperature, respectively. Beyond their excellent thermoelectric properties, the $Cu_{0.1}Bi_2Se_3$ nanoplate/PVDF fabrics also showed high mechanical durability, with only a 13% decrease in thermoelectric performance after 5000 bending cycles.

#### 3.4.1.2 Organic Thermoelectric Materials Related

Compared with traditional inorganics, organic thermoelectric materials, especially conducting polymers such as polypyrroles (PPy), polythiophenes (PTh), polyanilines (PANi), poly(3,4-ethylenedioxythiophene) (PEDOT), and metal-coordination polymers, are more suitable for the fabrication of cheap, flexible, and lightweight TEGs. Since the state-of-the-art performance of organic thermoelectric materials is lagging far behind their inorganic counterparts, at the moment they are mainly in the research stage instead of on the market. Although research on flexible organic TEGs is now focused on the materials part instead of the devices, there are still a few examples of flexible organic related TEGs. Because of the topic of this chapter, we will only show the readers some preliminary results on flexible organic TEGs instead of going into details of the development of various organic thermoelectric material systems in recent years. If the readers want to learn more about the materials, there are some related reviews to be referred to [4, 35, 37, 53, 54].

When Te nanorods were encapsulated with the excellent polymer thermoelectric material poly(3,4-ethylenedioxythiophene):polystyrene sulfonate (PEDOT:PSS), the power factor of the composite could be tuned by $H_2SO_4$ solution treatment. The optimized power factor could reach as high as 284 $W/(m\,K^2)$, as reported by Bae et al. [55]. By printing the Te-PEDOT:PSS composites, a flexible planar-type TEG could be embedded in textiles. The resulting TEG consisting

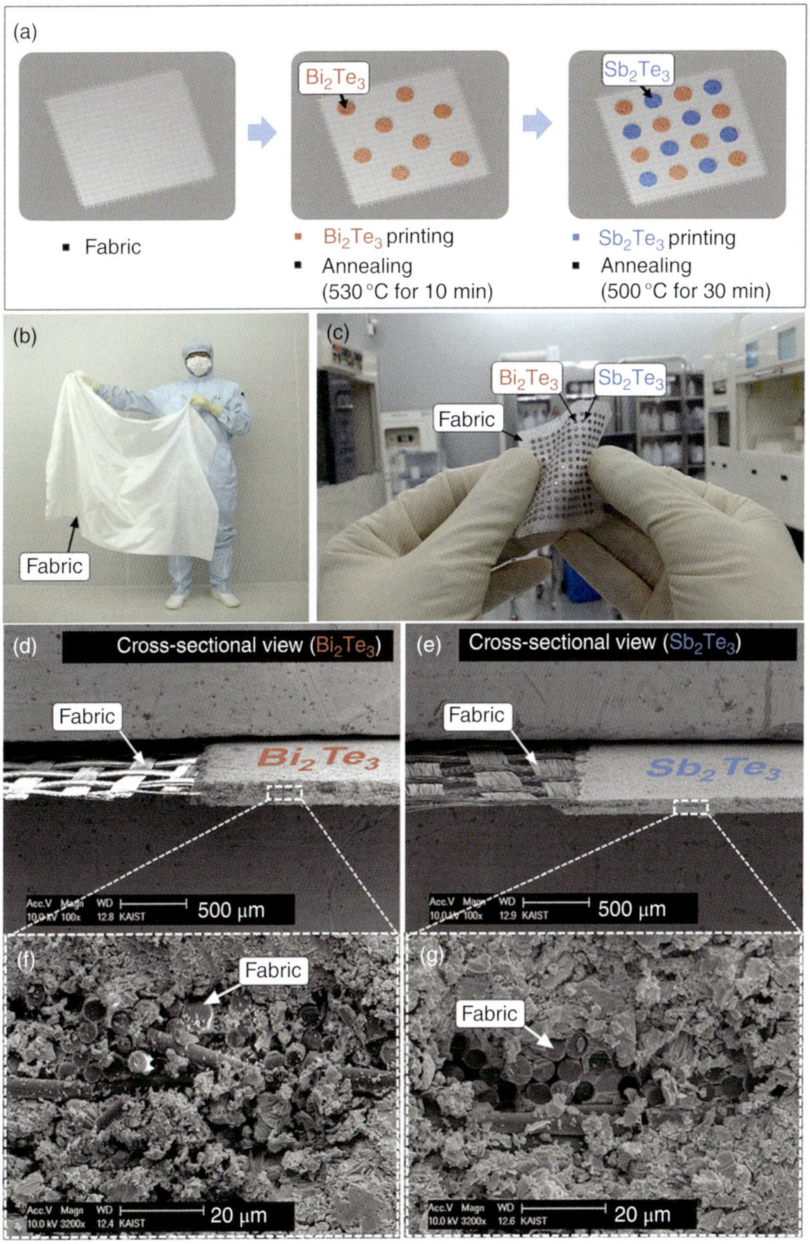


**Figure 3.6** (a) Schematic illustration of the fabrication process of the screen-printed $Bi_2Te_3$ and $Sb_2Te_3$ thick films on a glass fabric. (b) Photograph of a commercial glass fabric. (c) Photograph of $Bi_2Te_3$ and $Sb_2Te_3$ dots on a glass fabric of 40 mm × 40 mm. (d, e) Scanning electron microscope (SEM) images of the screen-printed $Bi_2Te_3$ and $Sb_2Te_3$ thick films on a glass fabric, respectively. (f, g) High magnification SEM images of cross-sectional view of the $Bi_2Te_3$ and $Sb_2Te_3$ thick films, respectively. Source: Kim et al. 2014 [49]. Copyright 2014, Reprinted with permission of the Royal Society of Chemistry.

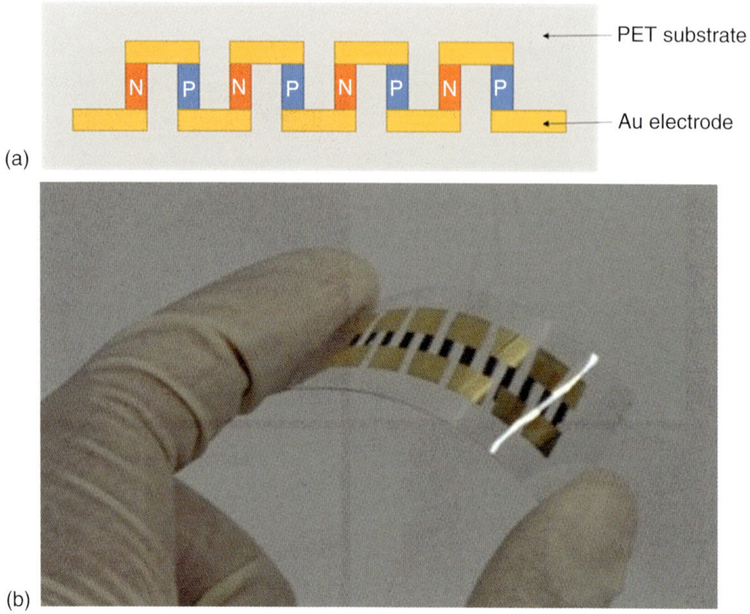


**Figure 3.7** (a) Schematic and (b) image of a six-leg inkjet-printed composite TE device on a flexible PET substrate. Source: Jiao et al. 2014 [56]. Copyright 2014, Reprinted with permission of the Royal Society of Chemistry.

of 32 legs in two rows showed a maximum power output of 10.59 nW when it was used to harvest human body heat with a temperature difference of 10 K.

Metal coordination polymers with 1,1,2,2-ethenetetrathiolate as linkers (poly(M-ett)) are excellent thermoelectric materials, but they are insoluble and infusible [22]. In 2014, Jiao et al. prepared both n- and p-type thermoelectric composites based on poly(M-ett) via a wet ball-milling method [56]. Through optimization of different parameters, a maximum PF at 300 K of 0.43 and 0.86 μW/(m K$^2$) was reached for poly[$K_x$(Ni-ett)]/PVDF (N) and poly[$Cu_x$(Cu-ett)]/PVDF (P), respectively. By inkjet printing, a flexible device on polyethylene terephthalate (PET) substrate with six thermocouples was fabricated, with a maximum power output of 45 nW at a temperature difference of 25 K (Figure 3.7).

Qu and coworkers demonstrated the concept of using poly(3-hexylthiophene) (P3HT) and silver paste coated cotton thread for the fabrication of flexible TEGs [57]. The coated thermoelectric cotton thread could be sewed on flexible fabrics and the temperature gradient was in the cross-plane direction. The maximum output power of the TEG with 13 p–n legs was 1.15 μW at a temperature gradient of 50 K.

#### 3.4.1.3 Carbon-Based Thermoelectric Materials Related

In recent years, carbon-based materials such as carbon nanotubes and graphene and their corresponding composites have been extensively investigated and showed promising thermoelectric properties [20, 58–60]. Thus, the possibility

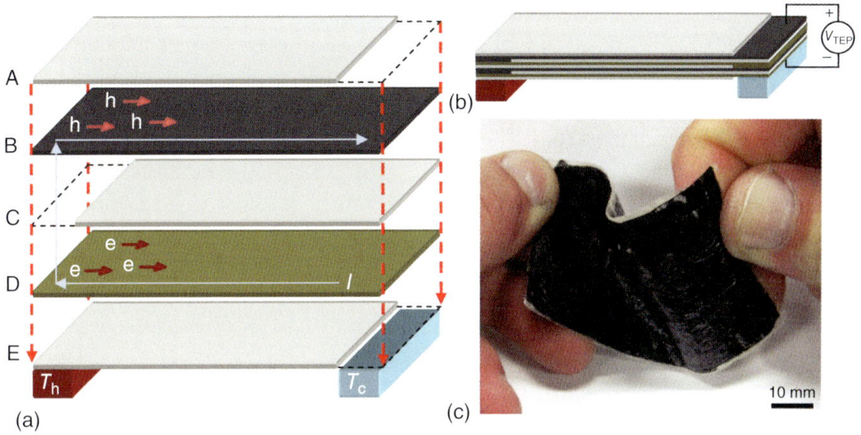


**Figure 3.8** (a) Layer arrangement for the multilayered fabric. Carbon nanotubes (CNTs)/PVDF conduction layers (B, D) are alternated between PVDF insulation layers (A, C, E). Every other conduction layer contains p-type CNTs (B), while the others contain n-type CNTs (D). The shorter insulating layers allow for alternating p/n junctions when the stack is pressed and heated to the polymer melting point of 450 K to bond the layers. Layers A–D can be repeated to reach the desired number of conduction layers N. When the film is exposed to a temperature gradient $\Delta T$, charge carriers (holes h, or electrons e) migrate from $T_h$ to $T_c$ resulting in a thermoelectric current $I$. (b) The resulting thermoelectric voltage $V_{TEP}$ can be read across the ends of the first and last conduction layers. (c) The thermoelectric fabric remains flexible and lightweight. Source: Hewitt et al. 2012 [62]. Copyright 2012, Reprinted with permission of American Chemical Society.

of carbon-based thermoelectric materials to be used as flexible TEGs has also been studied.

Carbon nanotube paper composites have been reported with promising p-type thermoelectric performance; however, it is difficult to obtain air-stable n-type properties. Doping with both polyethyleneimine (PEI) and sodium borohydride ($NaBH_4$), Yu et al. successfully changed the p-type carbon nanotube paper fabric to excellent n-type thermoelectric materials with a Seebeck coefficient as large as 80 μV/K [61]. Thermoelectric modules made of three p–n couples connected in series were fabricated, with an output voltage of 6 mV and power output of 25 nW, respectively, when a temperature gradient of 22 K was applied.

In 2012, Hewitt and coworkers demonstrated the possibility of individual composite films of multiwalled carbon nanotubes (MWCNTs)/polyvinylidene fluoride (PVDF) fabric as a realistic alternative for wearable TEGs [62]. The arrangement and photograph of a 72-layer fabric is shown in Figure 3.8. The thermoelectric voltage generated by the fabric TEG was the sum of the contributions of each layer, with an open circuit voltage of 26 mV at $\Delta T = 50$ K. At a matched internal resistance of 1270 Ω, the maximum power generation of 137 nW was reached.

Thermoelectric performance of bulk paper composite of single-walled carbon nanotubes with rubber polymer [63] and carbon nanotube forest/PEDOT:PSS composite [64] and their potential to be used as flexible TEGs were also investigated.

Besides carbon nanotubes and their composites, there are also some reports about flexible TEGs based on graphene. For example, Gao et al. fabricated reduced graphene oxide (rGO) and tellurium nanowire hybrid films via vacuum filtration, and when measured at 40 °C, an optimized power factor (PF) of 80 μW/(m $K^2$) was reached [65]. The same year, Guo and coworkers also reported a flexible and thermostable TEG based on porous all-graphene films, the output power of which was up to 0.43 μW at a temperature gradient of 75 K [66].

### 3.4.2 Fiber and Textile Related Thermoelectrics

Since there will be some problems in the integration of flexible TEGs, it is more attractive to fabricate the devices on fibers, threads, fabrics, and clothes. The low cost and safety (nontoxic) requirements wished for wearable TEGs may also be solved by this method. A remarkable progress on fiber-, fabric-, or textile-based flexible TEGs has been demonstrated thus far.

In 2016, Beeby and coworkers reported a flexible TEG based on bismuth tellurium ($Bi_{1.8}Te_{3.2}$) and antimony tellurium ($Sb_2Te_3$), which was fabricated by screen-printing technology on glass fiber textile substrate [67]. The formulations, curing conditions, and printing parameters of the screen-printable thermoelectric pastes were optimized to enable them to be successfully deposited on flexible textile substrates. With a temperature difference of 20 K, a maximum power output of 2 μW was reached.

Using commercial polyester fabric as substrate and commercial conducting polymer PEDOT:PSS as thermoelectric material, Du et al. fabricated a flexible TEG, which is shown in Figure 3.9 [68]. The authors demonstrated that PEDOT:PSS coated polyester fabric had very stable thermoelectric properties from 300 to 390 K. The output voltage of the fabric generator was 4.3 mV at a temperature difference of 75.2 K. Furthermore, the fabric TEG was also air-permeable, increasing its potential to be integrated with garments for

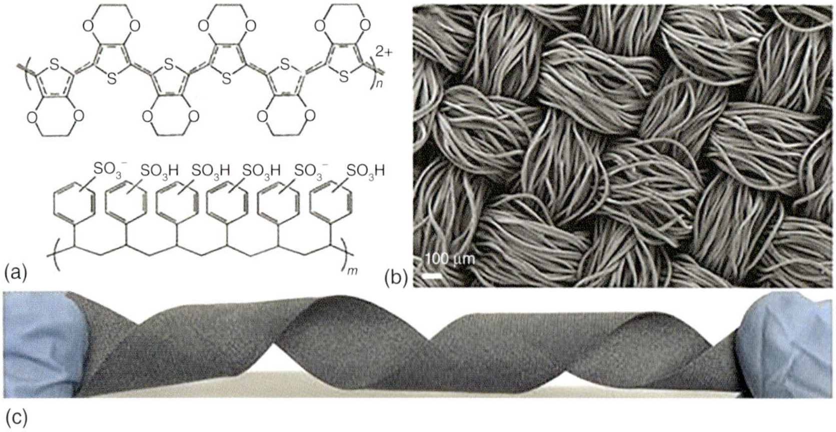


Figure 3.9 (a) Chemical structure of PEDOT:PSS, (b) SEM image, and (c) digital photograph of polyester fabric after coating treatment. Source: Du et al. 2015 [68]. Copyright 2015, Reprinted with permission of Springer Nature.

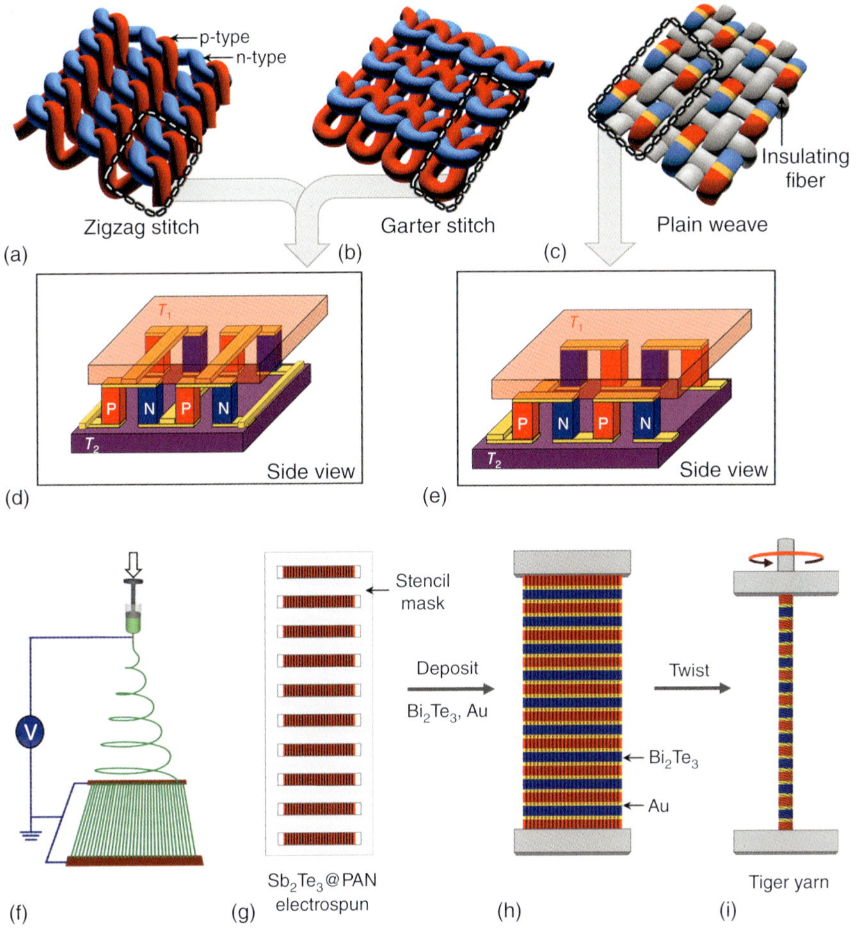


**Figure 3.10** Thermoelectric yarns and textiles and their fabrication. (a–c) Illustration of thermoelectric textiles utilizing a zigzag stitch, a garter stitch, and a plain weave, respectively. Red and blue yarns and yarn segments contain polymer nanofibers that are over coated with n-type and p-type thermoelectrics, respectively. In (c), n and p nanofiber segments are separated by gold-coated segments (yellow) and the grey yarns are insulating. (d) Illustration of the series–parallel electrical connectivity of thermoelectric elements for the highlighted areas of (a) and (b). (e) Illustration of the series electrical connectivity of thermoelectric elements for the highlighted area of (c). The yellow bars in (d) and (e) are electrical interconnects between the pictured n and p thermoelectric elements. (f) Illustration of the electrospinning process used to make PAN nanofiber sheets for conversion into a yarn. A PAN solution in the syringe is extruded through the metal needle. A high voltage supply is connected to the spinning tip and to two counter electrodes, so that highly aligned PAN nanofiber sheets are collected. (g, h) Illustration of the stencil-mask-based method used to provide alternating p-Au-n segments for tiger yarns by sequentially sputtering $Sb_2Te_3$-gold-$Bi_2Te_3$ on both sides of a PAN nanofiber sheet. (i) Illustration of the twist spinning process used to convert the nanofiber sheet into a tiger yarn. Source: Lee et al. 2016 [70]. Copyright 2016, Reprinted with permission of John Wiley & Sons.

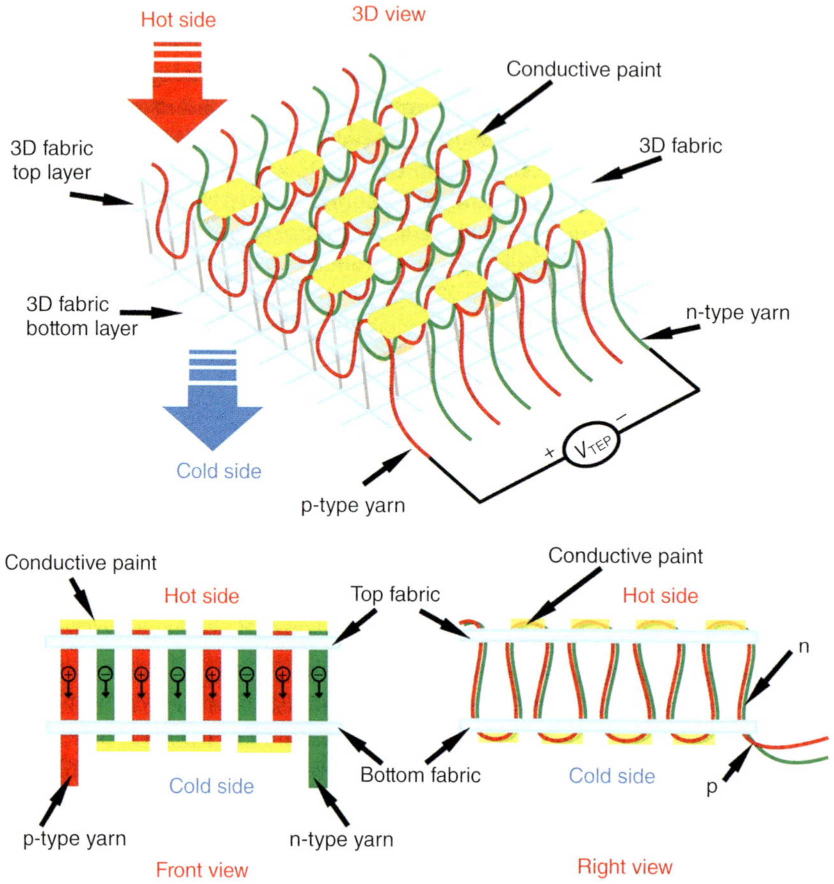


**Figure 3.11** Structure design of the 3D fabric TEG. Source: Wu and Hu 2017 [72]. Copyright 2017, Reprinted with permission of Institute of Physics.

the development of human comfortable wearable electronics. Later on, using PEDOT:PSS coated cotton fabric, a flexible TEG with five thermoelectric units was successfully fabricated. Under a temperature difference of 74.3 K, the achieved output voltage and power were 18.7 mV and 212.6 nW, respectively [69].

Lee et al. demonstrated the first example of flexible, woven, and knitted textile TEG based on electrospun polyacrylonitrile (PAN) nanofiber cores coated with $Bi_2Te_3$ and $Sb_2Te_3$ [70]. Nanofibers were twisted into flexible yarns, and the thermoelectric textiles were fabricated by knitting or weaving n- and p-type thermoelectric yarns, which were suitable for through-thickness thermoelectric power generation, as illustrated in Figure 3.10. For a temperature difference of 200 K, a high power output of up to 8.56 W/m$^2$ was reached in the textile thickness direction.

Later on, Choi et al. reported a flexible and ultralight TEG based on carbon nanotube yarn (CNTY) [71]. The as-prepared CNTY was alternatively doped with PEI and $FeCl_3$ into n and p type, respectively. At temperature differences of 5 and

40 K, the flexible all-carbon TEG with 60 pairs of CNTY reached a maximum power density of 10.85 and 697 μW/g, respectively.

To solve the wearing problem of flexible film TEGs, Wu and Hu designed a novel 3D fabric TEG structure [72]. Nonionic waterborne polyurethane (NWPU)/PEDOT:PSS/MWCNT composite was used for the p legs, while NWPU/N-doped MWCT composite was used for the n legs. The prototype TEG had a sandwich structure, which allowed power generation in the fabric thickness direction, as illustrated in Figure 3.11. The performance of this 3D fabric TEG was not high, however, which needs to be further enhanced in the future.

## 3.5 Prospects and Challenges

With the efforts of decades, great achievements have been made in flexible TEGs and thermoelectric textiles for smart clothing. Some of the reported results on wearable TEGs that are mentioned in this chapter are summarized in Table 3.3.

For the further development of wearable TEGs, there are many challenges. First, the thermoelectric performance of state-of-the-art TEGs is not very high, although it is efficient enough for some applications; the higher the $zT$ value, the wider their applications. Second, the integration of wearable TEGs with garments

**Table 3.3** Summary of some of the reported flexible TEGs.

| Device | Voltage output | Power (density) output | Temperature difference (K) | References |
|---|---|---|---|---|
| Coiled up device with antimony and bismuth | 1.6 V | 1.6 μW | 5 | [46] |
| Planar device with $Bi_2Te_3$ and $Sb_2Te_3$ | 430 mV | 32 nW | 40 | [47] |
| $Bi_2Te_3$ and $Sb_2Te_3$ on glass fabric | 90 mV | 3.8 mW/cm$^2$ | 50 | [49] |
| Planar device with Ag and PEDOT:PSS-Te | 12.75 mV | 10.59 nW | 10 | [55] |
| poly[$K_x$(Ni-ett)] and poly[$Cu_x$(Cu-ett)] composites on PET | 15 mV | 45 nW | 25 | [56] |
| Cotton fabric with P3HT and Ag | 7 mV | 1.15 μW | 50 | [57] |
| Carbon nanotube paper fabric | 6 mV | 25 nW | 22 | [61] |
| MWCNT/PVDF fabric | 26 mV | 137 nW | 50 | [62] |
| Porous all-graphene film | 22 mV | 0.43 μW | 75 | [66] |
| $Bi_{1.8}Te_{3.2}$ and $Sb_2Te_3$ on glass fiber fabric | | 2 μW | 20 | [67] |
| PEDOT:PSS coated cotton fabric | 18.7 mV | 212.6 nW | 74.3 | [69] |
| Textile with $Bi_2Te_3$/PAN and $Sb_2Te_3$/PAN fibers | | 856 μW/cm$^2$ | 200 | [70] |
| Device with carbon nanotube yarn | | 4.2 μW | 5 | [71] |
| 3D fabric with MWCT composite | 800 μV | 2.6 nW | 66 | [72] |

and possible factors that will affect their performance need to be studied. Third, as for the temperature differences for wearable TEGs are usually very small, it is extremely important to minimize the thermal resistances at the skin/TEG and TEG/ambient interfaces. Fourth, except for power generation, other functions of thermoelectric devices such as cooling and heating deserve to be investigated. Fifth, comfort, stability, as well as safety should be taken into consideration. Last but not the least, modeling for wearable TEGs needs to be developed at the same time to guide their future design.

## References

1 Rowe, D.M. (2005). *Thermoelectrics Handbook: Macro to Nano*. CRC Press.
2 Maciá, E. (2015). *Thermoelectric Materials: Advances and Applications*. CRC Press.
3 Kanatzidis, M.G., Hogan, T., and Mahanti, S. (2012). *Chemistry, Physics, and Materials Science of Thermoelectric Materials: Beyond Bismuth Telluride*. Springer Science & Business Media.
4 Bubnova, O. and Crispin, X. (2012). *Energy Environ. Sci.* 5: 9345.
5 Brand, O., Fedder, G.K., Hierold, C. et al. (2017). *Thermoelectric Energy Conversion: Basic Concepts and Device Applications*. Wiley.
6 Katz, H.E. and Poehler, T.O. (2016). *Innovative Thermoelectric Materials: Polymer, Nanostructure and Composite Thermoelectrics*. World Scientific.
7 Zhao, W., Liu, Z., Wei, P. et al. (2017). *Nat. Nanotechnol.* 12: 55.
8 Pei, Y., Tan, G., Feng, D. et al. (2017). *Adv. Energy Mater.* 7: 1601450.
9 Yamini, S.A., Mitchell, D.R.G., Gibbs, Z.M. et al. (2015). *Adv. Energy Mater.* 5: 1501047.
10 Wu, D., Zhao, L.-D., Tong, X. et al. (2015). *Energy Environ. Sci.* 8: 2056.
11 He, Y., Lu, P., Shi, X. et al. (2015). *Adv. Mater.* 27: 3639.
12 Zhao, L.-D., Lo, S.-H., Zhang, Y. et al. (2014). *Nature* 508: 373.
13 Wu, H.J., Zhao, L.D., Zheng, F.S. et al. (2014). *Nat. Commun.* 5: 4515.
14 Rogl, G., Grytsiv, A., Rogl, P. et al. (2014). *Acta Mater.* 63: 30.
15 Korkosz, R.J., Chasapis, T.C., Lo, S.-h. et al. (2014). *J. Am. Chem. Soc.* 136: 3225.
16 He, Y., Day, T., Zhang, T. et al. (2014). *Adv. Mater.* 26: 3974.
17 Liu, H., Yuan, X., Lu, P. et al. (2013). *Adv. Mater.* 25: 6607.
18 Biswas, K., He, J., Blum, I.D. et al. (2012). *Nature* 489: 414.
19 Kim, G.H., Shao, L., Zhang, K., and Pipe, K.P. (2013). *Nat. Mater.* 12: 719.
20 Zhao, W.Y., Fan, S.F., Xiao, N. et al. (2012). *Energy Environ. Sci.* 5: 5364.
21 Qu, S.Y., Yao, Q., Wang, L.M. et al. (2016). *NPG Asia Mater.* 8: e292.
22 Sun, Y.M., Sheng, P., Di, C.A. et al. (2012). *Adv. Mater.* 24: 932.
23 Sun, Y.H., Qiu, L., Tang, L.P. et al. (2016). *Adv. Mater.* 28: 3351.
24 Toshima, N. and Jiravanichanun, N. (2013). *J. Electron. Mater.* 42: 1882.
25 Yao, Q., Wang, Q., Wang, L.M., and Chen, L.D. (2014). *Energy Environ. Sci.* 7: 3801.
26 Toshima, N., Oshima, K., Anno, H. et al. (2015). *Adv. Mater.* 27: 2246.
27 Chen, Y.N., He, M.H., Liu, B. et al. (2017). *Adv. Mater.* 29: 1604752.

28 Fukumaru, T., Fujigaya, T., and Nakashima, N. (2015). *Sci. Rep.* 5: 7951.
29 Zhao, W., Tan, H.T., Tan, L.P. et al. (2014). *ACS Appl. Mater. Interfaces* 6: 4940.
30 Wang, H., Hsu, J.H., Yi, S.I. et al. (2015). *Adv. Mater.* 27: 6855.
31 Yang, L., Chen, Z.-G., Dargusch, M.S., and Zou, J. (2018). *Adv. Energy Mater.* 8: 1701797.
32 Ren, P., Liu, Y., He, J. et al. (2018). *Inorg. Chem. Front.* 5: 2380.
33 He, J. and Tritt, T.M. (2017). *Science* 357: eaak9997.
34 Tan, G., Zhao, L.-D., and Kanatzidis, M.G. (2016). *Chem. Rev.* 116: 12123.
35 Russ, B., Glaudell, A., Urban, J.J. et al. (2016). *Nat. Rev. Mater.* 1: 16050.
36 Chen, Y., Zhao, Y., and Liang, Z. (2015). *Energy Environ. Sci.* 8: 401.
37 Zhang, Q., Sun, Y., Xu, W., and Zhu, D. (2014). *Adv. Mater.* 26: 6829.
38 Petsagkourakis, I., Tybrandt, K., Crispin, X. et al. (2018). *Sci. Technol. Adv. Mater.* 19: 836.
39 Champier, D. (2017). *Energy Convers. Manage.* 140: 167.
40 Haras, M. and Skotnicki, T. (2018). *Nano Energy* 54: 461.
41 Du, Y., Xu, J., Paul, B., and Eklund, P. (2018). *Appl. Mater. Today* 12: 366.
42 Wu, H., Huang, Y., Xu, F. et al. (2016). *Adv. Mater.* 28: 9881.
43 Bahk, J.-H., Fang, H., Yazawa, K., and Shakouri, A. (2015). *J. Mater. Chem. C* 3: 10362.
44 Flipse, J., Bakker, F.L., Slachter, A. et al. (2012). *Nat. Nanotechnol.* 7: 166.
45 Leonov, V., Fiorini, P., Sedky, S., et al. (2005). Thermoelectric MEMS generators as a power supply for a body area network. *The 13th International Conference on Solid-State Sensors, Actuators and Microsystems*, 2005. Digest of Technical Papers. TRANSDUCERS '05 (5–9 June 2005).
46 Weber, J., Potje-Kamloth, K., Haase, F. et al. (2006). *Sens. Actuators, A* 132: 325.
47 Carmo, J.P., Goncalves, L.M., Wolffenbuttel, R.F., and Correia, J.H. (2010). *Sens. Actuators, A* 161: 199.
48 Francioso, L., De Pascali, C., Farella, I. et al. (2010). *2010 IEEE Sensors*: 747.
49 Kim, S.J., We, J.H., and Cho, B.J. (2014). *Energy Environ. Sci.* 7: 1959.
50 Dun, C., Hewitt, C.A., Huang, H. et al. (2015). *Nano Energy* 18: 306.
51 Dun, C., Hewitt, C.A., Huang, H. et al. (2015). *ACS Appl. Mater. Interfaces* 7: 7054.
52 Dun, C., Hewitt, C.A., Huang, H. et al. (2015). *Phys. Chem. Chem. Phys.* 17: 8591.
53 Culebras, M., Choi, K., and Cho, C. (2018). *Micromachines* 9: 30513632.
54 Chen, G., Xu, W., and Zhu, D. (2017). *J. Mater. Chem. C* 5: 4350.
55 Bae, E.J., Kang, Y.H., Jang, K.-S., and Cho, S.Y. (2016). *Sci. Rep.* 6: 18805.
56 Jiao, F., Di, C.-a., Sun, Y. et al. (2014). *Phil. Trans. R. Soc. A* 372: 20130008.
57 Qu, S., Chen, Y., Shi, W. et al. (2018). *Thin Solid Films* 667: 59.
58 MacLeod, B.A., Stanton, N.J., Gould, I.E. et al. (2017). *Energy Environ. Sci.* 10: 2168.
59 Kim, J.Y. and Grossman, J.C. (2015). *Nano Lett.* 15: 2830.
60 Blackburn, J.L., Ferguson, A.J., Cho, C., and Grunlan, J.C. (2018). *Adv. Mater.* 30: 1704386.
61 Yu, C., Murali, A., Choi, K., and Ryu, Y. (2012). *Energy Environ. Sci.* 5: 9481.

62 Hewitt, C.A., Kaiser, A.B., Roth, S. et al. (2012). *Nano Lett.* 12: 1307.
63 Nakano, M., Nonoguchi, Y., Nakashima, T., and Kawai, T. (2015). *Jpn. J. Appl. Phys.* 54: 04dn03.
64 Yusupov, K., Stumpf, S., You, S. et al. (2018). *Adv. Funct. Mater.* 28: 1801246.
65 Gao, J., Liu, C., Miao, L. et al. (2016). *RSC Adv.* 6: 31580.
66 Guo, Y., Mu, J., Hou, C. et al. (2016). *Carbon* 107: 146.
67 Cao, Z., Tudor, M.J., Torah, R.N., and Beeby, S.P. (2016). *IEEE Trans. Electron Devices* 63: 4024.
68 Du, Y., Cai, K., Chen, S. et al. (2015). *Sci. Rep.* 5: 6411.
69 Du, Y., Cai, K.F., Shen, S.Z. et al. (2017). *RSC Adv.* 7: 43737.
70 Lee, J.A., Aliev, A.E., Bykova, J.S. et al. (2016). *Adv. Mater.* 28: 5038.
71 Choi, J., Jung, Y., Yang, S.J. et al. (2017). *ACS Nano* 11: 7608.
72 Wu, Q. and Hu, J. (2017). *Smart Mater. Struct.* 26: 045037.

# Part II

# Energy

# 4

# Textile Triboelectric Nanogenerators for Energy Harvesting

*Xiong Pu*

*Chinese Academy of Sciences, Beijing Institute of Nanoenergy and Nanosystems, No. 30 Xueyuan Road, Haidian District, Beijing 100083, China*
*University of Chinese Academy of Sciences, School of Nanoscience and Technology, No. 19(A) Yuquan Road, Shijingshan District, Beijing 100049, China*

## 4.1 Introduction

Smart or electronic textile (e-textile), integrated with functional electronics, holds great promise for applications as fashionable/stylish wearable electronics [1]. It is appealing to integrate functional electronics/optoelectronics in textiles for fashion/aesthetic purposes, or to incorporate various sensors (temperature, pressure, heartbeat, pulse, etc.) in textiles for providing sports, health care, or medical monitoring/guidance to subjects wearing them. One crucial challenge is to find suitable power devices for the e-textile. It would be ideal if the power devices could be designed into fibers or fabrics, with minimum sacrifice of attributes of the original textiles, such as flexibility, comfort, lightweight, aesthetic appearance, and even washing capability. Today's portable/wearable electronics rely on high-energy batteries, which are, however, bulky, rigid, heavy, and hard to be incorporated into a highly flexible fiber or textile. Furthermore, safety concerns and frequent recharging are also limiting issues. Therefore, research has been conducted on textile-based power devices, which can harvest the environmental energies and generate electricity to power smart textile in real time.

The human body is rich in mechanical energies. For example, the footstep or arm movement of a 68 kg person can deliver about 67 and 60 W kinetic power, respectively [2]. Therefore, it is possible to design a fabric or fiber into a generator to convert human motion energies into electricity for powering various electronics integrated in smart textiles. Various mechanisms, such as piezoelectric nanogenerator (PENG), electromagnetic generator (EMG), and triboelectric nanogenerator (TENG), have been developed for generating electricity from different types of mechanical energy, including vibration [3], water flux/waves [4, 5], wind [6], sound [7], and human motions [8]. Conventional EMGs have been widely applied in today's industry. However, EMGs are not good candidates for powering smart textiles because of the rigid magnet used and the low voltage output for low-frequency mechanical motions (<5 Hz). The recently invented

*Flexible and Wearable Electronics for Smart Clothing*, First Edition.
Edited by Gang Wang, Chengyi Hou, and Hongzhi Wang.

TENGs, known for their high outputs at low frequencies and their versatility in structure designs and materials choices, are ideal power devices for smart textiles [9].

This chapter will overview the fundamentals of the TENG, followed by a summary of the progress in textile TENGs. The aspects of material choices, fabrication processes, design structures, and washing capabilities of state-of-the-art textile TENGs will be discussed. A brief introduction on self-charging power textiles with integrated energy-harvesting and storage functionalities will be provided. Lastly, a short summary and perspectives will be given.

## 4.2 Fundamentals of Triboelectric Nanogenerators (TENGs)

The nanogenerator was first proposed by Prof. Wang and Song in 2006 to harvest mechanical energies, by utilizing the piezoelectric and semiconducting properties of aligned ZnO nanowires (NWs) [10]. In 2012, Prof. Wang's group further invented the TENGs based on the two effects of contact electrification and electrostatic induction [11]. Since then, intensive research has been devoted to the TENG, leading to its more profound fundamental understandings, higher output performances, and wider range of applications.

### 4.2.1 Theoretical Origin of TENGs

The energy generation of TENG occurs in two steps. First, the triboelectrification or contact-electrification process will lead to the formation of immobile or static charges on the contact interfaces of two different dielectric materials. Second, the relative motions of the two electrified surfaces induce free charges flowing through external circuits, i.e. the electrostatic induction process [12]. A TENG with contact-separation mode has a typical structure as shown in Figure 4.1. Two different dielectric films display relative contact-separation motion in the vertical direction repeatedly. Electrostatic charges generated at the surfaces when they are in contact will induce an electric potential difference between the two electrodes at the backsides. This potential difference will drive free charges flowing through the external load. When the two surfaces display contact-separation motion repeatedly, AC output can be obtained. Inside the TENG, the variation of the polarization field (this filed is built by the surface static charges) causes the Maxwell's displacement current; outside the TENG, it is the capacitive conduction current that flows through the load [5].

The theoretical foundation of TENG is traced back to the Maxwell's displacement current. The Maxwell's displacement current can be described as [5, 13]

$$\boldsymbol{J}_{\mathrm{D}} = \frac{\partial \boldsymbol{D}}{\partial t} = \varepsilon \frac{\partial \boldsymbol{E}}{\partial t} + \frac{\partial \boldsymbol{P}_{\mathrm{s}}}{\partial t} \tag{4.1}$$

where $\boldsymbol{J}_{\mathrm{D}}$ is the displacement current, $\boldsymbol{D}$ is the electric displacement field, $\varepsilon$ is the permittivity of the dielectrics, $\boldsymbol{E}$ is the electric field, and $\boldsymbol{P}_{\mathrm{s}}$ is the polarization field induced by surface polarization charges. The first term refers to the

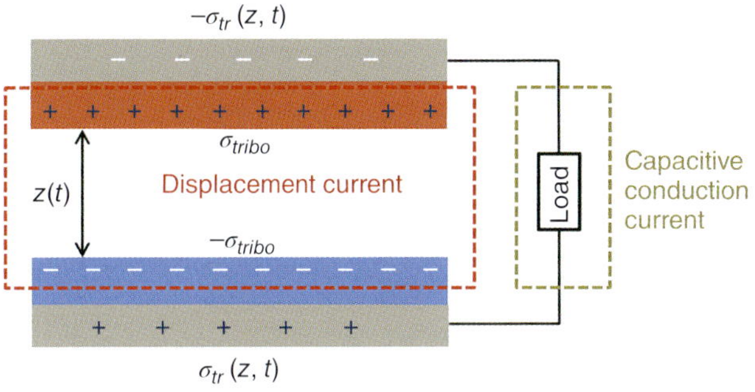


**Figure 4.1** Fundamentals of the TENG. Inside the TENG, it is the Maxwell's displacement current based on the time-variation of polarization field. The polarization field is generated by the triboelectrified surface charges. Outside the TENG, it is the capacitive conduction current through external load.

time-varying electric field and is the origin of electromagnetic waves, while the second term is induced by the time-varying surface polarization field and is the origin of nanogenerators. For the TENG, physical contact of two dielectric materials will generate opposite electrostatic charges at the surfaces. The electrostatic field built drives the electron to flow through the external circuit. The internal circuit of the TENG is related to the displacement current, and the output current in the external circuit is the capacitive conduction current [5]. A recent study demonstrated that the TENG has much better output performance than EMGs for harvesting mechanical energies at low frequency, especially <5 Hz [9]. This unique attribute of TENG makes it the most competitive technology for harvesting irregular, low-frequency mechanical energies.

### 4.2.2 Four Working Modes

Although various structures of TENGs have been designed for harvesting different types of mechanical energies, they are based on four basic working modes, as shown in Figure 4.2 [14]: (i) vertical contact-separation mode, as discussed in Figure 4.1. (ii) Lateral sliding mode – the dielectric films move in the lateral direction to have contact-sliding motions. The sliding can also be cylinder rotation or disc rotation. Interdigitated electrodes can be designed to improve the performances. (iii) Single-electrode mode – the backside electrode is grounded (or connected to a reference electrode). The electrified dielectric film will induce free charges flowing through the external load between the ground and the electrode. The relative motion can be either vertical or lateral. (iv) Freestanding mode – two similar electrodes are placed under a dielectric film with electrified static charges. The dielectric film moving at the top has the same size and shape as the one on the bottom electrode. The electrified static charges of top dielectric film can last for a long period of time, so the top dielectric film does not need to contact the bottom electrodes. The oscillating motions of the dielectric film induce asymmetric charge distribution in the two bottom electrodes, which will then drive the electron flowing back and forth through the external circuit. With this

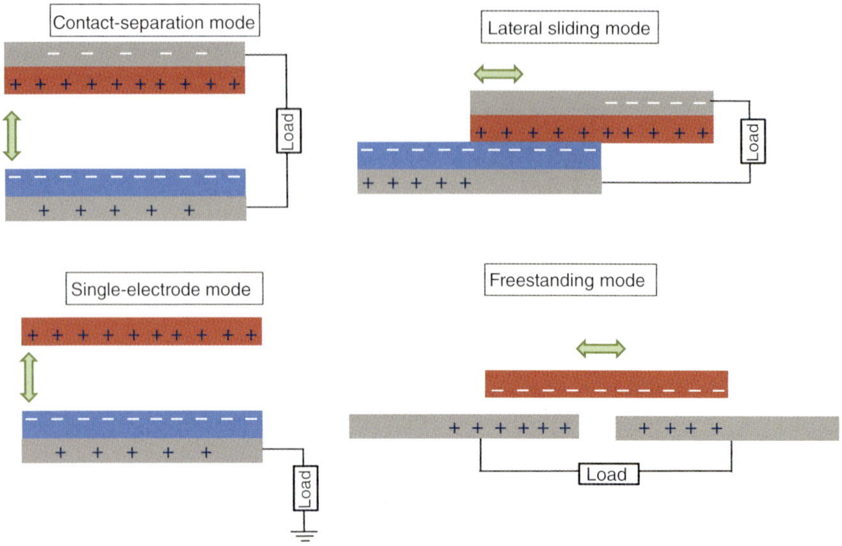


**Figure 4.2** The four fundamental working modes of TENGs.

mode, the potential mechanical wearing damage of materials can be avoided for better durability.

Based on the four basic working modes, a variety of structures have been developed for TENGs to harvest different types of mechanical energies: motional energies of rotation, sliding, pressing and bending, vibrational energies, water drops or waves, wind, acoustic sound, and heart beating, to name a few [15].

### 4.2.3 Materials for TENGs

Nearly all materials have the triboelectrification effect, such as metals, ceramics, polymers, or natural leathers, wools, cottons, cellulose, and silks. Therefore, the choice for electrifying materials is abundant. A recent report even used p- and n-type semiconductors for electrification [16]. But, insulating dielectric polymers are frequently used as electrifying materials, because the charges generated in conductive materials can be dissipated more easily. The tendency of two materials brought into contact to generate net positive or negative charges has been evaluated. The triboelectric series table first summarized was proposed by John Carl Wilcke in 1757, with some recent updates (see Table 4.1) [17]. The accurate and quantitative evaluation is hard as most of these materials are insulating. The larger the difference between the two contacting materials in triboelectric tables, the higher the amount of static charges generated. The capability of a material in capturing or donating electrons can be tuned by introducing surface functional molecules or grafting groups. Surface micro- or nanoscale morphologies also lead to higher effective contacting areas, as also the generated static charges.

As for the electrode, any conductive material can be utilized, such as metal, conductive polymer, carbonaceous materials, indium-doped tin oxides (ITOs), or even water and hydrogels [18]. Considering that the triboelectrification occurs

**Table 4.1** The triboelectric series table.

| | | | |
|---|---|---|---|
| Positive | Aniline-formol resin | Polyvinyl alcohol | |
| | Polyformaldehyde 1.3–1.4 | Polyester (Daeron) (PET) | |
| | Etylcellulose | Polyisobutylene | |
| | Potyamide 11 | Polyuretane flexible sponge | |
| | Potyamide 6–6 | Polyethylene terephthalate | |
| | Melanime formol | Polyvinyl butyral | |
| | Wool, knitted | Formo-phenolique, hardened | |
| | Silk, woven | Polychlorobutadiene | |
| | Polyethylene glycol succinate | Butadiene-acrylonitrile copolymer | Negative |
| | Cellulose | Nature rubber | |
| | Cellulose acetate | Polyacrilonitrile | |
| | Polyethylene glycol adipate | Acrylonitrile-vinyl chloride | |
| | Polydiallyl phthalate | Polybisphenol carbonate | |
| | Cellulose (regenerated) sponge | Polychloroether | |
| | Cotton, woven | Polyvinylidine chloride (Saran) | |
| | Polyurethane elastomer | Poly(2,6-dimethyl polyphenyleneoxide) | |
| | Styrene-acrylonitrile copolymer | Polystyrene | |
| | Styrene-butadiene copolymer | Polyethylene | |
| | Wood | Polypropylene | |
| | Hard rubber | Polydiphenyl propane carbonate | |
| | Acetate, rayon | Polyimide (Kapton) | |
| | Potymethyl methacrylate (Lucite) | Polyethylene terephtalate | |
| | Polyvinyl alcohol | Polyvinyl chloride (PVC) | |
| | (continued) | Polytrifluorochloroethylene | |
| | | Polytetrafluoroethylene (Teflon) | |

Source: Wang 2013 [17]. Copyright 2013, Reproduced with permission of American Chemical Society.

only in the contacting interfaces, both the electrification layer and the electrode layer can be very thin. By properly choosing the electrification and electrode materials, the TENG can be thin, lightweight, self-healable, flexible, stretchable, or transparent [18, 19].

## 4.3 Progresses in Textile TENGs

Recent progresses in TENG for various applications have been summarized in some review papers [13, 15, 17, 20]. Among these applications, TENGs for smart textiles are of special interest, since they can offer electric power at minimum sacrifices of the comfort of textiles, compared to all other energy solutions, such as solar cells, electrochemical energy storage, EMGs, and thermoelectrics. Many reports indicate that TENGs have been realized in polymeric fibers, yarns, or fabrics.

### 4.3.1 Materials for Textile TENGs

Triboelectrification is nearly a universal phenomenon in all kinds of materials. Most of the materials used in textiles are excellent candidates for generating tribo-electrostatic charges. According to the tribo-series table, nylon, silk, wool, leather, cellulose, and cotton tend to donate electrons to obtain net positive static surface charges. Other artificial fiber materials tend to capture electrons to achieve net negative static surface charges, such as polyester, acrylic fibers, and polyurethane fibers. To achieve a stronger triboelectric effect, two materials with larger difference in tribo-series should be coupled to have contact-separation motions. Therefore, the choices of electrification dielectric materials for fabricating a textile TENG are abundant.

The challenge comes from the electrical conductivity and comfort required of the fiber/textile electrodes. Most natural and artificial fibers are insulating. Metal fibers can be directly used as electrodes in a textile TENG; for example, several studies reported textile TENGs utilizing stainless steel (SS) fiber electrodes. [21, 22]. Chen et al. simply attached Cu foil on a dielectric textile with adhesive tape for use as electrodes [23]. Nevertheless, the metal is heavy, rigid, and uncomfortable. Carbon fibers, spun carbon nanotube (CNT) fibers, or spun reduced graphene oxide (rGO) fibers are better choices in terms of flexibility and lightweight, but the cost is also higher [24]. Alternatively, methods have also been developed to coat conductive materials on common insulating fibers/textiles. CNTs, rGO, or Ag nanowires are frequently reported to be coated on common fibers/textiles for use as electrodes in a textile TENG [25, 26]. The percolated networks provide pathways for electron transfers, but the conductivity is limited. Coating a thin metal film on polymeric fiber surfaces is another approach. As fabrics feature a wavy surface and porous morphology, the metal layer produced by most metal coating methods (such as e-beam evaporation, thermal evaporation, and sputtering) is not continuous and the treated fabric is not conductive. Electroless deposition in wet chemical bath can produce a continuous and conformal metal film (Al, Cu, Ni, etc.) on cylinder fibers and porous fabrics [27, 28]. The thin metal-coated fibers/fabrics are ideal electrodes for textile TENGs, considering that high conductivity, lightweight, flexibility, and low-cost are simultaneously achieved.

### 4.3.2 Fabrication Processes for Textile TENGs

The fabrication of textiles has a long history, tracking back to ancient times. The modern industrial revolution also started with textile industry. To achieve a comfortable and high-performance textile TENG, the fundamentals of fabric manufacturing should be considered. Nearly all the textile manufacturing techniques can be applied to a textile TENG. By far, textile TENGs fabricated by weaving, knitting, and sewing processes have all been reported.

Weaving is the most common technique for textile manufacturing, and many reported textile TENGs are fabricated through this process [26–29]. Most reports simply utilize the structure of plain weaving. Chen et al. tried plain, satin, twill weaving structures and their combinations showed that their output performances could be different [23]. Other than 2D weaving, 3D weaved textile TENG has also been reported [30, 31].

Knitted textile TENGs have also been reported. Kwak et al. demonstrated the effect of knitting structures on the performances of a stretchable textile TENG [32]. The authors fabricated textile TENGs with weft-knitted plain, double, and rib structures, and showed that the rib-structured textile TENG has the largest effective contacting area and thus highest output voltage. 3D knitting has also been applied to textile TENG [31]. 3D knitted fabrics are produced by the spatial formation of 2D warp- or weft-knitted fabrics.

The treated fiber electrode can also be sewn into an existing fabric to produce a textile TENG. A stretchable textile TENG has been reported by using the sewing process [22]. A silicone-coated stainless steel fiber was sewn in a stretchable fabric with a serpentine configuration. Then, the final fabric can still be stretched up to about ~100% tensile strain. This textile TENG worked on a single-electrode mode to generate electricity. The peak values of voltage and current reached 200 V and 200 μA, respectively.

### 4.3.3 Structures of Textile TENGs

#### 4.3.3.1 1D Fiber TENGs

1D fibers or yarns are building blocks of a textile or fabric. Therefore, TENGs with 1D configuration have been reported, so that electricity can be generated when stretching, bending, or pressing the fibers/yarns. For example, a two-ply yarn consisting of two fiber electrodes was reported to be a fiber-based TENG by Zhong et al. as illustrated in Figure 4.3 [25]. One cotton thread was coated with CNT; another thread was coated with CNT and an extra top layer of polytetrafluoroethylene (PTFE). These two cotton threads were entangled into one composite yarn. Both the CNT and PTFE were coated through a dip-coating process. The CNT-coated cotton yarn is conductive, with a resistance of about ~0.664 kΩ/cm. When stretching the two modified cotton threads, the PTFE layer and CNT coating are brought into contact, and electrification occurs at the interface, with net negative charges generated at the PTFE surface of one thread and net positive charges at the CNT coating of another thread (Figure 4.3b). By releasing the two-ply yarns, the two threads are separated, and the induced charges flow through the external load from one CNT-cotton electrode to another. If the yarn is stretched again, the induced current flows in the reverse direction. The repeated stretching–releasing will then yield an external alternating current. This process is similar to the contact-separation mode of TENG discussed in the above Section 4.2.2. A 9.0 cm long fiber TENG was reported to output a pulsed AC current of 11.22 nA with a tensile strain of 2.15%. When attaching it on a finger, the maximum power generated by finger motions can reach ~0.91 μW. This

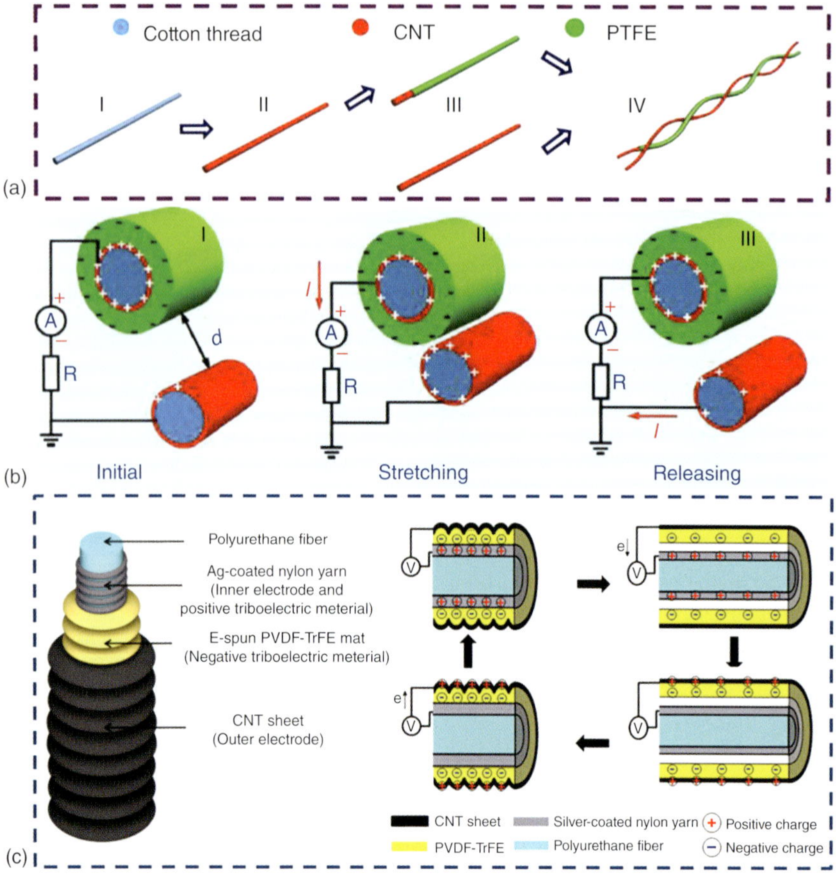


**Figure 4.3** A fiber TENG: (a) fabrication and (b) working mechanism of the fiber TENG. (c) Structure and working mechanism of the core–shell fiber TENG. Source: (a, b) Zhong et al. 2014 [25]. Copyright 2014, Reproduced with permission of American Chemical Society; (c) Reproduced with permission of Sim et al. 2016 [33]. https://www.nature.com/articles/srep35153#rightslink. http://creativecommons.org/licenses/by/4.0/. Licensed under CCBY 4.0.

fiber TENG is demonstrated to power a liquid crystal display (LCD) and charge a 2.2 μF capacitor to 2.4 V within 27 s.

Another possible configuration for 1D fiber/yarn TENG is the core–shell structure. If a core fiber is wrapped in a shell tube with certain void space, contact-separation motion between the core and shell layers can be realized by stretching/pressing the fiber. To achieve this structure, both core fiber and shell layer should be stretchable. Sim et al. designed such a fiber-based TENG, as shown in Figure 4.3c [33]. The inner fiber is a polyurethane fiber entangled by silver-coated nylon yarn; and the shell layers are wrinkled electrospun PVDF-TrFE layer and wrinkled CNT sheet coated through a prestraining–buckling–releasing process. A 1D TENG working in contact-separation mode is then constructed. In the meantime, both the CNT shell electrode and core silver electrode are stretchable. Therefore, electricity is generated when the fiber is reversibly stretched to 50% tensile strain. Other similar structures were also reported [34].

The characteristic of 1D fiber TENG is that each single fiber/yarn is an electricity-generating TENG. Although the 1D fiber/yarn TENGs can be easily woven or knitted into a garment or fabric, the major drawback is that their outputs are low. For example, for the core–shell fiber TENG, the output voltage is only about 24 mV at 50% strain [33]. The major reasons for low outputs are as follows: first, the contact interface area of two cylinder fibers is low, as is the generated static charge density; second, although the separation motion between core and shell fibers or two-ply fibers is realized, the separated distance is limited.

#### 4.3.3.2 2D Fabric TENGs

Since the outputs of TENGs are dependent on the amount of electrostatic charges generated at the interface, 2D fabrics can expose more surface areas for the electrification process and then better performances are expected. A 2D fabric-based TENG can generally be fabricated by two approaches: (i) bottom-up process starting with fiber/yarn electrodes and (ii) directly treating 2D fabrics for use as electrodes. Typically, a conductive layer and a dielectric electrification layer have to be coated successively on a common 1D fiber or a 2D fabric. Then two such electrodes with different top coatings will be combined in certain configuration to realize contact-separation or relative sliding motions.

As for the first approach, Pu et al. designed a plain-woven TENG fabric by weaving Ni-coated polyester yarn as wefts and parylene/Ni-coated polyester yarns as warps [27]. The conductive Ni coating on polyester is realized by electroless deposition process, while the top parylene layer is coated through a chemical vapor deposition (CVD) method. The weft and warp fiber electrodes were connected through an external circuit (Figure 4.4a). Electricity generation through different types of mechanical motions has been demonstrated, for example, rubbing two TENG fabrics with each other, pressing the TENG fabric by human skin, or contact-separation between one TENG fabric and another insulating fabric. This versatility of the TENG fabric allows energy generation from nearly every moving part of human bodies. The authors demonstrated the electricity generation by attaching the TENG fabric in the hand, under the foot, under the arm, and at the elbows. Commercial capacitors and even a flexible Li ion battery can be charged. The working mechanism of the textile TENG is shown in Figure 4.4b.

Upon contact, net negative charges are generated on the surface of parylene. Induced charges in the Ni layers will then be driven flowing through the external circuit when repeating the reaching-releasing motions between Ni-coated and paraylene-Ni-coated fibres, electrostatically induced charges in the Ni layers of the two fibres will flow back-and-forth through the external circuits. The typical output of short-circuit current is of ~4 μA amplitude and open-circuit voltage is of ~50 V amplitude (Figure 4.4c,d). Similar strategies were then adopted by several other researchers but with different coating layers. For example, Zhao et al. reported a textile TENG with Cu coatings and polyimide coatings on polyethylene terephthalate yarns as electrodes and electrification layers, respectively [28].

The major drawback of the coating methods is that it is hard to form uniform coating layers, especially for the top dielectric polymer coatings. The coating method for the conductive metal layer (Ni, Cu) by electroless deposition has been well established and commercialized, whereas it is difficult to achieve

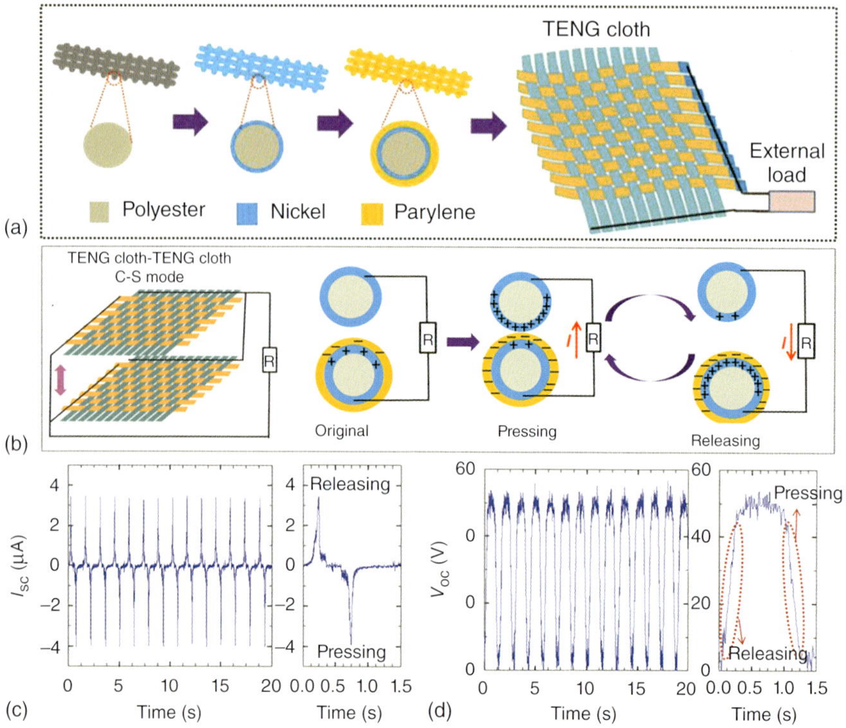


**Figure 4.4** A woven textile TENG: (a) fabrication process, (b) working mechanism, (c) current, and (d) voltage output. Source: Pu et al. 2015 [27]. Copyright 2015, Reproduced with permission of John Wiley and Sons.

continuous, thin, and uniform top dielectric coatings. To address this issue, Yu et al. proposed a textile TENG by using core–shell yarn electrodes as the building blocks [21]. The core is made up of two parallel stainless steel fibers, and the shell is made up of covering fibers (polyurethane, i.e. Spandex) twined tightly around the core fibers, as shown in Figure 4.5. The diameter of each core SS fiber is about ~30 μm, and the diameter of each covering fiber is about 20 μm, resulting in a final core–shell yarn with a diameter of about ~300 μm. With this route, the shell dielectric layers are uniform, and the flexibility of the final yarn is still excellent. Then, a textile TENG can be realized by any fabrication route, such as weaving, knitting, or sewing. The authors then demonstrated the versatility of the textile TENG to generate electricity in the single-electrode mode, contact-separation mode, or even in-plane sliding mode. This TENG textile can also work even after being cut and sewn back together.

As for the second approach, common 2D fabrics, after being treated with conductive coatings and/or dielectric films, can be directly used in textile TENGs without extra weaving, knitting, or sewing processes [31, 35–38]. Seung et al. reported a textile TENG with a configuration as shown in Figure 4.6 [35]. Ag-coated textile was used as the conductive electrode. To enhance the outputs, a nanopatterned surface was designed by growing aligned ZnO nanorods

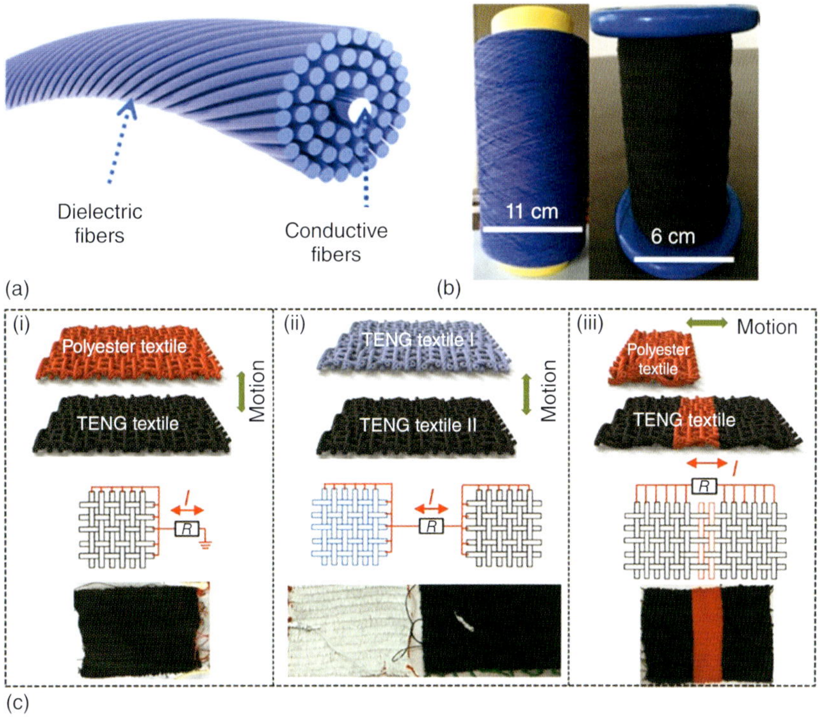


**Figure 4.5** The textile TENG woven from core–shell fibers: (a) scheme and (b) photographs of the core–shell fibers; (c) scheme of the textile TENGs fabricated for different working modes. Source: Yu et al. 2017 [21]. Copyright 2017, Reproduced with permission of American Chemical Society.

on an Ag-coated textile followed with a top polydimethylsiloxane (PDMS) layer grown through a dip-coating process. The nanopattern morphology was maintained in the final PDMS textile. Therefore, higher effective contacting area can be expected for the triboelectrification process upon contact-separation motion between the PDMS-coated textile and another Ag-coated textile. When working in the two-electrode contact-separation mode, the textile TENG with nanopattern can output a voltage of 120 V and current of 65 μA, whereas, the flat counterpart without nanopattern can only output a voltage of 30 V and current of 20 μA. Finally, self-powered operations of an LCD, light-emitting diodes (LEDs), and a keyless vehicle entry system by the output power only from the textile TENG were demonstrated.

Textile TENGs featuring a stacked structure were also reported by attaching conductive fabrics on common fabrics, but the resulting devices were generally not air breathable due to the use of polymer sealing agents [36]. Zhang et al. synthesized a stacked textile TENG by sewing dielectric and conductive textiles together [38]. Conductive fabrics were at the backsides of a nylon and cotton fabric, respectively. A so-called silanization treatment was further carried out to introduce functional groups on cotton fabrics. This process makes the cotton fabric strongly tribo-negative due to the grafted fluoroalkylated siloxane film.

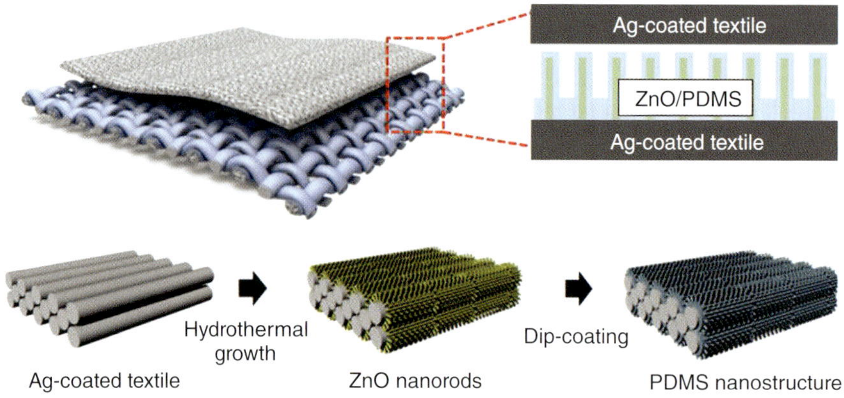


**Figure 4.6** Structure illustration of a nanopatterned textile TENG. Source: Seung et al. 2015 [35]. Copyright 2015, Reproduced with permission of American Chemical Society.

Because nylon is strongly tribo-positive, the contact of modified cotton fabric and nylon fabric can generate more electrostatic charges. Compared with the cotton fabric without surface modification, the final textile TENG with surface functionalization showed a fivefold increase in voltage and threefold increase in current density. According to the above two studies, it is also demonstrated that nanoscale surface morphology and surface chemical modifications are two effective ways to improve the outputs of textile TENGs.

Generally, sliding-mode TENGs with interdigitated electrodes can output higher current, compared to TENGs with contact-separation mode. Therefore, Pu et al. have reported a sliding-mode textile TENG (Figure 4.7) [39]. Grating-structured electrodes were fabricated on a textile substrate by a mask-aided electroless deposition of Ni. A top layer of parylene was then coated by a CVD process that functions as a triboelectrification layer (Figure 4.7a). A photograph of a pair of final TENG textiles is shown in Figure 4.7b. When wearing the TENG textiles underneath the arm as shown in Figure 4.7c, the electricity generated by swinging the arm is enough to power LED bulbs and LCD screens. Reducing the finger electrode width can increase the output current. When the finger electrode width is 1 mm, the achieved power density can be 3.2 W/m$^2$ at a sliding speed of 0.75 m/s. This textile TENG is also capable of converting low-frequency human motion energy into high-frequency current outputs, as shown by the rectified current in Figure 4.7d, where one swing of the arm can result in tens of current pulses. The authors also demonstrated that this TENG fabric was flexible, breathable, washable, and capable of integrating with common textiles.

#### 4.3.3.3 3D Fabric TENGs

Complex 3D weaving/knitting techniques have also been reported for fabricating textile TENGs. Dong et al. developed a textile TENG with 3D orthogonal woven (3DOW) structure (Figure 4.8) [30]. The 3DOW textiles were fabricated from three types of yarns: 3-ply-twisted stainless steel/polyester fiber blended yarn (warp yarn), PDMS-coated energy-harvesting yarn (weft yarn), and binding

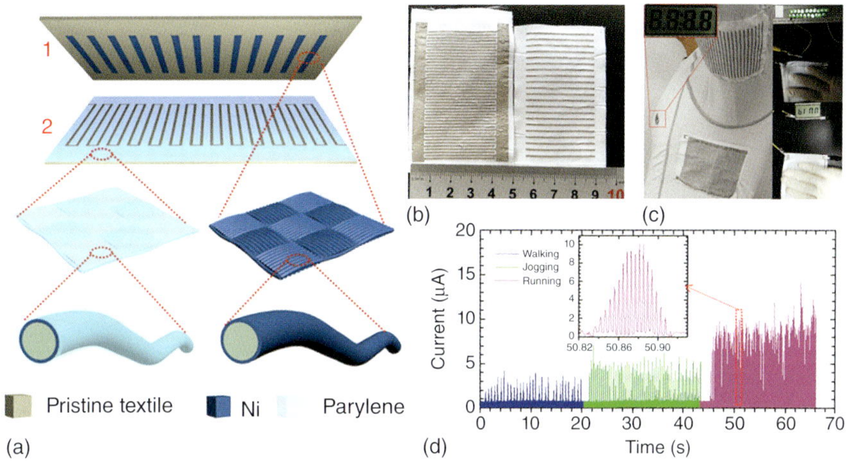


**Figure 4.7** A sliding-mode textile TENG with interdigitated electrodes: (a) structure illustration, (b) photograph of the prototype, (c) a textile TENG worn under the arm to power LED bulbs and LCD screens, and (d) rectified current output at different human motions. Source: Pu et al. 2016 [39]. Copyright 2016, Reproduced with permission of John Wiley and Sons.

yarns in the thickness direction (Z-yarn). Figure 4.8a shows the scheme of fabrication procedures of the 3DOW textile. Figure 4.8b shows the schematic texture of the 3DOW textile. Different types of circuit connections have been compared. Basically, d1 is a 2D woven textile; d2 is similar to Figure 4.8a, but woven without the middle warp conductive yarns; d31, d32, and d33 have the same structure as shown in Figure 4.8b. The d32 is a two-electrode mode textile TENG (the circuit is as shown in Figure 4.8c), but other four types work in single-electrode mode. As shown in Figure 4.8d,e, the two-electrode d32-type textile TENG shows the highest output voltage and current. The maximum peak power density of two-electrode 3DOW textile can reach 263.36 mW/m$^2$ under the tapping frequency of 3 Hz. The authors demonstrated that the 3D textile TENG can charge capacitors, power a digital watch, and serve as a self-powered active motion sensor.

Zhu et al. also reported a 3D spacer fabric for TENG applications [31]. This textile TENG was based on a 3D weft-knitted structure, consisting of three layers: two parallel outer fabric layers and a spacer yarn layer in between. The top fabric is nylon, with a top coating of graphene ink for using as the electrode. The bottom fabric is coated with commercially available PTFE coatings. So, the triboelectrification occurs when the PTFE layers contact the nylon fabrics, and AC electricity can be generated when the 3D textile is repeatedly pressed and released.

### 4.3.4 Washing Capability

It is important for textile TENGs to retain the flexibility, comfort, and washing capability of the original common textiles. In order to be able to be washed by a machine, a textile TENG should be mechanically durable, able to withstand

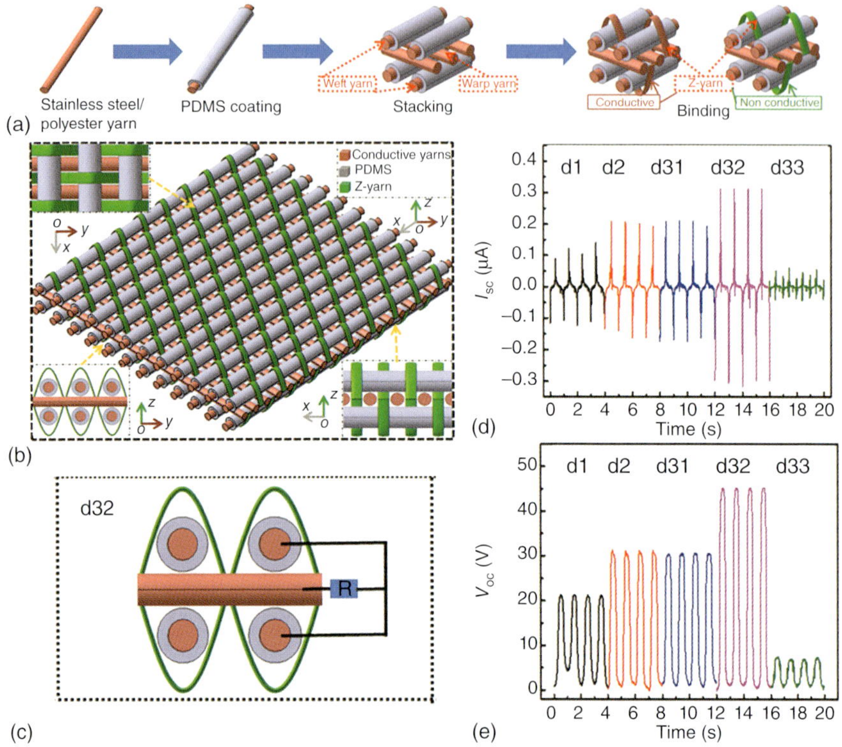


**Figure 4.8** A 3D woven textile TENG. (a) Fabrication process and (b) structure scheme of the textile TENG. (c) The structure of a d32 type TENG textile. (d) Current and (e) voltage output of TENG textiles with different weaving structures. Source: Dong et al. 2017 [30]. Copyright 2017, Reproduced with permission of John Wiley and Sons.

detergents, washing, drying, or even ironing. For most of the textile TENGs discussed above, the mechanical flexibility is maintained. Some reported textile TENGs are even stretchable. Most of the textile TENGs also showed no performance degradation after being bended or twisted for many cycles. However, a few washing experiments have been conducted to demonstrate the washing durability of textile TENGs [21, 28]. As for the conductive coatings of electrodes, it is expected that metal wires and electroless-coated metal coatings are durable to washing. But, the dip-coated CNTs, rGO, or metal nanowires are possibly detached from the substrate during harsh washing with detergents.

The textile TENGs woven from core–shell fibers (SS core fiber and Spandex shell fibers) are demonstrated to be able to withstand 120 washes when placed in a laundry bag [21]. No obvious change in the fiber morphology or the device performance is observed after the machine laundry. Textile TENGs made from fiber electrode with electroless-deposited thin metal coatings are also washable. The sliding-mode textile designed by Pu et al. can be washed with a household laundry for 20 minutes and ironed to be flat again [39]. Zhao et al. also reported a textile TENG with electroless-deposited Cu electrodes, which can withstand 20 times of machine washes [28]. The output current of the textile TENG drops after the first three times of washes (from 13.78 to 3.52 mA/m$^2$), but then maintains 92% after subsequent washing cycles.

### 4.3.5 Self-charging Power Textiles

The output of a textile TENG is pulsed electric voltage/current. In the meantime, the human motion is irregular, as is the electricity generated by textile TENGs. Therefore, based on these two reasons, the electricity of a textile TENG has to be managed and stored for stable DC power supply to electronics. Pu et al. proposed a self-charging power unit by integrating TENG cloths with a belt-shaped flexible lithium-ion battery (LIB) [27]. The textile TENG can be placed under the arm for harvesting human motion energy. The AC current of the TENG was converted into DC current by a rectifier. The authors demonstrated that the LIB belt, charged by the textile TENG, can power a heartbeat monitor. As for the LIB belt, $LiFePO_4$ cathode and $Li_4Ti_5O_{12}$ anode were coated on the fabric substrate as well. The LIB belt can withstand 180° bending without damage or performance degradation. The voltage of the LIB belt rises quickly to the operational voltage (≈1.9 V) when charging it by the textile TENG with a 0.7 Hz contact-separation motion. When charging the LIB belt for 4, 9, and 14 hours, the corresponding discharge capacities are 1.3, 2.8, and 4.4 mAh/m$^2$, respectively.

Later, a self-charging power textile, as shown in Figure 4.9, was further proposed with yarn supercapacitors and textile TENGs as energy storage and energy-harvesting devices, respectively [29]. The Ni-coated polyester yarn with top rGO coatings was used as the capacitive electrode in a symmetric yarn supercapacitor. Two such rGO-coated yarns were bound by solid-state gel electrolyte in parallel. The yarn-supercapacitor achieved high capacitance (13.0 mF/cm, 72.1 mF/cm$^2$). The supercapacitor yarns can be sewed, knitted, or woven in common fabrics, so a single textile can possess two functions of both energy harvesting and storage. With a rectifier, three supercapacitors (connected in series) were charged by a textile TENG to 2.1 V.

## 4.4 Conclusions and Perspectives

In summary, many prototypes of textile TENGs have been demonstrated to be viable to harvest human motion energies for potentially powering smart textiles. Almost all the materials used for today's textile can be applied in a textile TENG, due to the universal triboelectrification effect. The textile TENGs are reported to be generally flexible, wearable, or even stretchable and machine washable. Considering that human motion is of low frequency, textile TENGs are probably the best choice of kinetic energy-harvesting device. Nanoscale surface morphology, surface chemical modifications, structure optimization, and textile weaving/knitting/sewing process can all be optimized for improving the output performances. Textile TENGs with 1D fiber configuration, or 2D and 3D fabric configuration have been proposed to power small electronics, such as

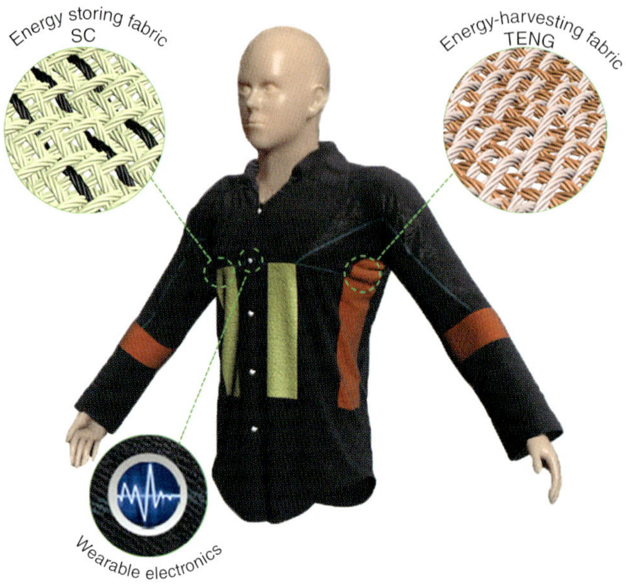


**Figure 4.9** The scheme of a self-charging power textile that integrates energy-harvesting, energy storage, and functional electronics. Source: Pu et al. 2016 [29]. Copyright 2016, Reproduced with permission of John Wiley and Sons.

LEDs, electronic watches, or health-care electronics. Preliminary demonstration on self-charging power textiles has also been conducted for stable power supply.

Despite these progresses, there are still some challenges. First, the output performances have to be improved to meet the power requirements of wearable electronics. Otherwise, the practical application of textile TENGs is still not viable. Second, power management has to be designed specifically for textile TENGs, so that stable DC output can be supplied for smart textiles or wearable electronics. It is also a challenge to make the management circuit flexible or wearable. A possible alternative way is to design management circuit into a miniature module. Third, the material choices, structure designs, and synthesis processes have to be optimized for scale-up fabrication at low cost. Most current coating methods used for carbonaceous electrode materials and top electrification polymers are hard to be scaled. Last, the fabrication of textile TENGs needs to be compatible with current textile industry, and consideration needs to be paid to their aesthetic/stylish designs. One disadvantage of most current textile TENGs is that their appearances are not pleasing or appealing after multiple surface treatments. Frequently utilized coatings or sealing adhesive agents may also make them not breathable. Nevertheless, in light of the growing advancements in this field and more attention received from the research community, it can be optimistically expected that textile TENGs would become viable for practical applications in the near future.

## References


1 (a) Yetisen, A.K., Qu, H., Manbachi, A. et al. (2016). *ACS Nano* 10: 3042. (b) Stoppa, M. and Chiolerio, A. (2014). *Sensors* 14: 11957.
2 Qi, Y. and McAlpine, M.C. (2010). *Energy Environ. Sci.* 3: 1275.
3 Zhang, H., Yang, Y., Su, Y. et al. (2014). *Adv. Funct. Mater.* 24: 1401.
4 Wang, Z.L. (2017). *Nature* 542: 159.
5 Wang, Z.L., Jiang, T., and Xu, L. (2017). *Nano Energy* 39: 9.
6 (a) Zhao, Z., Pu, X., Du, C. et al. (2016). *ACS Nano* 10: 1780. (b) Bae, J., Lee, J., Kim, S. et al. (2014). *Nat. Commun.* 5: 4929.
7 Yu, A., Jiang, P., and Lin Wang, Z. (2012). *Nano Energy* 1: 418.
8 Bai, P., Zhu, G., Lin, Z.-H. et al. (2013). *ACS Nano* 7: 3713.
9 Zi, Y., Guo, H., Wen, Z. et al. (2016). *ACS Nano* 10: 4797.
10 Wang, Z.L. and Song, J. (2006). *Science* 312: 242.
11 Fan, F.R., Tang, W., and Wang, Z.L. (2016). *Adv. Mater.* 28: 4283.
12 (a) Lowell, J. and Rose-Innes, A. (1980). *Adv. Phys.* 29: 947. (b) Duke, C.B. and Fabish, T.J. (1978). *J. Appl. Phys.* 49: 315.
13 Wang, Z.L. (2017). *Mater. Today* 20: 74.
14 Wang, Z.L. (2014). *Faraday Discuss.* 176: 447.
15 Wang, Z.L., Chen, J., and Lin, L. (2015). *Energy Environ. Sci.* 8: 2250.
16 Zhang, Q., Xu, R., and Cai, W. (2018). *Nano Energy* 51: 698.
17 Wang, Z.L. (2013). *ACS Nano* 7: 9533.
18 Pu, X., Liu, M., Chen, X. et al. (2017). *Sci. Adv.* 3: e1700015.
19 Sun, J., Pu, X., Liu, M. et al. (2018). *ACS Nano* 12: 6147.
20 (a) Hinchet, R., Seung, W., and Kim, S.-W. (2015). *ChemSusChem* 8: 2327. (b) Zhu, G., Peng, B., Chen, J. et al. (2014). *Nano Energy* 14: 126.
21 Yu, A., Pu, X., Wen, R. et al. (2017). *ACS Nano* 11: 12764.
22 Lai, Y.-C., Deng, J., Zhang, S.L. et al. (2017). *Adv. Funct. Mater.* 27: 1604462.
23 Chen, J., Huang, Y., Zhang, N. et al. (2016). *Nat. Energy* 1: 16138.
24 (a) Xu, Z. and Gao, C. (2015). *Materials Today* 18: 480. (b) Weng, W., Chen, P., He, S. et al. (2016). *Angew. Chem. Int. Ed.* 55: 6140.
25 Zhong, J., Zhang, Y., Zhong, Q. et al. (2014). *ACS Nano* 8: 6273.
26 Zhou, T., Zhang, C., Han, C.B. et al. (2014). *ACS Appl. Mater. Interfaces* 6: 14695.
27 Pu, X., Li, L., Song, H. et al. (2015). *Adv. Mater.* 27: 2472.
28 Zhao, Z., Yan, C., Liu, Z. et al. (2016). *Adv. Mater.* 28: 10267.
29 Pu, X., Li, L., Liu, M. et al. (2016). *Adv. Mater.* 28: 98.
30 Dong, K., Deng, J., Zi, Y. et al. (2017). *Adv. Mater.* 29: 1702648.
31 Zhu, M., Huang, Y., Ng, W.S. et al. (2016). *Nano Energy* 27: 439.
32 Kwak, S.S., Kim, H., Seung, W. et al. (2017). *ACS Nano* 11: 10733.
33 Sim, H.J., Choi, C., Kim, S.H. et al. (2016). *Sci. Rep.* 6: 35153.
34 (a) Yu, X., Pan, J., Zhang, J. et al. (2017). *J. Mater. Chem. A* 5: 6032. (b) Kim, K., Song, G., Park, C., and Yun, K.-S. (2017). *Sensors* 17: 2582.
35 Seung, W., Gupta, M.K., Lee, K.Y. et al. (2015). *ACS Nano* 9: 3501.

**36** (a) Ko, Y.H., Nagaraju, G., and Yu, J.S. (2015). *RSC Adv.* 5: 6437. (b) Li, S., Zhong, Q., Zhong, J. et al. (2015). *ACS Appl. Mater. Interfaces* 7: 14912.
**37** Cui, N., Liu, J., Gu, L. et al. (2015). *ACS Appl. Mater. Interfaces* 7: 18225.
**38** Zhang, L., Yu, Y., Eyer, G.P. et al. (2016). *Adv. Mater. Technol.* 1: 1600147.
**39** Pu, X., Song, W., Liu, M. et al. (2016). *Adv. Energy Mater.* 6: 1601048.

# 5

# Flexible and Wearable Solar Cells and Supercapacitors

*Kai Yuan, Ting Hu, and Yiwang Chen*

*Nanchang University, Department of Chemistry, College of Chemistry/Institute of Polymers and Energy Chemistry, 999 Xuefu Avenue, Nanchang 330031, China*

## 5.1 Introduction

Energy is a basic necessity for the existence and growth of human civilization. The ever-growing energy crisis and environmental problems arising from burning fossil fuels has stimulated global efforts to explore renewable and clean energy sources. The sun is the most abundant energy source with no running cost and leaves no pollution or residual waste to meet this demand. However, solar energy is diffuse and intermittent, owing to the unreliability of nature. For example, energy is mainly captured during daylight for solar power. These pending hurdles have stimulated the research on efficient energy conversion and storage devices. Energy conversion and storage devices with high comprehensive performance, including solar cells, lithium-ion batteries, and supercapacitors are very promising technologies for mitigating the worldwide energy challenge. Owing to the rapid development of wearable electronics, portable, flexible, and wearable energy conversion and storage devices are in urgent demand. However, conventional energy conversion and storage devices, for example, the commercial silicon-based solar cells, show high energy conversion efficiency while suffering from a complex fabrication process. Particularly, most conventional energy conversion and storage devices, including dielectric capacitors, silicon-based solar cells, lead-acid batteries, and lithium-ion batteries, are generally made into rigid devices, which are unable to meet the requirements for flexible and wearable electronics. Therefore, it is desirable to develop energy conversion and storage devices with competitive performances and flexibility that allows their fabrication and integration in wearable electronics. In this chapter, the development of flexible and wearable energy conversion and storage devices is presented, focusing on solar cells (such as dye-sensitized solar cells (DSSCs), polymer solar cells (PSCs), perovskite solar cells (PVSCs)), supercapacitors, and their integrated devices.

*Flexible and Wearable Electronics for Smart Clothing*, First Edition.
Edited by Gang Wang, Chengyi Hou, and Hongzhi Wang.

## 5.2 Flexible and Wearable Solar Cells

The ever-growing innovations and developments in the field of energy research make the flexible and wearable energy devices become increasingly smaller with compact dimensions. For wearable applications, flexible and wearable solar cells play a very important role in the devices for which energy harvesting is necessary [1, 2]. Since inorganic solar cells, such as silicon-based solar cells, are not amenable to bending for meeting the wearable applications, most of the developed flexible and wearable solar cells are based on organic or hybrid materials.

### 5.2.1 Flexible and Wearable Dye-Sensitized Solar Cells

The conventional planar-shaped DSSCs with sandwich-like configuration are developed by Grätzel and coworker [3]. The cell includes three primary parts, a working electrode, a counter electrode, and an electrolyte to separate the two electrodes placed face to face. The working electrode is made of transparent glass with a thin transparent conductive layer, such as fluorine-doped tin dioxide (FTO) or indium-doped tin oxide (ITO), and serves as the substrate for depositing mesoporous semiconductor oxide (usually is titanium dioxide, $TiO_2$) with high surface area by sintering. The counter electrode is made by depositing platinum, carbon materials, etc. as a catalytic layer. After the working electrode is immersed into a photosensitive dye solution for soaking and bonding the dye onto the surface of the $TiO_2$, the two electrodes are sealed followed by injecting redox electrolytes such as $I^-/I^{3-}$.

The working process in the DSSCs for the energy harvesting and conversion is achieved through a photoelectrochemical process with light-induced redox reactions (Figure 5.1). When illuminated the dye molecules harvest photons, and the electrons are excited from the highest occupied molecular orbital (HOMO) to the lowest unoccupied molecular orbital (LUMO). The excited dye molecules then inject electrons into the conduction band of the nearby semiconductor because the LUMO for dye molecules is above the conduction band edge of the semiconductor layer. The oxidized dye molecules are regenerated by accepting the electrons released from the $I^{3-}$ reduced into $I^-$ in the electrolyte.

Based on the conventional planar structure, DSSCs can be also made into flexible and wearable configurations [5–7], such as twisted and coaxial fiber-shaped structures [8–11]. The conception of FDSSCs, or "solar cells in a fiber form," as they are called, was first reported by Baps et al. as early as 2001 [12]. In their case, a steel wire coated with a layer of dye-absorbed $TiO_2$ nanoparticles was used as the working electrode, and a conductive polymer coated inside a transparent polymer tube was used as the counter electrode of the tube-shaped solar cell. The transparent tube was sealed, followed by injection of liquid electrolyte between the two electrodes. The as-fabricated tube-shaped solar cell that is 10 cm in length could deliver a voltage of 0.3–0.35 V. However, the current is of only a few microamperes due to the low conductivity of the polymer electrode as well as the corrosive electrolyte. Moreover, liquid electrolytes also limit the application of these flexible fibers in wearable devices unless it is completely sealed and well packaged. The premiere of the "solar cells in a fiber form" may not be recognized as a success,

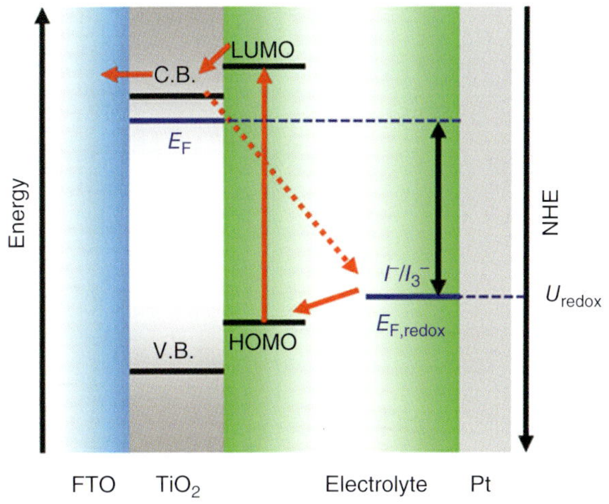


**Figure 5.1** Schematic illustration of the working mechanism of a DSSC. Source: Peng 2015 [4]. Copyright 2015, Reprinted with permission of Springer Nature.

but it is not a failure either. This visionary work paved the first stepping stone toward its application to wearable electronic devices, and inspired the scientists to develop wearable DSSCs toward high performance.

Until 2008, to avoid the use of transparent conducting oxides, Zou and coworkers made a flexible and wearable DSSC fiber on twisted fiber-like electrodes [13]. The FDSSC was made by twisting two metal electrodes in a helical structure, as shown in (Figure 5.2). A stainless steel wire coated with dye-sensitized $TiO_2$ and a platinum wire were saved as the working electrode and the counter electrode, respectively. The surface-coated $TiO_2$ as a barrier layer prevented the short circuit between the two electrodes. Such twisted FDSSCs are available to receive substantial light from an arbitrary angle. The thickness of the $TiO_2$ layer plays a critical role in the performance of the FDSSCs. After the optimization, with 3.5 μm thick porous $TiO_2$ layer, the FDSSC could give short-circuit current density ($J_{sc}$), open circuit voltage ($V_{oc}$), and fill factor ($FF$) of 0.06 mA/cm$^2$, 0.61 V, and 0.38, respectively. The final energy conversion efficiency was recorded as 0.27%. In order to improve the energy conversion efficiency, the porous titania-modified steel wire was replaced by a titanium wire coated with a layer of titania nanoparticles as working electrode. A much enhanced energy conversion efficiency of 5.41% was obtained [14]. They also fabricated a novel fiber solar cell housed in a parabolic-shaped reflector, which can effectively harvest diffuse light from all directions. The maximum efficiency of the FDSSCs had been further increased to 7.02% [15].

Peng and coworkers have developed a new and general method to fabricate novel stretchable, wearable DSSC textiles based on elastic, electrically conducting fibers [16]. In their work, multiwalled carbon nanotube (MWCNT) sheets were firstly winding aligned on rubber fibers to prepare fiber electrodes. Then, modified Ti wire was twisted onto the elastic MWCNT fiber as a working electrode. Photoactive materials were coated on the electrodes to fabricate a wire-shaped DSSC. The wire-shaped DSSCs could weave into wearable and stretchable photovoltaic textile (Figure 5.3). The energy conversion efficiency of the wire-shaped

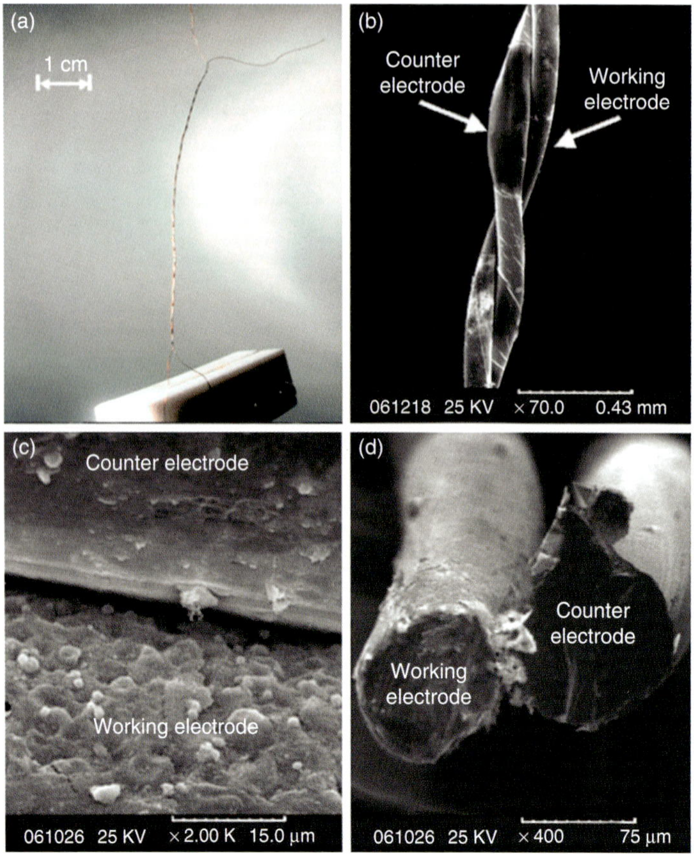


**Figure 5.2** (a) Optical photograph of a twisted wire-shaped flexible dye-sensitized solar cells (WSF-DSSC) (Uncut; radius: ~0.2 mm); (b, c) scanning electron microscopic (SEM) photograph of a WSF-DSSC (top view); (d) SEM photograph of a WSF-DSSC (sectional view); Source: Fan et al. 2008 [13]. Copyright 2008, Reprinted with permission of John Wiley & Sons.

DSSC can reach as high as 7.13%, and also can remain at this level even under stretched condition.

Wang and coworkers developed a grating-structured triboelectric nanogenerator fabric and integrated it with fiber-shaped dye-sensitized solar cells (FDSSCs) so as to achieve a whole textile-based energy-harvesting system [17]. In the FDSSCs, mesoporous $TiO_2$ layer coated Ti wire served as the working electrode, and twisted Pt wire was used as the counter electrode. The electrolyte was filled and sealed by a transparent polytetrafluoroethylene (PTFE) tube. When bending with a radius of ≈2 cm, the FDSSCs showed a slight decrease in the performance. The power conversion efficiency (PCE) retention of the FDSSCs remained at about 90% after bending for 100 times. The $J_{sc}$ and $V_{oc}$ of the FDSSCs increased with the increase in the light intensity. Under an AM 1.5 solar (100 mW/cm$^2$), the FDSSCs displayed an average $J_{sc}$ of 10.6 mA/cm$^2$

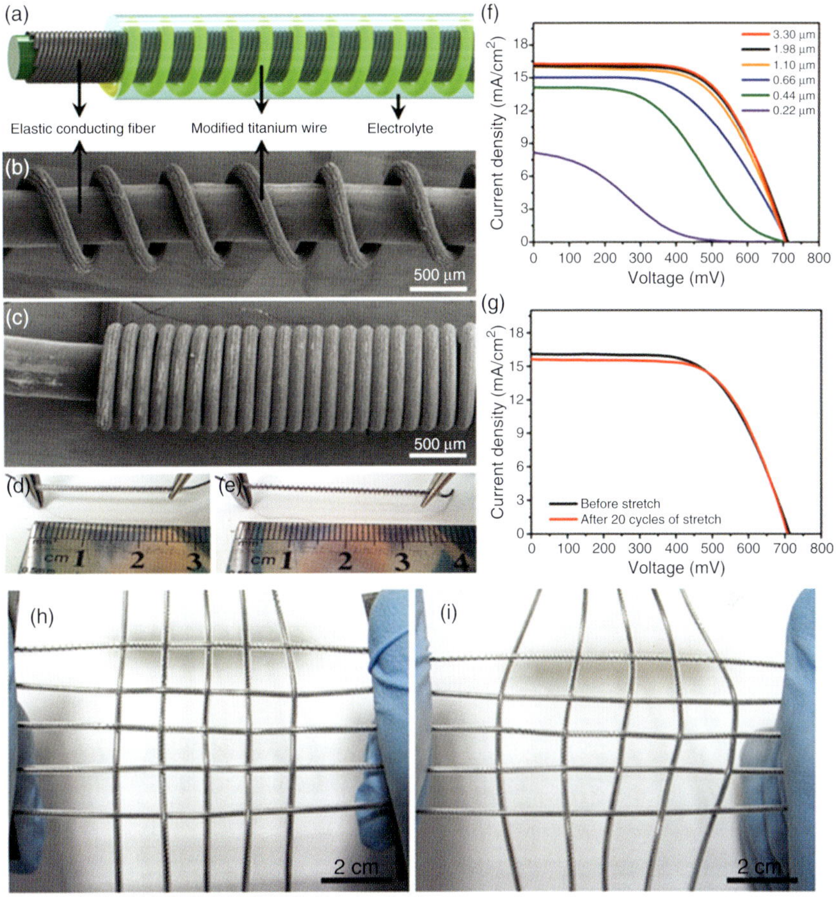


**Figure 5.3** (a) Schematic illustration of a stretchable DSSC. (b, c) SEM images of a stretchable DSSC with pitch distances of 560 and 164 μm, respectively. (d, e) A stretchable DSSC before and after stretch by 30%, respectively. (f) *J–V* curves of DSSCs using elastic conducting fibers with MWCNT sheets of different thicknesses as the counter electrodes and the same Ti wire impregnated with 30 μm long $TiO_2$ nanotubes as the working electrode. (g) *J–V* curves of a stretchable DSSC before and after stretch for 20 cycles at a strain of 30%. (h, i) Photographs of a flexible and stretchable photovoltaic textile before and after stretch, respectively. Source: Yang et al. 2014 [16]. Copyright 2014. Reprinted with permission of John Wiley & Sons.

and $V_{oc}$ of 0.68 V, with an average PCE of 6%. Moreover, the devices showed long-term stability and durability.

Soni and coworkers used a smart thermoreversible water-based polymer gel [poly(ethylene oxide)-poly(propylene oxide)-poly(ethylene oxide) (PEO-PPO-PEO)] electrolyte to fabricate environment-friendly quasi-solid-state DSSC (Figure 5.4) [18]. The water-based polymer gel electrolyte shows self-healing characteristics at a low temperature (below 0 °C). The quasi-solid-state DSSC displays superior flexibility with self-healing behavior. When the quasi-solid-state DSSC is subjected to bending, shrinking, or twisting, the performance can recover close to the initial level by a simple cooling treatment

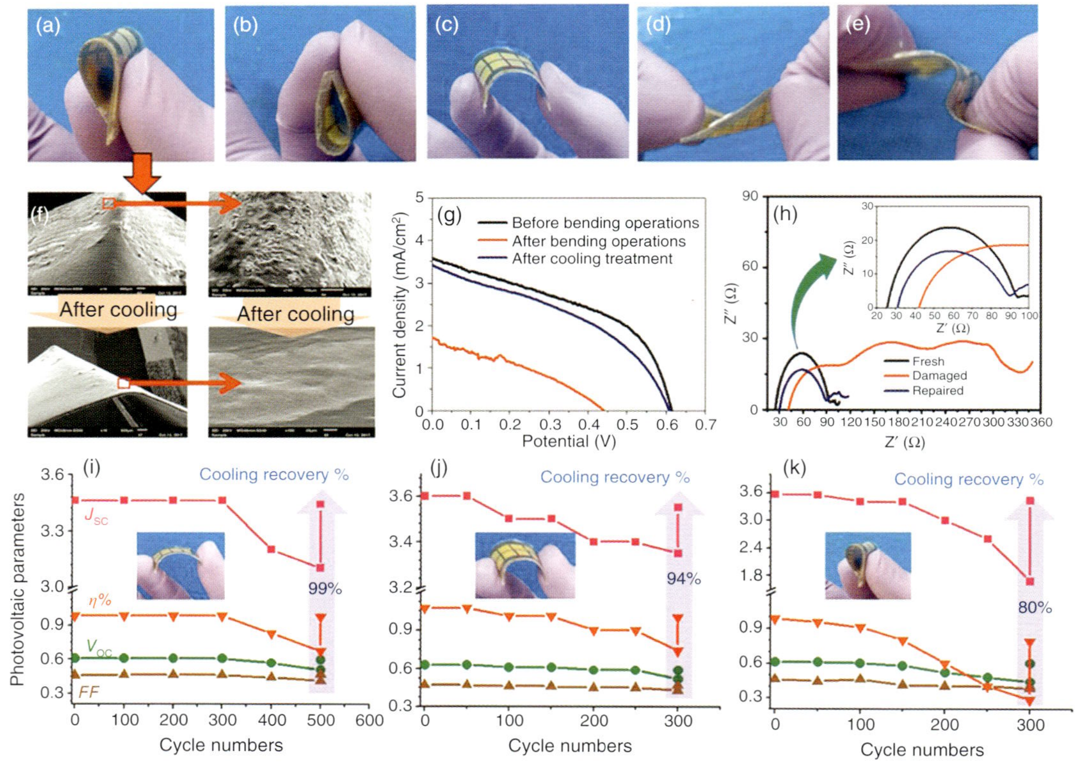


**Figure 5.4** Digital photograph of model device under different deformation modes: (a) bending from anode side, (b) bending from cathode side, (c) 90° bending, (d) twisting, and (e) shrinking. (f) SEM image of the top view for damaged interface between polymer gel electrolyte (PGE)photoanode and repaired interface after cooling through self-healing. (g) *J–V* curves represent the fresh to damaged device after 300 strong bending operations at 180° bend angle and recovered curve after cooling treatment. (h) Electrochemical impedance spectroscopy (EIS) characteristics of model device in dark condition (inset: magnification at high frequency). Cooling recovery efficiency of device performance with respect to deformation cycle numbers at (i) 45°, (j) 90°, and (k) 180° bending operations. Source: Sonigara et al. 2018 [18]. Copyright 2018, Reprinted with permission of John Wiley & Sons.

to regenerate the electrode/electrolyte interface. After 300 cycles of 90° bending, the quasi-solid-state DSSC can maintain its performance at as high as 94% of the initial level. This is the first aqueous DSSC device fabricated with a smart thermoreversible polymer gel as electrolyte, and shows self-healing behavior, which provides a perspective in flexible and wearable solid-state photovoltaic devices.

### 5.2.2 Flexible and Wearable Polymer Solar Cells

PSCs are an attractive choice for flexible and wearable solar cells because of the use of flexible and lightweight polymers as the active materials [19–21]. The most commonly developed PSC is a bulk heterojunction (BHJ) solar cell with mixed electron donor and acceptor materials to form a bicontinuous interpenetration network as active layer sandwich between the hole blocking layer and the electron blocking layer (Figure 5.5a) [22]. The working mechanism of a PSC can be described as follows: when a PSC device is illuminated, the

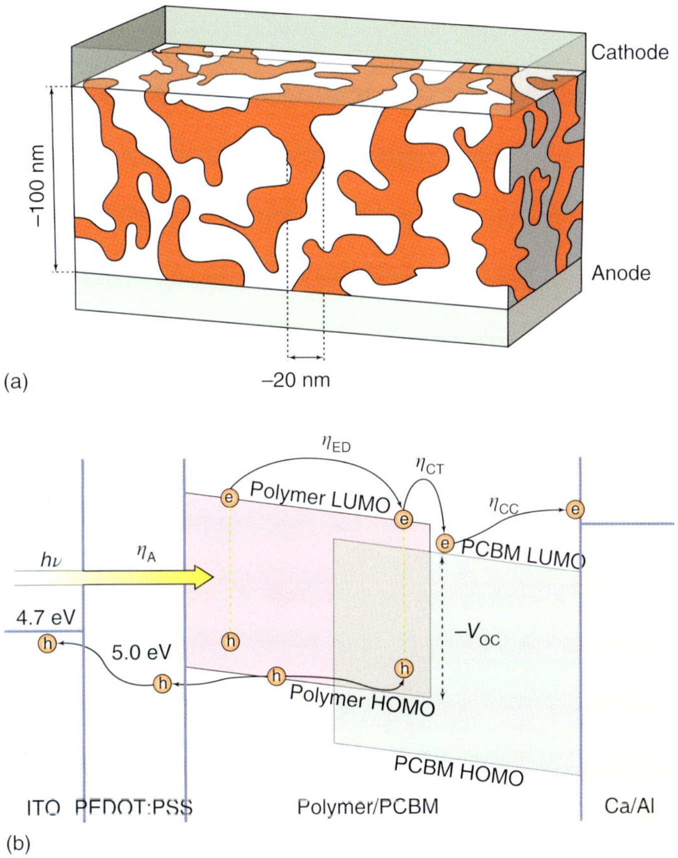


**Figure 5.5** (a) Conceptual morphology model with bicontinuous interpenetration network of the polymer and the acceptor. (b) Schematic illustration of the working mechanism of a PSC. Source: Li et al. 2012 [22]. Copyright 2012, Reprinted with permission of Springer Nature.

donor material absorbing light is excited and generates electron–hole pairs (excitons). The excitons diffuse to the donor–acceptor interface and dissociate into free-charge carriers, the electron is transferred to the LUMO of the acceptor and the hole to the HOMO of the donor molecule, which then transfer to the respective electrodes (Figure 5.5b).

Although the flexible DSSCs could deliver efficiencies up to 8% [23], the use of volatile electrolytes needs additional packaging, making them not ideal for wearable application. Ahead of the flexible and wearable DSSC, organic solar cells and PSCs were successfully made into flexible and wearable format and achieved a conversion efficiency of 0.5% and 0.6%, respectively [24, 25]. For developing high performing flexible and wearable solar cells, high transparency and good electrical conductivity for efficient light harvesting and charge collection are necessary. An all-solid-state, flexible, wearable, and lightweight PSC textile was reported by Peng and coworkers [1]. A Ti metal wire textile was used as the cathode. The Ti metal wires were covered with perpendicularly aligned $TiO_2$ nanotubes. Poly(3-hexylthiophene):phenyl-C61-butyric acid methyl ester (P3HT:PCBM) was dip coated on the textile cathode as active layer. Two transparent carbon nanotube (CNT) sheets with remarkable mechanical and electronic properties were coated onto both sides of the modified metal textile as anode to produce the desired PSC textile (Figure 5.6). The PSC textile displays the same energy conversion efficiencies irrespective of the side from which it is illuminated due to the designed sandwich structure. The PSC textiles are highly flexible and stable; after bending for more than 200 cycles the energy conversion efficiencies varied by less than 3%.

For practical applications such as electronic textiles, flexibility and stretchability are very important. In order to obtain bendable and stretchable solar cells, Peng and coworkers developed a spring-like architecture [26]. In this architecture, a Ti wire was wound around a flexible and bendable rod made of rubber to acquire a spring-like shape (Figure 5.7). $TiO_2$ nanotubes were grown by an electrochemical method perpendicular to the outer surface of the Ti spring. Then, P3HT:PCBM and poly(3,4-ethylenedioxythiophene):poly(styrenesulfonate) (PEDOT:PSS) were coated on the Ti spring as photoactive layer and hole-transport layer, respectively. Subsequently, a thin layer (c. 18 nm in thickness) of CNT sheet was wound over the Ti spring to finish the fabrication of the spring-like PSCs. The spring-like PSCs can weave into wearable PSC textiles without sealing (Figure 5.7). After 1000 cycles of bending, the PCE of the as-fabricated spring-like PSCs can maintain 90% of the initial value, which is superior to conventional planar PSCs. Moreover, after stretching by 30% for 50 cycles, the PCE values changed less than 10%. The limitation of the spring-like architectures is that a major part of the total area of the solar cell is not exposed to sunlight. Therefore, the total power output of the spring-like solar cell is inferior to that of a conventional structure.

Mechanical resilience, low weight, and large-area solar cells are designable for future specific applications such as electronic skin and textiles. Kaltenbrunner et al. demonstrate flexible PSCs on plastic foil substrates less than 2 μm thick [27]. The devices were fabricated by coating P3HT:PCBM active layer on a PEDOT:PSS-modified 1.4 μm thick PET substrate foil, and with 115 nm Ca/Ag as the top electrode (Figure 5.8). The as-fabricated PSCs exhibit extreme bending

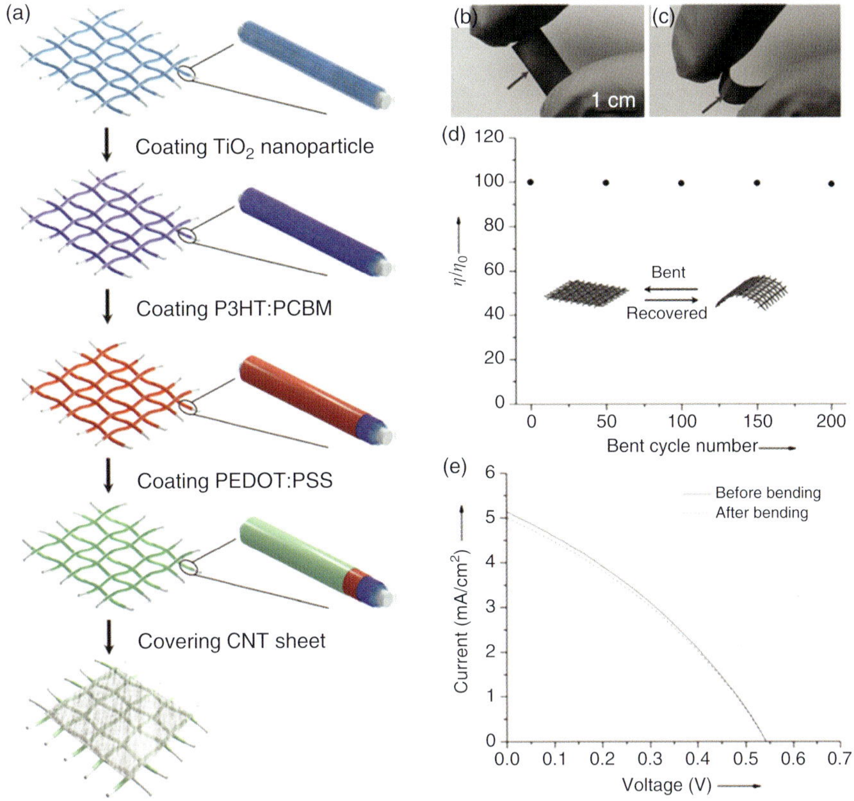


**Figure 5.6** (a) Fabrication of the PSC textile. Photographs of the PSC textile before (b) and after (c) bending. (d) Dependence of the energy conversion efficiency of the PSC textile on the bending cycle. $\eta_0$ and $\eta$ correspond to the energy conversion efficiencies before and after bending, respectively. (e) *J–V* curves of the PSC textile before and after bending for 200 cycles. Source: Zhang et al. 2014 [1]. Copyright 2014, Reprinted with permission of John Wiley & Sons.

flexibility and can wrap around a human hair with a radius of 35 μm. Moreover, these devices can reversibly withstand extreme mechanical deformation and have unprecedented solar cell-specific weight. Particularly, the device performs with $J_{SC}$ of 11.9 mA/cm$^2$, open circuit potential ($V_{OC}$) of 580 mV, and fill factor (FF) of 61% and gave a PCE of 4.2% under one sun illumination. These results are comparable to devices fabricated on ITO-coated glass substrates. The device also shows good durability during stretching and compression. The repeated stretching and compression cycling test (from original size to 50% quasi-linear compression and back) indicated a gradual decrease in $J_{SC}$, FF, and PCE, resulting in a 27% reduction in PCE after 22 cycles. The processing methods are standard, which could be used to fabricate ultrathin transistors, light-emitting diodes, and capacitors to achieve ultrathin organic electronics.

Recently, Someya and coworkers fabricated ultrathin and ultra-flexible organic solar cells to power organic electrochemical transistors as sensors on a 1 μm thick ultra-flexible substrate [28]. The structure of the devices is illustrated in Figure 5.9. For the device fabrication, a solution-processable

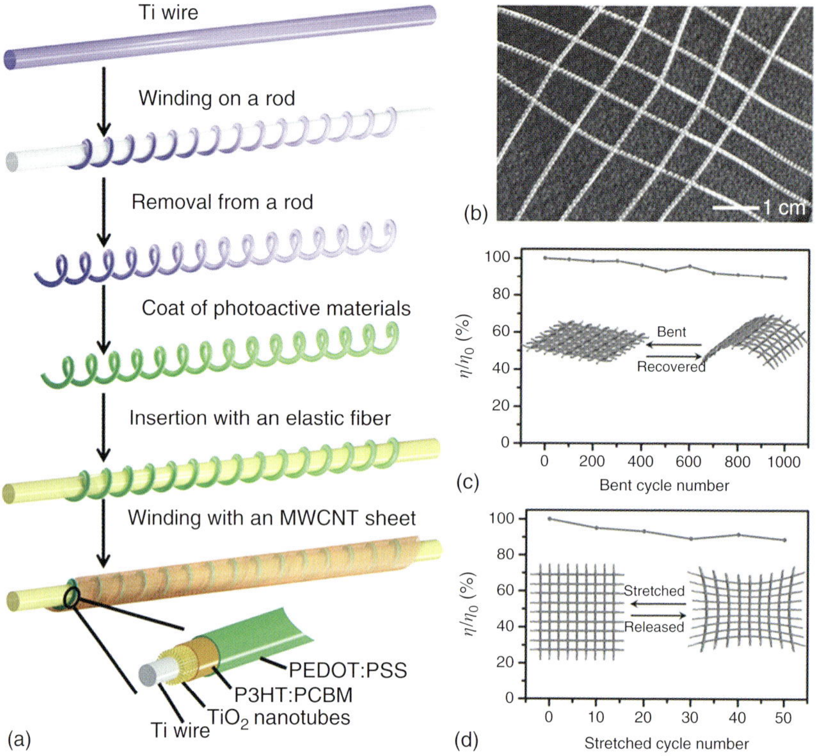


**Figure 5.7** (a) Schematic illustration of the fabrication of a stretchable fiber shaped PSC. (b) A stretchable PSC textile being integrated into clothes. (c) Dependence of energy-conversion efficiency on bending cycle number. $\eta_0$ and $\eta$ correspond to the energy-conversion efficiencies before and after bending, respectively (the inserted image schematically shows the bending process). (d) Dependence of energy-conversion efficiency on the stretching cycle number. $\eta_0$ and $\eta$ correspond to the energy-conversion efficiencies before and after stretching for different cycles, respectively (the inserted image schematically shows the stretching process). Source: Zhang et al. 2015 [26]. Copyright 2015, Reprinted with permission of John Wiley & Sons.

ZnO nanoparticle layer was used as the electron-transporting layer, and poly[4,8-bis(5-(2-ethylhexyl)thiophen-2-yl)benzo [1,2-*b*;4,5-*b*′] dithiophene-2, 6-diyl-*alt*-(4-octyl-3-fluorothieno[3,4-*b*] thiophene)-2-carboxylate-2-6-diyl] (PBDTTT-OFT) and [6,6]-phenyl-C71-butyric acid methyl ester ($PC_{71}BM$) BHJ was used as the photoactive layer. The obtained device is 3 μm thick and is ultra-flexible and ultra-lightweight (36.6 μg per unit device). The flexible solar cell exhibited unprecedented photovoltaic parameter values, including $V_{OC}$ of 0.785 V, $J_{SC}$ of 19.17 mA/cm$^2$, and FF of 0.697, giving a high PCE of 10.49% and a high power-per-weight of 11.46 W/g under illumination (AM 1.5 G), superior to that of other flexible organic photovoltaics (OPVs). These results indicate that the ultrathin and ultra-flexible devices are sufficient for application as a practical mobile and portable power source.

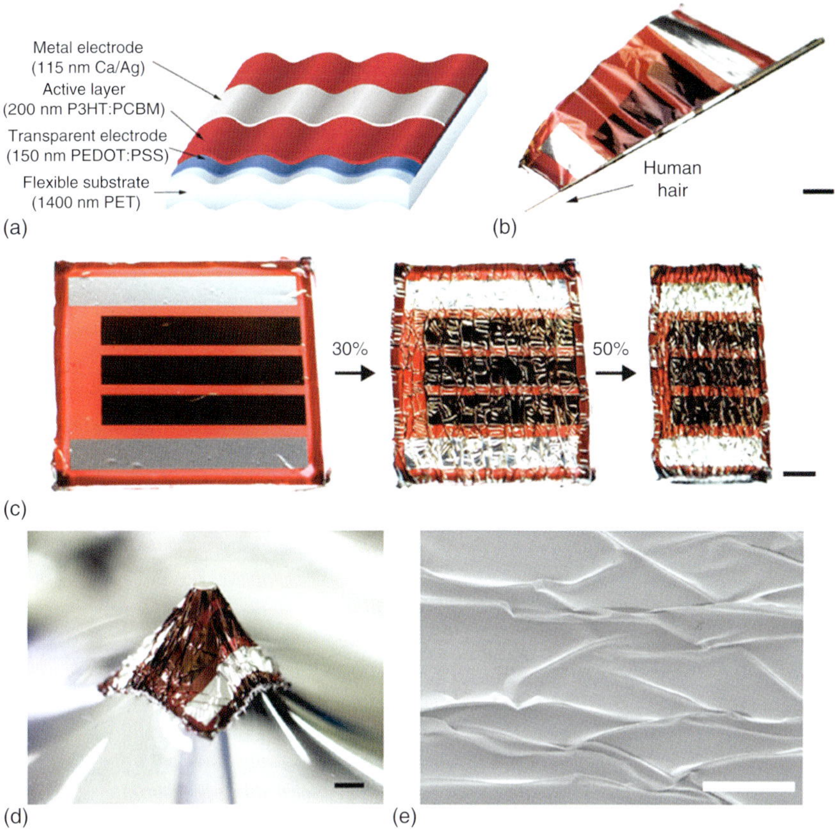


**Figure 5.8** (a) Scheme of the ultralight and flexible organic solar cell. (b) Extreme bending flexibility demonstrated by wrapping a solar cell around a human hair of 35 μm radius. Scale bar (also in c, d) 2 mm. (c) Stretchable solar cells made by attaching the ultrathin solar cell to a pre-stretched elastomer. They are shown flat (left) and at 30% (middle) and 50% (right) quasi-linear compression. (d) The device attached to the elastomeric support, under three-dimensional deformation by pressure from a 1.5 mm diameter plastic tube. (e) SEM image of the PET surface of the solar cell in compressed state. The radius of curvature for the shown wrinkles is estimated to be on the order of 10 μm. Scale bar 500 μm. Source: Kaltenbrunner et al. 2012 [27]. Copyright 2012, Reprinted with permission of Springer Nature.

One of the distinguishing features for PSC is the possibility of using roll-to-roll printing methods to print the device directly onto a thin flexible substrate to fabricate large-scale flexible devices [29–34]. In our group, we demonstrated the fabrication of scalable and highly conductive PEDOT:PSS:CNTs transparent electrode by roll-to-roll printing (Figure 5.10) [35]. The as-fabricated PEDOT:PSS:CNTs-based transparent electrode showed a high transmittance of 80.7% with a sheet resistance of 17 Ω/sq. When using the PEDOT:PSS:CNTs electrodes to fabricate PSCs, ITO-free flexible devices with a PCE of 7.47% and high stability were achieved. We also fabricated a PEDOT:PSS:graphene hybrid transparent electrode with high flexibility by using roll-to-roll printing

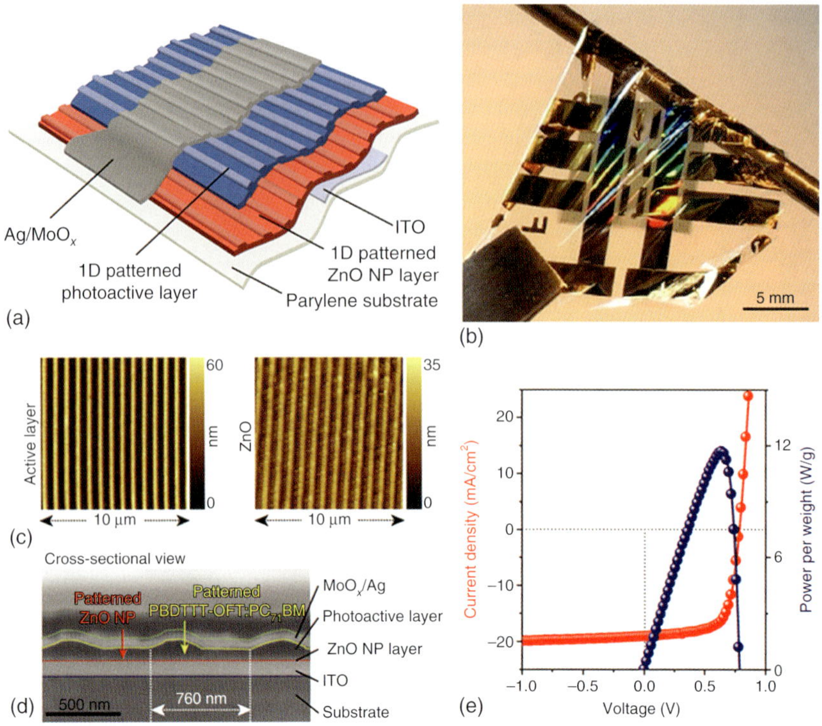


**Figure 5.9** (a) Structure of the OPV device. (b) Photograph of the OPV device wrapped over a spatula rod and pulled by tweezers, after delamination from the supporting glass substrate. (c) Atomic force microscopy (AFM) images of the grating-patterned surface of the bulk-heterojunction layer (left) and the ZnO NP layer (right). (d) High-resolution cross-sectional SEM image of the grating-patterned structure. (e) Current density–voltage (*J*–*V*; red) and power-per-weight–voltage (*P*–*V*; blue) curves of the freestanding devices. Source: Park et al. 2018 [28]. Copyright 2018, Reprinted with permission of Springer Nature.

[36]. Flexible PSCs based on the hybrid transparent electrode provided a high PCE of 9.4%. We also demonstrated roll-to-roll fabrication of flexible orientated graphene transparent electrodes by shear force and one-step reducing post-treatment [37]. Flexible PSCs based on the transparent electrode with a PCE of 8.32% was achieved with an effective device area up to 1.01 $cm^2$. By using the slot-die roll-to-roll technique, we demonstrated the fabrication of composite transparent electrodes composed of PEDOT:PSS and Ag grid/PET with optimized mechanical and optical properties [38]. The as-fabricated transparent electrode exhibited superior flexibility; after bending 2000 times under various bending situations, no obvious performance loss was found. When using the electrode to fabricate flexible PSCs, a maximal PCE of 8.08% was achieved.

### 5.2.3 Flexible and Wearable Perovskite Solar Cells

Currently, hybrid inorganic–organic PVSCs have attracted considerable attention due to their potential for low-cost solution fabrication processing and high

**Figure 5.10** (a) Roll-to-roll machine, (b) roll-to-roll process of PEDOT:PSS:CNTs printing, the wet PEDOT:PSS:CNTs film (dark blue), (c–e) picture of the gravure-printed PEDOT:PSS:CNTs electrode with superior flexibility and high transparency, the corresponding dried films (light blue), and (f) transparent PEDOT:PSS:CNTs electrode coated onto PET substrate showing the best sheet resistance of 17 (Ω/sq). Source: Hu et al. 2014 [35]. Copyright 2014, Reprinted with permission of American Chemical Society.

PCE (presently exceed 22%) [39–49]. The PVSC works like a PSC. Briefly, under illumination, the excitons are generated in the perovskite layer and dissociate into electrons and holes. The holes are transferred to the hole-transport layer and the electrons are transferred to the electron-transport layer from the perovskite layer, and the carriers are transferred to the respective electrodes.

PVSCs, endowed with high efficiency and solid state, have attracted increasing interest in transformation toward flexibility [43, 49, 50]. Much effort has been devoted to developing a solution process to fabricate flexible PVSCs. By using a low-temperature and solution-based process (electrodeposited ZnO as a compact layer and chemical bath deposition grown nanorods), Mathews and coworkers have fabricated flexible devices on the flexible substrate with promising performance (Figure 5.11a) [51]. The as-fabricated flexible devices show a PCE of 2.62%. Kelly and coworker using a thin film of ZnO nanoparticles as an electron-transport layer successfully constructed a high-efficiency flexible PVSC with conversion efficiencies of 10.3% (Figure 5.11b) [52]. Snaith and coworkers demonstrated that a device in planar heterojunction structure consisting of flexible ITO plastic foil, PEDOT:PSS, $CH_3NH_3PbI_{3-x}Cl_x$ perovskite absorber, PCBM, TiO*x*, and aluminum can deliver a PCE of 6.5% [54]. Similar

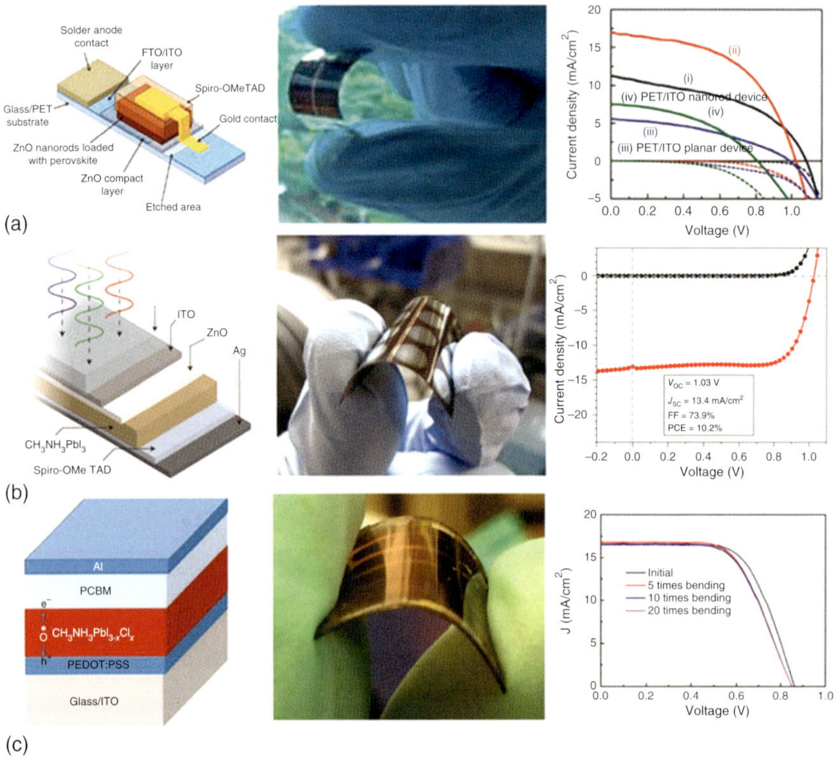


**Figure 5.11** (a) Flexible PVSC in a mesoscopic structure. (b) Planar structure and (c) compatible with PSC structure and modification materials. Source: (a) Kumar et al. 2013 [51]. Copyright 2013, Reprinted with permission of Royal Society of Chemistry; (b) Liu and Kelly 2013 [52]. Copyright 2014, Reprinted with permission of Springer Nature. (c) You et al. 2014 [53]. Copyright 2014, Reprinted with permission of American Chemical Society.

work was also reported by Yang and coworkers [53]. The device structure was PET/ITO/PEDOT:PSS/$CH_3NH_3PbI_{3-x}Cl_x$/PCBM/Al (Figure 5.11c). The flexible device shows $J_{SC}$ of 16.5 mA/cm$^2$, $V_{OC}$ of 0.86 V, FF of 64%, and achieves a PCE of 9.2%. The flexible device also tolerates repeated mechanical deformation. After mechanical bending up to 20 times, the device still maintains its performance.

Stretchable electronics may offer unrestricted conformability with biological tissues, which will require multidimensional deformability. Kaltenbrunner et al. reported planar flexible and stretchable perovskite solar arrays with a total thickness of 3 μm, including substrate, electrodes, and a protective encapsulating layer [55]. The cells displayed a stabilized PCE of 12% after being operated for days under normal environmental conditions. The devices were fabricated at low temperatures from low-cost materials at high yield. PET foils of 1.4 μm thickness served as the substrate, and PEDOT:PSS was used as the transparent hole selective electrode. An alkylammonium lead trihalide $RNH_3PbX_3$ perovskite served as the active layer. N,N′-dimethyl-3,4,9,10-tetracarboxylic perylene diimide (PTCDI) or 6,6-phenyl C61 butyric acid methylester (PCBM) were used as electron-transport layers. A chromium layer with accompanying $Cr_2O_3$

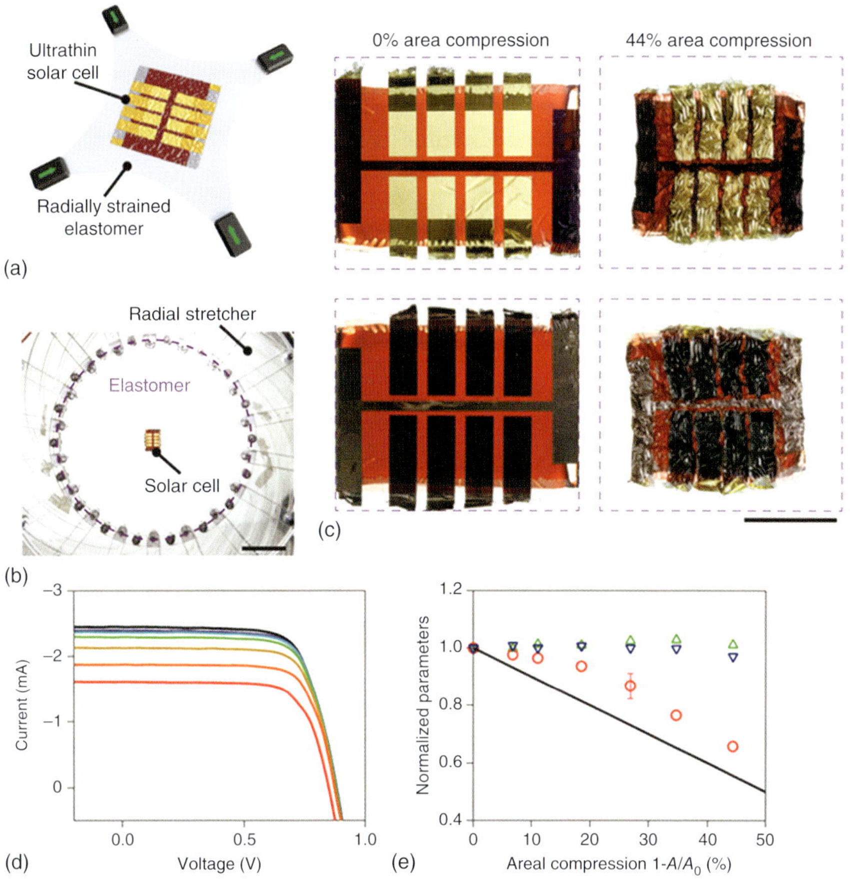


**Figure 5.12** (a) Schematic of radially stretchable solar cells. (b) Photograph of a solar panel mounted on a radial extender. Scale bar, 5 cm. (c) Close-up images of a solar foil undergoing radial compression down to 44% in area. Scale bar, 1 cm. (d) *I–V* characteristics measured while the device is radially compressed. Starting from a flat configuration (black solid trace), the cell is compressed down to a 44% decrease in area (red solid trace). (e) Solar cell performance metrics as a function of radial compression. $V_{OC}$ (blue downward triangles), FF (green upward triangles), $J_{SC}$ (red circles), and device area (solid black line). Source: Kaltenbrunner et al. 2015 [55]. Copyright 2015, Reprinted with permission of Springer Nature.

stabilizes the metal top contact for operation in ambient air. Low-resistivity metals, for example, gold, copper, and aluminum, complete the device. Optionally, polyurethane (PU) serves as a 1 μm thick capping layer for mechanical protection when transferring the ultrathin perovskite solar foils onto a radially pre-stretched elastomer. On relaxation of the tape, a complex microstructure of folds and wrinkles forms, allowing subsequent multidirectional deformation (Figure 5.12). When the device is compressed, the $V_{OC}$ and FF remained virtually unchanged. The $J_{SC}$ decreased with reduction of illuminated active area. The decrease in $J_{SC}$ was less than expected based on the shrinkage of device area.

The ultrathin perovskite solar foils endure up to 44% areal compression with changeless photovoltaic performance.

Using roll-to-roll process for large-scale production of PVSCs, each layer of the device requires solution operation possibility, flexibility, and full printability [56, 57]. However, perovskite is a crystalline material, and hence cracks may occur upon bending. Thus, the practical application of roll-to-roll process is limited to fabricating large-scale PVSCs. Bolink and coworkers fabricated a flexible solar cell on a PET substrate and investigated the bending endurance of the device [58]. After bending for 50 times, no obvious performance deterioration was found, and the devices retained a PCE of 7%, suggesting the compatibility of perovskite with roll-to-roll processing. Chen and coworkers found that incorporating elastomer PU into the perovskite precursor solution can retard the crystallization rate of perovskite and improve its bendability [59]. The device delivered an impressive PCE of 18.7%. After bending 200 times, no obvious cracks were detected on the perovskite films, due to the denseness and high elasticity created by PU, which forms a cross-linked network at the grain boundaries. Kelly and coworker suggested that the flexibility of PVSC is limited by the flexibility of the PET/ITO substrate and not by the perovskite layer [52]. Song and coworkers designed a nanocellular scaffold as an interfacial layer to construct a mechanics buffer layer for wearable PVSCs [60]. The nanocellular scaffold releases mechanical stresses during flexural experiences and significantly enhances the mechanical stability of the perovskite films. Thus, these flexible PVSCs were practically fabricated in modules as a wearable solar-power source.

van Hest and coworkers demonstrated highly crystalline, large-scale, smooth, and uniaxially oriented perovskite films printed in the ambient environment at room temperature [61]. In their case, an all blade-coated metal halide perovskite cell with $V_{OC}$ of 1.124 V, $J_{SC}$ of 22.86 mA/cm$^2$, and FF of 76.17% resulted in a PCE up to 19.6% (Figure 5.13). When $MAPbI_3$ films were coated on the $SnO_2$/ITO/glass substrate with slot-die coating, the resulting devices show $V_{OC}$ of 1.111 V, $J_{SC}$ of 22.4 mA/cm$^2$, and FF of 69.5%, with PCE of 17.31%. Slot-die coating is perfectly implementable with R2R coating. They also tested the slot-die coating method on the R2R printing of perovskite on a flexible Corning Willow Glass substrate. The resulting R2R slot-die coated flexible glass-based devices (device area: 0.15 cm$^2$) with $V_{OC}$ of 1.076 V, $J_{SC}$ of 21.24 mA/cm$^2$, and FF of 61.7% gave an efficiency of 14.12%. The developed printing method can be applied to various perovskite compositions and pave the way for the future R2R printing of highly efficient perovskite cells.

With the increasing attention on portable and wearable devices, conventional planar flexible energy devices could hardly fulfill the demand. Electronic textiles are highly desired in diverse fields. Combining high efficiency and solid state, PVSCs are promising in fiber-shaped energy devices [8]. Peng and coworkers demonstrated a novel coaxial fiber-shaped PVSC with low cost and high flexibility [62]. In the coaxial fiber-shaped PVSC (Figure 5.14), a stainless steel fiber with a compact mesoporous $TiO_2$ layer was used as the anode, and a perovskite $CH_3NH_3PbI_3$ layer was used as active layer, and a semi-transparent MWCNT sheet wound over the inner fiber as the cathode. This all-solid-state fiber-shaped solar cell provided an energy conversion efficiency of 3.3% ($J_{SC}$ = 10.2 mA/cm$^2$,

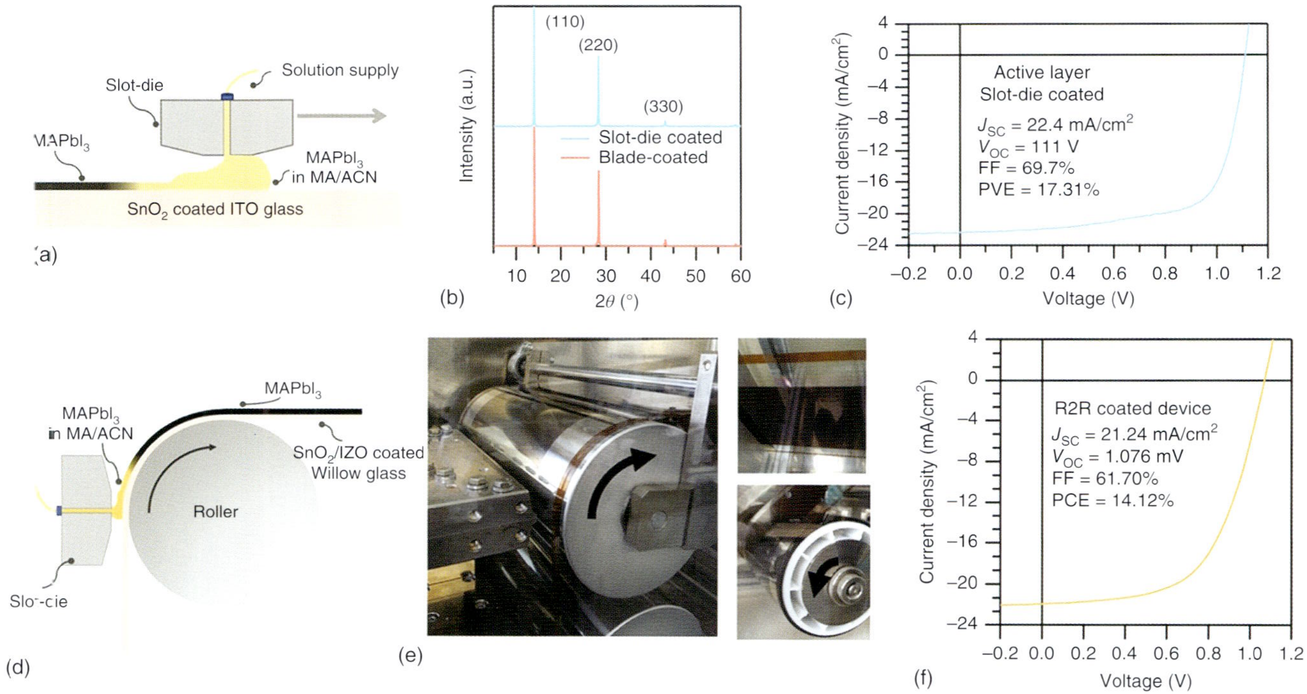


**Figure 5.13** (a–c) Schematic representation of slot-die coating on rigid glass (a), film X-ray diffraction (XRD) (b), and device performance (c). (d) Schematic representation of slot-die coating on flexible glass with a R2R coater. (e) Photograph of R2R coating in action (left), after annealing (top right), and the rewinding process (bottom right). (f) *J–V* curve of the champion R2R-coated device. Source: Dou et al. 2018 [61]. Copyright 2018, Reprinted with permission of American Chemical Society.

$V_{OC}$ = 0.664 V, FF = 0.487) even after 50 bending cycles. When the angle of illumination was varied from 0–180° the photovoltaic parameters for the coaxial fiber-shaped PVSC remain unchanged. Moreover, the fiber-shaped PVSC can further be woven into flexible textiles. Peng and coworkers developed a cathodic deposition technique to fabricate fiber-shaped PVSCs [63]. By using this method, perovskite layers can provide uniformity and high coverage on titanium (Ti) wires. When transparent aligned CNT sheet is used to cover perovskite layer modified Ti wire to produce a coaxial fiber-shaped PVSC, a high open-circuit voltage of 0.85 V with a high PCE of 7.1% has been achieved. Peng and coworkers also developed elastic fiber-type PVSC by using a stretchable aligned CNT based fiber and a modified spring-like Ti wire as electrodes [64]. These elastic fiber-type PVSCs provided energy conversion efficiencies ranging from 4.81% to 5.22%. These devices also showed good stretchability. Even after 250 stretching cycles, the devices maintained 90% performance, making them feasible candidates for practical applications. Li et al. developed a double-twisted fiber-shaped PVSC by using CNT electrodes and packing in a transparent polymer film [65]. The as-prepared device provided a PCE of 3.03% and a bending stability for more than 1000 cycles. Zou and coworkers fabricated a fiber-shaped PVSC with the structure of Ti/c-$TiO_2$/meso-$TiO_2$/perovskite/spiro-OMeTAD/Au. The fiber-shaped PVSCs exhibited a PCE of 5.3% and 8.4% under AM 1.5 illumination and in the diffuse model, respectively [66].

### 5.2.4 Flexible and Wearable Supercapacitors

Owing to the rapid development of flexible and wearable electronic devices, development of high-performance wearable energy conversion and storage devices is urgently desired [9, 67]. Supercapacitors and rechargeable batteries are two typical kinds of energy storage devices [68–73]. However, the lower power densities of batteries limits their application. Compared with batteries, supercapacitors show many advantages such as ultrafast charge/discharge rates, higher power densities, and superb cycling stability. Consequently, the development of flexible and wearable supercapacitors is considered to be an efficient approach to meet the high demand for wearable power sources. Moreover, their long cycle lifetime renders supercapacitors more suitable than batteries. According to the capacitive storage mechanisms, supercapacitors can be mainly divided into electric double-layer capacitors (EDLCs) and pseudocapacitors [74, 75]. The working mechanisms of two kinds of supercapacitors and batteries as well as their differences are outlined in Figure 5.15. EDLCs store energy through ion adsorption/desorption, and thus display fast charge/discharge rates and superior cycling stability [76–78]. The charge storage of pseudocapacitors by means of fast reversible redox reactions usually shows higher energy density than EDLCs [79, 80].

The supercapacitor device performance is highly dependent on electrode materials. The requirements of electrode materials for high-performance wearable supercapacitors include lightweight, flexibility, high energy density, high power density, safety, durability, and so on. Recently, numerous efforts have

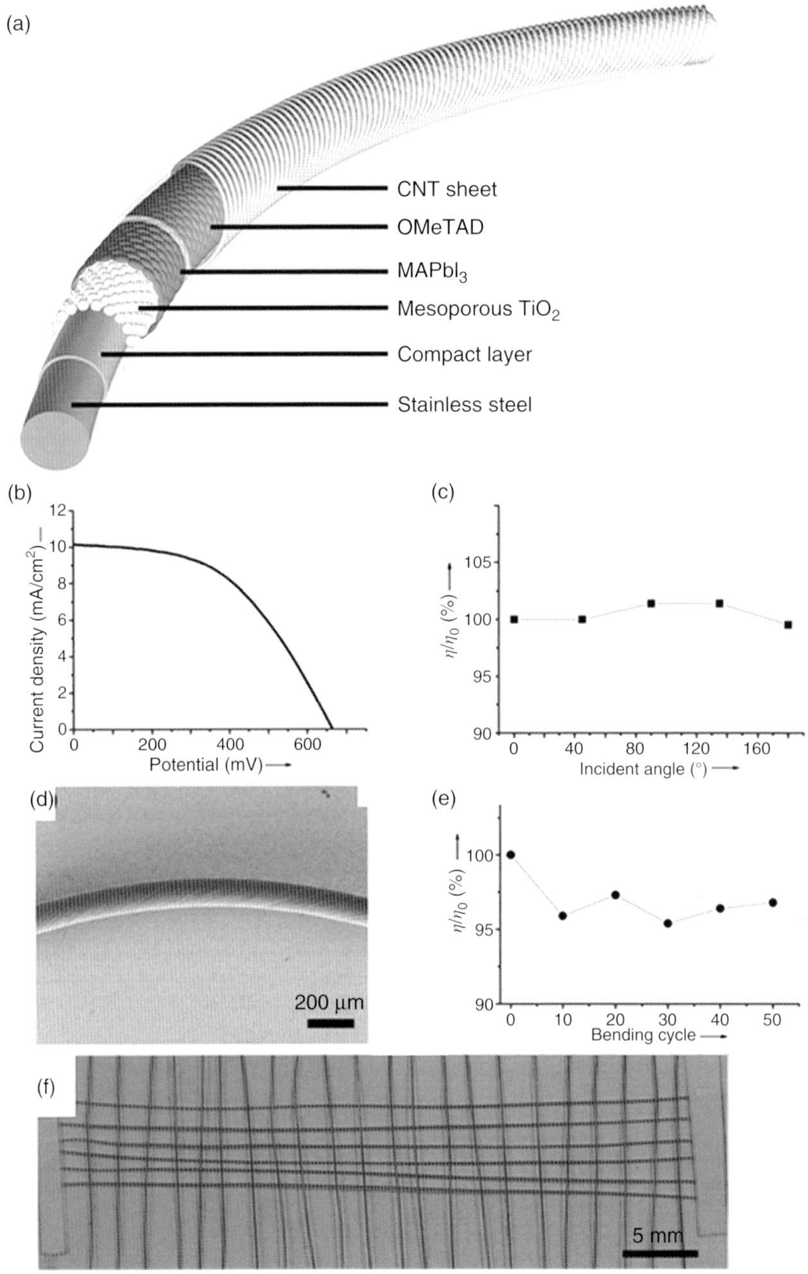


**Figure 5.14** (a) Structure of the fiber-shaped PVSC. (b) *J–V* curve of a coaxial fiber-shaped PVSC. (c) Dependence of the energy conversion efficiency on the angle of incident light. Here, $\eta_0$ and $\eta$ correspond to the energy conversion efficiencies at 0° and the other angle, respectively. (d) SEM image of a bent fiber-shaped PVSC. (e) Dependence of the energy conversion efficiency on the bending cycle number. Here, $\eta_0$ and $\eta$ correspond to the energy conversion efficiencies before and after bending, respectively. (f) Photograph of a textile. Source: Qiu et al. 2014 [62]. Copyright 2014, Reprinted with permission of John Wiley & Sons.

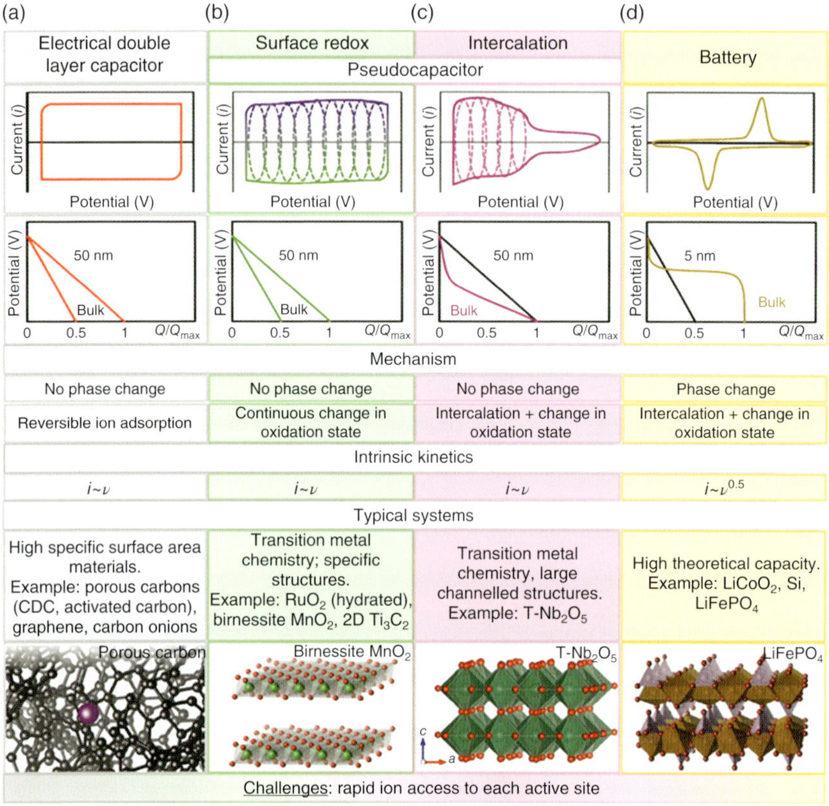


**Figure 5.15** Faradaic and capacitive energy storage. Summary of the characteristic metrics such as cyclic voltammetry, galvanostatic profiles, key mechanism descriptions, and typical systems that are known to utilize the mentioned charge storage mechanisms: (a) double-layer capacitor, (b) surface redox pseudocapacitance due to adsorption and/or fast intercalation of ions, (c) intercalation pseudocapacitance, and (d) batteries. Source: Lukatskaya et al. 2016 [75]. Copyright 2016, Reprinted with permission of Springer Nature. https://www.nature.com/articles/ncomms12647#rightslink. http://creativecommons.org/licenses/by/4.0/. Licensed under CCBY 4.0

been devoted to developing high-performance electrode materials for supercapacitors. More and more materials are employed to fabricate supercapacitors, as shown in Figure 5.16. Nanocarbons (such as CNTs and graphene) [78, 81–87], conducting polymers (such as PANi, PPy, and PEDOT) [88–93], and transition metal oxide/hydroxides/sulphides [94–103] have been widely investigated as electrode materials for supercapacitors. In addition, several new electrode materials, such as MXenes (transition metal carbides, nitrides, and carbonitride, e.g. $Ti_3C_2$, $Ti_4N_3$, and $Ti_3CN$) [104–109], metal organic frameworks (MOFs) [110–115], covalent organic frameworks (COFs) [116–123], polyoxometalates (POMs) [124], and black phosphorus (BP) [125], have been introduced recently due to their many advantages in meeting the requirements of high-performance supercapacitors.

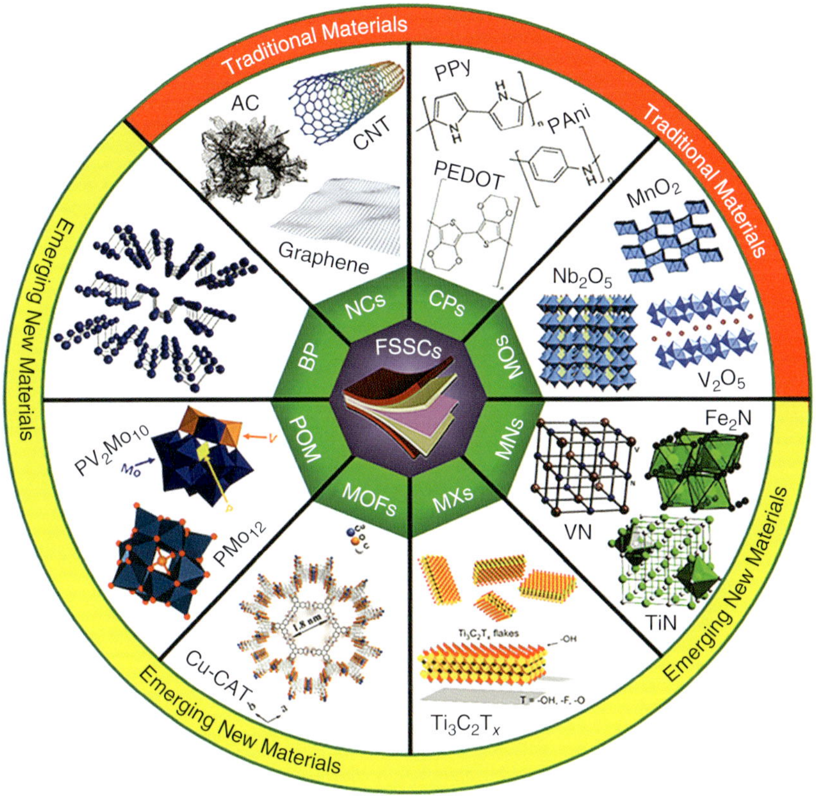


**Figure 5.16** An overview of promising electrode materials used so far for flexible and wearable supercapacitors. The materials are divided into two subsections such as traditional materials (NCs – nanocarbons, CPs – conducting polymers, MOs – metal oxides) and emerging new materials (MNs – metal nitrides, MXs – MXenes, MOFs – metal organic frameworks, POMs – polyoxometalates, BP – black phosphorus). Source: Dubal et al. 2018 [69]. Copyright 2018, Modified and reprinted with the permission of Royal society of Chemistry.

Supercapacitor configuration is also an important factor defining device performance. In order to be easily incorporated in wearable electronics, it is highly desirable that supercapacitors have a compact design and flexibility. Therefore, except for the electrode materials, further investigations must focus on breakthroughs for new designs of supercapacitor configuration. Several innovative supercapacitor designs for flexible/bendable/wearable electronics are shown in Figure 5.17. Generally speaking, there are three main cell designs including sandwich-type, planar, and fiber-shaped supercapacitors. Fiber-shaped supercapacitors can be divided into wire-type, fiber on plane, and cable-type supercapacitors. Sandwich-type supercapacitors have the most common device design since these can be fabricated easily. Planar supercapacitors are of high volumetric capacitances and considered to be promising power sources for implantable medical devices. Cable-type and wire-shaped supercapacitors are

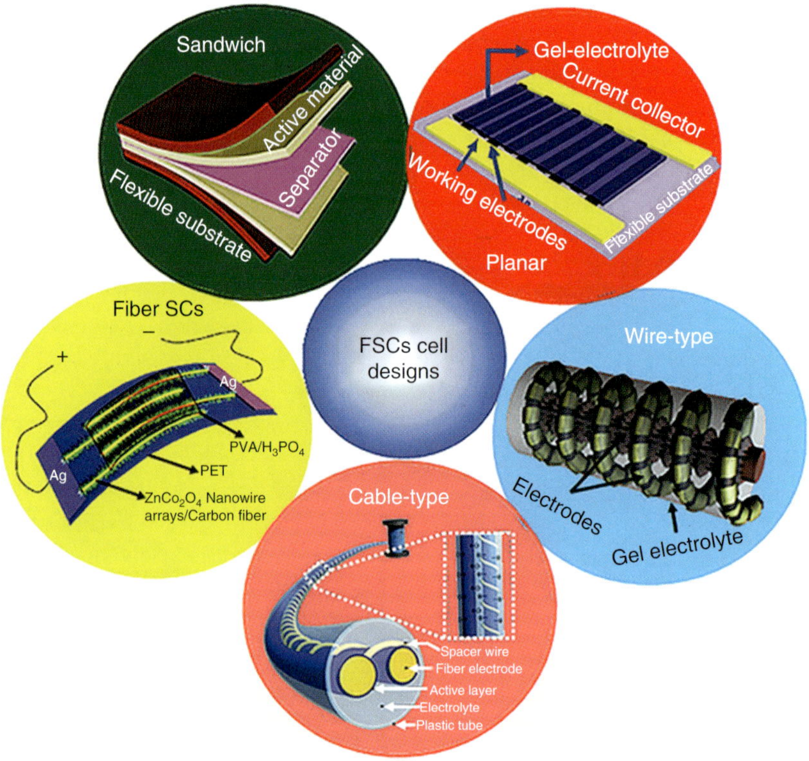


**Figure 5.17** Recently developed innovative flexible supercapacitor (FSCs) device configuration designs: conventional sandwich type, planar, fiber, wire-type, and cable-type. Source: Dubal et al. 2018 [69]. Copyright 2018, Reprinted with the permission of Royal Society of Chemistry.

suitable for wearable energy textiles and can be designed to stretch. Thus, all the three device configurations are important depending on the application perspective. Therefore, flexible and wearable supercapacitors for integration into wearable electronics require to be designed to store sufficient amount of energy, have good mechanical flexibility, and be safe to satisfy the application requirements. More essentially, flexibility is frequently mentioned as a strict requirement for flexible and wearable electronics.

### 5.2.5 Flexible and Wearable Electric Double-Layer Capacitors (EDLCs)

Wearable electronics requires wearable energy storage devices with high flexibility, high energy density and retention, and feasibility for integrating into different wearable systems. EDLCs using nanocarbon materials (such as CNTs and graphene) as electrode materials are promising supercapacitors (SCs) to satisfy the electrochemical requirement because of the high electrical conductivity and stable electrochemical properties suitable for high charging speed, long life cycle, and good safety. Müllen and coworkers directly printed in-plane supercapacitors on paper and ultrathin polyethylene terephthalate

(PET) substrates using electrochemically exfoliated graphene as active materials [126]. The fabricated supercapacitors on a paper substrate provided a significant areal capacitance of 5.4 mF/cm$^2$, and showed an excellent rate capability of 75% when operated from 10 to 1000 mV/s. Moreover, use of a PET substrate resulted in an ultrathin supercapacitor that exhibited ultra-flexibility, thus making it suitable for flexible electrochemical energy storage devices. Kaner and coworkers reduce graphite oxide films to graphene by using the laser of a standard LightScribe DVD optical drive [127]. The as-prepared graphene films show high electrical conductivity (1738 S/m), robust mechanical strength, and high specific surface area (1520 m$^2$/g). When directly using the graphene films as electrodes to fabricate supercapacitors in different electrolytes without binders and current collectors, the devices afford ultrahigh energy density values, high power density, and excellent cycle stability (96.5% after 10 000 cycles). Moreover, these supercapacitors also showed excellent flexibility with only 5% loss after 1000 bending cycles, and are thus promising for flexible electronics. Jiang and coworkers developed a stretchable supercapacitor by using sinusoidal single-walled carbon nanotube (SWCNT) macrofilms as stretchable electrodes with a polymeric separator and an organic electrolyte [128]. The as-fabricated stretchable supercapacitors exhibited good energy and power densities. Moreover, the stretchable supercapacitors showed unchanged electrochemical performance even under 30% applied tensile strain. Yu and coworkers prepared a foldable porous graphene film with engineered microvoids by reduction of graphene oxide film under confinement [129]. Even after thermal annealing at a high temperature of 1300 °C, the folding performance of the porous graphene film is not compromised. When the porous graphene film is used as electrode material to fabricate solid-state supercapacitor, the device also shows good foldability. After 2000 cycles of single folding followed by another 2000 cycles of double folding, the device displays nearly identical capacitance performance (Figure 5.18). Xie and coworkers reported a graphene-based solid-state supercapacitor with a zwitterionic gel electrolyte, manifesting in superior electrochemical performance [130]. The flexible solid-state supercapacitor achieves a high volume capacitance of 300.8 F/cm$^3$ at 0.8 A/cm$^3$ with a rate capacity of 85.1% when the current density increases from 0.8 to 20 A/cm$^3$.

Fiber-based or wire-shaped supercapacitors can be integrated into stretchable fabrics or yarns easily to meet a more practical requirement of wearable energy storage in our daily life [131–136]. Wallace and coworkers reported a superelastic wet-spun hybrid CNT graphene fiber to achieve a high-performance fiber-based supercapacitor [131]. When using the hybrid CNT graphene fibers to prepare spring-like coiled fiber and coated with an elastic polymer to fabricate spring-like supercapacitor, the device provides a specific capacitance of 138 F/g and shows an excellent elasticity capable of 800% strain. The elastic rubber coating enables remarkable stretchability. After thousands of cycles with up to 500% strain, the performance of the device shows no significant change. Multiple devices can be easily assembled in parallel or series to fulfill specific energy and power needs. Zheng and coworkers report a hierarchical graphene–metallic textile composite electrode concept to fabricate one-dimensional flexible supercapacitor yarns with high energy and power density [137]. The hierarchical composite electrodes

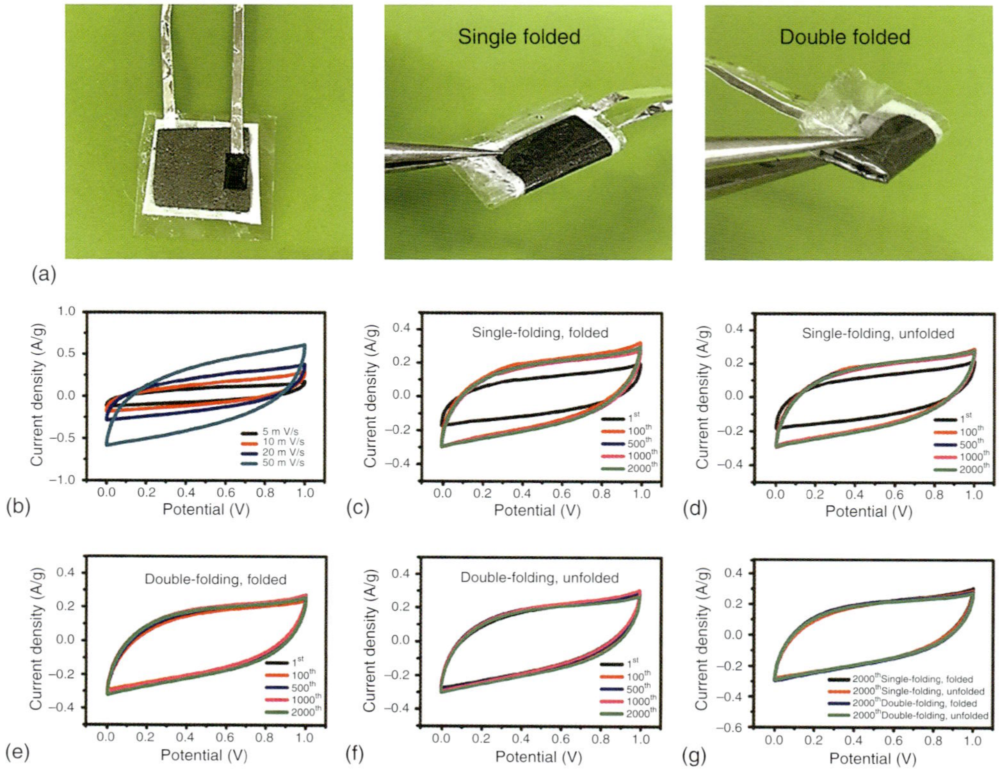


**Figure 5.18** (a) Optical images of the flexible rGO film supercapacitor with the unfolding (left), single-folding (middle), and double-folding situation (right). (b) *CV* curves of the supercapacitor at different scan rates, *CV* curves of the supercapacitor with (c) single folding (in the folded state) and (d) single folding (in the unfolded state). Data are shown after the 1st, 100th, 500th, 1000th, and 2000th folding/unfolding cycle. Corresponding results for (e) double folding (in the folded state) and (f) double folding (in the unfolded state), after the 1st, 100th, 500th, 1000th, and 2000th cycle of folding and unfolding. (g) Comparison of single vs. double folds of the supercapacitor after 2000 folding/unfolding cycles at a scan rate of 5 mV/s. Source: Huang et al. 2018 [129]. Copyright 2018, Reprinted with permission of John Wiley & Sons.

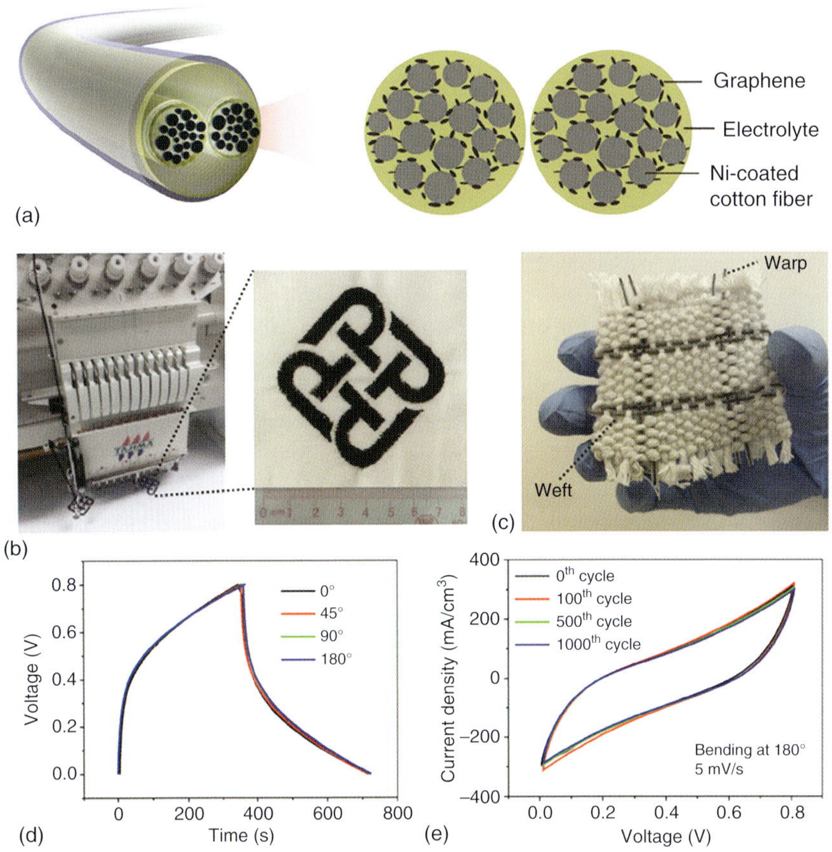


**Figure 5.19** Wearable applications of solid-state SC yarns. (a) Schematic illustration of the structure of one SC yarn. (b) Digital images of embroidery logos of The Hong Kong Polytechnic University using the composite electrode yarns. (c) A woven fabric made with solid-state SC yarns. (d) Galvanostatic charge-discharge (GCD) curves of solid-state SC yarns at different bending angles. (d) CV curves of the device at different bending (180° bending angle) cycles. Source: Liu et al. 2015 [137]. Copyright 2015, Reprinted with permission of Springer Nature. https://www.nature.com/articles/ncomms8260#rightslink. http://creativecommons.org/licenses/by/4.0/. Licensed under CCBY 4.0

are fabricated by highly scalable electrochemical deposition of graphene and electroless deposition of Ni on commercial cotton yarns. When using one pair of these composite electrodes to fabricate all solid-state supercapacitor yarn, the devices showed high power density and volumetric energy density of 1400 mW/cm$^3$ and 6.1 mWh/cm$^3$, respectively (Figure 5.19). In addition, the supercapacitor yarn is highly flexible, lightweight, strong, durable in bending fatigue and life cycle tests, and can be integrated into different wearable electronic devices.

### 5.2.6 Flexible and Wearable Pseudocapacitor

Pseudocapacitors store energy by fast reversible redox reactions and show higher energy density than EDLCs. Conducting polymers (such as polypyrrole [PPy], polyaniline [PAni], and polythiophene [PT]) and metal oxides (such as $MnO_2$, $RuO_2$, $Nb_2O_5$, and $Fe_3O_4$) are commonly used as active materials for pseudocapacitors [80, 138–140]. Zhu et al. fabricated a freestanding, highly flexible supercapacitor electrode by hybridizing MXene (l-$Ti_3C_2$) with polypyrrole chains (PPy/l-$Ti_3C_2$) [141]. Flexible supercapacitors assembled with the freestanding electrode afford a high capacitance of 203 mF/cm$^2$ (406 F/cm$^3$). Moreover, the devices based on PPy/l-$Ti_3C_2$ exhibited superior cycling stability and excellent deformation tolerance. After 20 000 charging/discharging cycles, no obvious capacitance loss was observed. This study provides an approach to improve the capacitance and cycling stability of conducting polymer-based electrodes, making them more attractive for practical applications. Cho and coworkers used an assembly approach (ligand-mediated layer-by-layer assembly) to minimize the contact resistance between neighboring metal and/or metal oxide nanoparticles to produce metallic cellulose paper-based supercapacitor electrodes [142]. Using this method, insulating paper can be converted into highly porous metallic paper. The areal capacitance and rate capability of supercapacitors can be easily tailored by alternating the metal and pseudocapacitive nanoparticles on the metallic papers. The maximum energy and power density of the metallic paper-based flexible supercapacitors are estimated to be 267.3 μWh/cm$^2$ and 15.1 mW/cm$^2$, respectively. Zhou and coworkers developed a strategy to prepare layered highly conductive thick PEDOT:PSS [143]. Using the PEDOT:PSS films to fabricate all solid-state flexible symmetric supercapacitors, a high volumetric energy density of 6.80 mWh/cm$^3$ was achieved at a power density of 100 mW/cm$^3$.

Ma and coworkers fabricated polyaniline–polyvinyl alcohol hydrogel through dynamic boronate bond [144]. The as-prepared conductive hydrogels had remarkable mechanical and electrochemical properties. Flexible solid-state supercapacitors based on the conductive hydrogels show remarkable electrochemical capacitance (306 mF/cm$^2$ and 153 F/g). The device exhibits excellent mechanical stability and remarkable cyclic lifetime. After 1000 mechanical folding cycles the device maintained 100% capacitance retention. The robustness and high capacitance performance enable the as-fabricated supercapacitor as a promising power source for flexible and wearable electronics.

Kang and coworkers fabricated breathable and wearable supercapacitors by using highly flexible paper electrodes (Figure 5.20) [145]. The paper electrodes were prepared by depositing $MnO_2$ and CNTs as electrochemically active materials on flexible air-laid paper supports. The electrodes possess exceptional rate capability and cycling stability. In addition, the electrodes exhibit superior flexibility, can be repeatedly bent, stretched, kinked, crumpled into a ball, freely cut (into wire-like, stripe-shaped, and round electrodes), and even folded into a paper crane. The high flexibility makes the electrodes show stable electrochemical performances after various deformations. Moreover, flexible solid-state supercapacitors assembled with the paper electrodes provide almost identical capacitance under different bending states. Breathable supercapacitors were obtained by making numerous through-holes on the supercapacitors. The breathable supercapacitors also show good flexibility and capacitance performance, with 94% of its initial capacitance being retained. After 200 repeated bending cycles from 0–180°, more than 91% of its initial capacitance is retained.

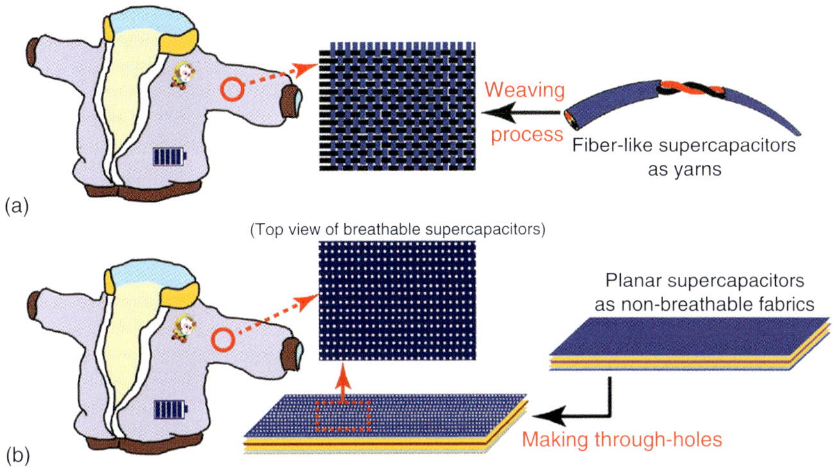


**Figure 5.20** Schematics for the production of breathable and smart garments through two strategies, i.e. (a) weaving flexible fiber-like supercapacitors into a textile, and (b) making numerous through-holes on flexible planar supercapacitors. Source: Dong et al. 2016 [145]. Copyright 2016, Reprinted with permission of John Wiley & Sons.

Fiber-shaped electrode plays an important role in flexible and wearable supercapacitors and their practical applications in portable and wearable electronics. Yu and coworkers fabricated a 3D polyaniline/graphene hydrogel via self-assembly of polyaniline hydrogels and graphene oxide macrostructures [146]. The as-prepared 3D hybrid hydrogel showed enhanced mechanical properties, outstanding flexibility, mouldability, and good electrochemical properties. All-gel-state fibrous supercapacitor based on the 3D hybrid hydrogel provided an excellent volumetric energy density of 8.80 mWh/cm$^3$ and can achieve a large strain up to ≈40%. Yao and coworkers fabricated zinc–nickel–cobalt oxide (ZNCO)@Ni(OH)$_2$ nanowire arrays (NWAs) on a carbon nanotube fiber (CNTF) as electrode for supercapacitors [98]. The as-prepared fiber-shaped electrode had an ultrahigh specific capacitance of 2847.5 F/cm$^3$ (10.678 F/cm$^2$) at a current density of 1 mA/cm$^2$. Hu and coworkers had grown hierarchical tectorum-like $\alpha$-$Fe_2O_3$/polypyrrole nanoarrays (T-$Fe_2O_3$/PPy NAs) on conductive carbon cloth as electrode for supercapacitors [147]. The electrode displayed a high areal capacitance of 382.4 mF/cm$^2$ at a current density of 0.5 mA/cm$^2$. When using T-$Fe_2O_3$/PPy NAs and $MnO_2$ as electrodes to fabricate a solid-state asymmetric supercapacitor, the device provided a high energy density of 0.22 mWh/cm$^3$. Chen and coworkers fabricated fiber-like MoS2-rGO/MWCNT electrodes by twisting a well-aligned MWCNT sheet incorporated with $MoS_2$ and reduced graphene oxide (rGO) nanosheets [148]. This fiber-based flexible, solid-state, asymmetric supercapacitor provided a wide potential window of 1.4 V with good rate and cycling stability and high energy density.

In our group, we designed and fabricated a hierarchical electrode by growing core-branch Al-doped cobalt sulfide nanosheets anchored on Ni nanotube arrays combined with carbon cloth (denoted as CC/H-Ni@Al-Co-S) [94]. Owing

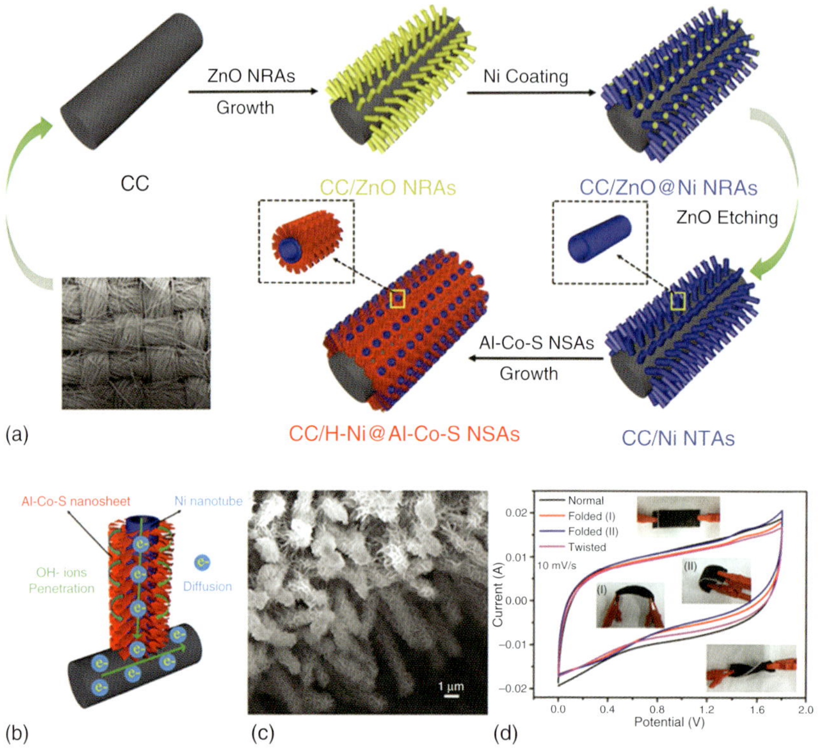


**Figure 5.21** (a) Schematic illustration of the fabrication of a hierarchical core-branch CC/H-Ni@Al-Co-S nanosheet electrode. (b) Schematic illustration displaying the merits of the core-branch CC/H-Ni@Al-Co-S electrode for energy storage. (c) SEM images of the CC/H-Ni@Al-Co-S nanosheet electrode. (d) *CV* curves of the flexible ASC under different bending conditions. The insets of (d) show the photographic images of the device under different bending states. Source: Huang et al. 2018 [94]. Copyright 2018, Reprinted with permission of American Chemical Society.

to the combination of structural and compositional advantages, the as-prepared self-standing electrode can serve as cathode for flexible asymmetric supercapacitors. The electrode showed superior rate capability and high specific capacitance of 2434 F/g at 1 A/g in a three-electrode configuration. When CC/H-Ni@Al-Co-S was used as cathode and multilayer graphene/CNT film as anode to fabricate all-solid-state asymmetric supercapacitors, the devices provided a high energy density up to 65.7 Wh/kg and superb cycling stability. Moreover, the asymmetric supercapacitors also showed good flexibility and stability under different bending conditions (Figure 5.21). Using the combination of coaxial electrospinning technology and a hydrothermal process, we fabricated a hierarchical structure of $NiCo_2S_4$ nanosheets aligned on a double capillary carbon nanofiber ($NiCo_2S_4$@DCCNF) cloth to produce a self-standing electrode [149]. Flexible all-solid-state supercapacitors based on NiCo2S4@DCCNF afford good electrochemical performance with a high capacitance of 166 F/g and a maximum energy density of 55.6 Wh/kg as well as remarkable electrochemical stability.

Electrolyte materials are also very important for flexible supercapacitors. Zhi and coworkers reported an intrinsic self-healable and highly stretchable electrolyte comprising polyacrylic acid dual cross-linked by hydrogen bonding and vinyl hybrid silica nanoparticles for flexible, self-healable, and stretchable supercapacitors [150]. Supercapacitors assembled by the use of PPy-deposited CNT paper electrodes with this polyelectrolyte exhibit the merits of self-healability and stretchability. After 20 cycles of breaking/healing, the devices retained the capacitance completely. Moreover, the supercapacitors can be stretched up to 600% strain with enhanced performance. When polyacrylamide was used to replace polyacrylic acid to prepare electrolyte, the new electrolyte was super-stretchable and compressible [151]. The supercapacitor with the new electrolyte can be stretched up to 1000% strain with enhanced performance, and compressed to 50% strain with good retention of the initial performance.

### 5.2.7 Integrated Solar Cells and Supercapacitors

Although photovoltaic technology is an effective way to utilize solar energy, because of the intermittent nature of the sunlight the output power of solar cells cannot be continuous and stable. Therefore, a dual-functional device that combines solar cell with supercapacitor with photoelectric conversion and energy storage capability is desired. The concept of a solar cell/supercapacitor integrated device (named as photo-capacitor) was first proposed by Miyasaka and Murakami [152]. Attributed to their potential applications in future portable and wearable devices, photo-supercapacitors have been developing rapidly and receiving more and more attention [2, 8, 153–155]. According to the type of division, DSSC, PSC, and PVSC type photo-supercapacitors have been reported one after another (Figure 5.22).

Flexible power sources are highly desirable for next-generation wearable electronic systems. Yu and coworkers developed a novel planar-structure electrochemical device for highly flexible semitransparent power generation/storage

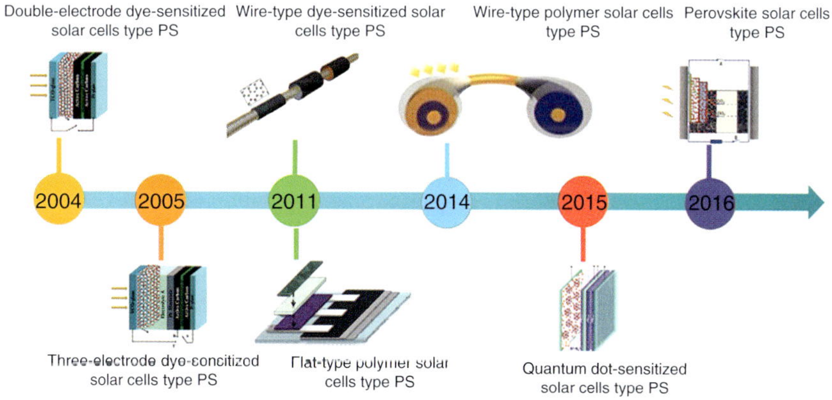


**Figure 5.22** A chronology of the development of photo-supercapacitors (PSs). Source: Sun and Yan 2017 [153]. Copyright 2017, Reprinted with permission of John Wiley & Sons.

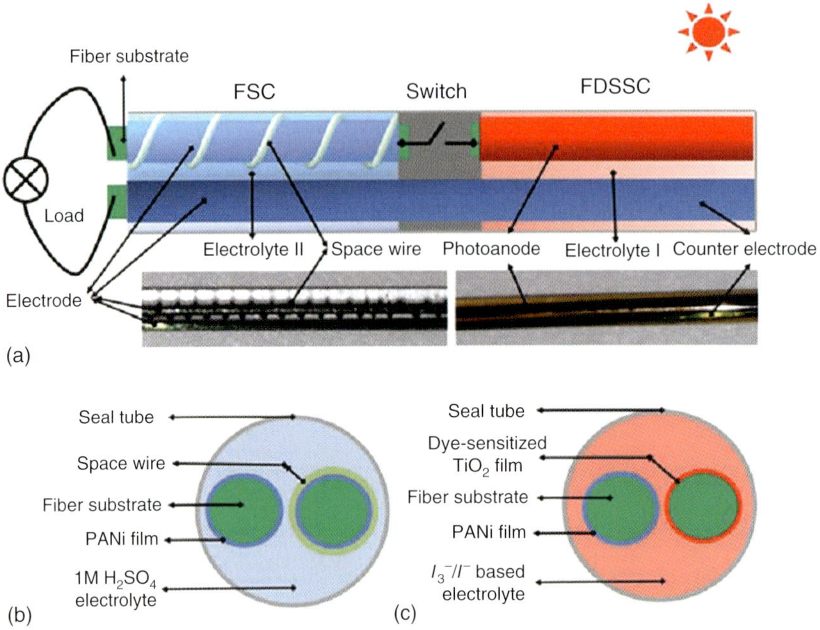


**Figure 5.23** (a) Schematic diagram and photograph of a parallel-like wire-type DSSC-photo-supercapacitors (DSSCPS). (b) Cross-section schematic diagram of the energy storage (ES) part. (c) Cross-section schematic diagram of the photoelectric conversion (PC) part. Source: Fu et al. 2013 [156]. Copyright 2013, Reprinted with permission of Royal Society of Chemistry.

sources [7]. In their device, two types of prototype DSSCs and supercapacitors have been fabricated as planar devices and with good light transparency, lightweight, high mechanical flexibility, and excellent multiple large bending tolerance. The combination of DSSCs and supercapacitors in this work provides a strategy toward devices for wearable energy management. Zou and coworkers fabricated an integrated power fiber for energy conversion and storage with the incorporation of a FDSSC and a fiber-shaped supercapacitor (Figure 5.23) [156]. For the FDSSC part, the dye-sensitized $TiO_2$-Ti wire was utilized as the photoanode, and polyaniline coated stainless steel wire was utilized as the counter electrode. For the fiber-shaped supercapacitor, the two polyaniline coated stainless steel wires were used as the working electrodes. The integrated flexible power fiber provided an overall energy conversion efficiency as high as 2.1%. Similarly, Peng and coworkers reported an "energy fiber" by coaxially integrating DSSC and electrochemical capacitor [157]. The "energy fiber" can simultaneously realize photoelectric conversion and energy storage. The "energy fiber" is very flexible, and can be woven into lightweight textiles and other structures to meet the portable facilities in electronics.

By introducing PSCs, Peng and coworkers developed an all-solid-state, flexible "energy fiber," which was able to function as an integrated photovoltaic energy converter and energy storage device [158]. For the PSC part, the energy fiber consisted of a planar PSC in the radial direction. For the supercapacitor part,

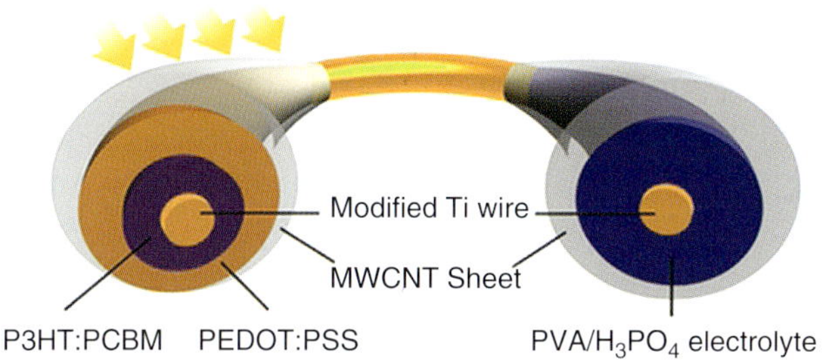


**Figure 5.24** Schematic illustration of the structure of all-solid-state, coaxial and integrated fiber device. The left and right sections correspond to the PC and ES parts, respectively. Source: Zhang et al. 2014 [158]. Copyright 2014, Reprinted with permission of John Wiley & Sons.

the energy fiber consisted of a supercapacitor in the coaxial architecture with a higher contact area (Figure 5.24). This configuration favors rapid charge transport. The superior flexibility exhibited by the energy fiber property enables it to be bent into various forms. After 1000 cycles of bending, the energy conversion and storage efficiency of the energy fiber was more than 90%. This efficient and stable performance is attributed to the use of transparent, flexible, conductive, and strong MWCNT sheets as well as the coaxial architecture of the device, greatly improving the flexibility, stability, photoelectric conversion, and energy storage.

Taking advantage of the high PCE of PVSCs, a supercapacitor and PVSCs integrated device was fabricated [159–161]. Wang and coworkers fabricated a power pack integrating a $CH_3NH_3PbI_3$-based PVSC with a polypyrrole-based supercapacitor [160]. Fan and coworkers developed an integrated energy conversion and storage device constructed with printable PVSC and PEDOT-based supercapacitor to achieve dual function [159]. The device delivers a high energy storage efficiency of 73.77% and a maximum overall energy conversion and storage efficiency up to 4.70%.

Recently, Thomas and coworkers have developed a bifunctional flexible device by integrating PVSC with supercapacitor on a copper ribbon, which is suitable for charging wearable devices [162]. Unlike the traditional method of using two separate devices, this flexible solar ribbon strategy makes it easy to use as a wearable device. When illuminated with simulated solar light, the supercapacitor provided an energy density of 1.15 mWh/cm$^{-3}$ and a power density of 243 mW/cm$^3$ (Figure 5.25). The lightweight, all-solid-state, energy-harvesting and storing (ENHANS) ribbon exhibits high flexibility. The flexible nature of the electrodes enables ENHANS ribbon to withstand external mechanical deformation. After a certain number of bending cycles, the discharge curve of the device is almost identical to a device without bending. The interlacing of ENHANS ribbons with cotton yarns into a textile provides added mechanical strength. To demonstrate the working of ENHANS ribbons as a fabric for simultaneous energy generation and storage, the matrix was exposed to a solar simulator (1 sun, AM 1.5 G) to photocharge for one minute and the stored charge used to light a light-emitting diode after ceasing the photocharging. The all-solid-state ribbon can be used as an excellent energy source for wearable applications.

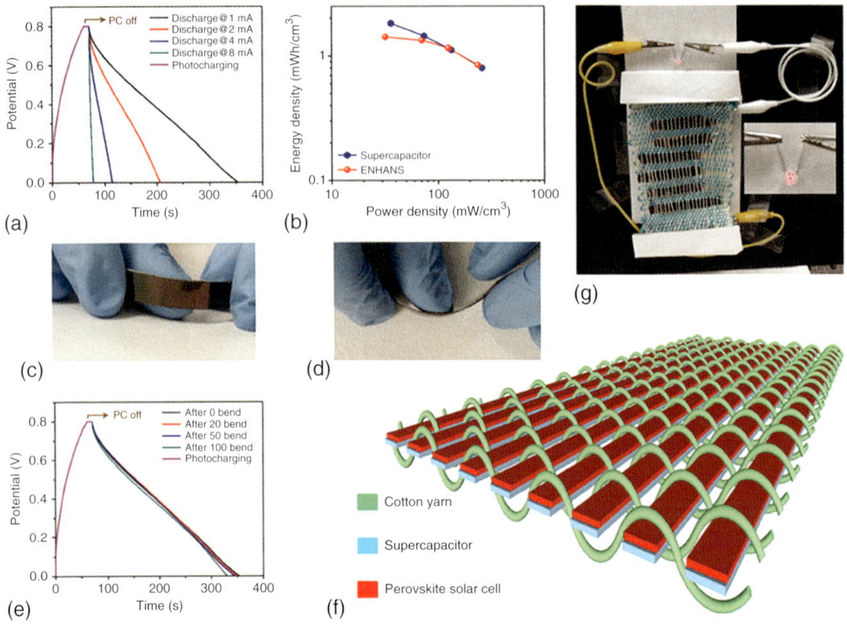


**Figure 5.25** (a) Charge–discharge profile of the energy-harvesting and storing (ENHANS) ribbon. (b) Ragone plots of an independent supercapacitor and an ENHANS ribbon; (c, d) photograph of the ENHANS ribbon being bent at different angles; (e) charge–discharge profile of the ENHANS ribbon after different bending cycles; (f) schematic illustration of the ENHANS ribbon after weaving with the cotton yarn to make a portable lightweight cloth and (g) ENHANS ribbons weaved with cotton thread to demonstrate the working of the lightweight fabric. The photograph shows the charge deliverability of weaved matrix as a result of one minute photocharging. Source: Li et al. 2016 [162]. Copyright 2016, Reprinted with permission of Springer Nature. https://www.nature.com/articles/ncomms13319#rightslink. http://creativecommons.org/licenses/by/4.0/. Licensed under CCBY 4.0

## 5.3 Conclusions and Outlook

With the rapid development of flexible and wearable smart devices and electronics, low cost, lightweight, mechanical flexibility, and less thickness wearable energy sources with high energy density and power density, long-term stability, and operational safety are in urgent need. Although considerable progress has been achieved for flexible and wearable energy conversion and storage devices (such as solar cells and supercapacitors), there are still several roadblocks to overcome before deploying them for real-world applications.

For flexible and wearable solar cells, the efficiency and durability need to improve further. The best efficiency attained for flexible and wearable solar cells is usually lower than that of the corresponding devices fabricated on glass substrate. Additionally, fabricating flexible and wearable solar cells with high efficiency on a commercial scale is not an easy task due to the different layers (active layer, electron/hole-transport layer et al.) that need to be deposited cautiously. To accomplish this goal, solution-processed organic/polymer/PVSCs are a promising technology for high power-conversion efficiency and in low-cost, flexible solar modules. However, the lifetime of the flexible and wearable solar cells based on these materials in an open environment is still a challenge. For example, the stability of PVSCs is usually less than 2000 hour under continuous simulated solar illumination. Thus, new encapsulation methods or device structures and other feasible strategies are desired to improve the overall efficiency and stability of solar cells.

Flexible supercapacitors can be provided with functions such as self-healing, stretchability, and so on, which greatly broadens their application in flexible and wearable electronics and devices. Although great achievements have been attained with flexible and wearable supercapacitors so far, there are still some challenges for this promising and hot research field. The performance of the all-solid-state flexible supercapacitors is relatively lower than that of conventional supercapacitors using liquid electrolytes. Therefore, much attention should be paid to fabricating and designing novel electrodes and devices. Moreover, the performance of flexible all-solid-state supercapacitors also greatly depends on the ionic conductivity of the electrolyte. Therefore, much effort should be devoted to designing and developing novel solid-state electrolytes with excellent ion and/or charge transport and diffusion property.

Integrated devices that simultaneously harvest (solar cells) and store (supercapacitors) energy represent a very interesting and promising research field, which may greatly expand the practical application areas of solar cells and supercapacitors. However, relatively little attention has been paid to this field. The overall performance of the integrated solar cell and supercapacitor devices, currently limited by their inconsistencies in the electrical and electrochemical performances, should be improved further. To obtain high performance, efficient electrodes should be designed and developed for integrated devices. The structures of the integrated devices should make the solar cell and supercapacitor parts match well with each other. Although challenges still remain, it is predicted that flexible and wearable energy sources (solar cells, supercapacitors, and their integrated devices) would herald the coming of a new generation of flexible and wearable electronic systems.

## Acknowledgments

This work is financially supported by the National Science Fund for Distinguished Young Scholars (51425304), the NSFC-DFG Joint Research Project (51761135114), the National Natural Science Foundation of China (21704038, 51763018), the Natural Science Foundation of Jiangxi Province (20171ACB21009, 2018ACB21021), the China Postdoctoral Science Foundation (2018M632599), and the National Postdoctoral Program for Innovative Talents (BX201700112).

## References


1 Zhang, Z., Li, X., Guan, G. et al. (2014). A lightweight polymer solar cell textile that functions when illuminated from either side. *Angew. Chem. Int. Ed.* 53: 11571–11574.
2 Zhang, N., Chen, J., Huang, Y. et al. (2016). A wearable all-solid photovoltaic textile. *Adv. Mater.* 28: 263–269.
3 O'Regan, B. and Grätzel, M. (1991). A low-cost, high-efficiency solar cell based on dye-sensitized colloidal $TiO_2$ films. *Nature* 353: 737.
4 Peng, H. (2015). *Fiber-Shaped Energy Harvesting and Storage Devices*. Berlin, Heidelberg: Springer-Verlag.
5 Xue, Z., Jiang, C., Wang, L. et al. (2014). Fabrication of flexible plastic solid-state dye-sensitized solar cells using low temperature techniques. *J. Phys. Chem. C* 118: 16352–16357.
6 Yoo, K., Kim, J.-Y., Lee, J.A. et al. (2015). Completely transparent conducting oxide-free and flexible dye-sensitized solar cells fabricated on plastic substrates. *ACS Nano* 9: 3760–3771.
7 Li, H., Zhao, Q., Wang, W. et al. (2013). Novel planar-structure electrochemical devices for highly flexible semitransparent power generation/storage sources. *Nano Lett.* 13: 1271–1277.
8 Varma, S.J., Sambath Kumar, K., Seal, S. et al. (2018). Fiber-type solar cells, nanogenerators, batteries, and supercapacitors for wearable applications. *Adv. Sci.* 5: 1800340.
9 Yi, F., Ren, H., Shan, J. et al. (2018). Wearable energy sources based on 2D materials. *Chem. Soc. Rev.* 47: 3152–3188.
10 Sun, H., Zhang, Y., Zhang, J. et al. (2017). Energy harvesting and storage in 1D devices. *Nat. Rev. Mater.* 2: 17023.
11 Chen, T., Qiu, L., Yang, Z., and Peng, H. (2013). Novel solar cells in a wire format. *Chem. Soc. Rev.* 42: 5031–5041.
12 Baps, B., Eber-Koyuncu, M., and Koyuncu, M. (2001). Ceramic based solar cells in fiber form. *Key Eng. Mater.* 206–213: 937–940.
13 Fan, X., Chu, Z.Z., Wang, F.Z. et al. (2008). Wire-shaped flexible dye-sensitized solar cells. *Adv. Mater.* 20: 592–595.
14 Lv, Z., Fu, Y., Hou, S. et al. (2011). Large size, high efficiency fiber-shaped dye-sensitized solar cells. *Phys. Chem. Chem. Phys.* 13: 10076–10083.
15 Fu, Y., Lv, Z., Hou, S. et al. (2011). Conjunction of fiber solar cells with groovy micro-reflectors as highly efficient energy harvesters. *Energy Environ. Sci.* 4: 3379–3383.
16 Yang, Z., Deng, J., Sun, X. et al. (2014). Stretchable, wearable dye-sensitized solar cells. *Adv. Mater.* 26: 2643–2647.
17 Pu, X., Song, W., Liu, M. et al. (2016). Wearable power-textiles by integrating fabric triboelectric nanogenerators and fiber-shaped dye-sensitized solar cells. *Adv. Energy Mater.* 6: 1601048.
18 Sonigara, K.K., Machhi, H.K., Vaghasiya, J.V. et al. (2018). A smart flexible solid state photovoltaic device with interfacial cooling recovery feature through thermoreversible polymer gel electrolyte. *Small* 14: 1800842.

19 Service, R.F. (2011). Outlook brightens for plastic solar cells. *Science* 332: 293.

20 Graetzel, M., Janssen, R.A.J., Mitzi, D.B., and Sargent, E.H. (2012). Materials interface engineering for solution-processed photovoltaics. *Nature* 488: 304–312.

21 Meng, L., Zhang, Y., Wan, X. et al. (2018). Organic and solution-processed tandem solar cells with 17.3% efficiency. *Science* 361: 1094–1098.

22 Li, G., Zhu, R., and Yang, Y. (2012). Polymer solar cells. *Nat. Photonics* 6: 153–161.

23 Liu, G., Peng, M., Song, W. et al. (2015). An 8.07% efficient fiber dye-sensitized solar cell based on a $TiO_2$ micron-core array and multilayer structure photoanode. *Nano Energy* 11: 341–347.

24 Liu, J., Namboothiry, M.A.G., and Carroll, D.L. (2007). Fiber-based architectures for organic photovoltaics. *Appl. Phys. Lett.* 90: 063501.

25 O'Connor, B., Pipe, K.P., and Shtein, M. (2008). Fiber based organic photovoltaic devices. *Appl. Phys. Lett.* 92: 193306.

26 Zhang, Z., Yang, Z., Deng, J. et al. (2015). Stretchable polymer solar cell fibers. *Small* 11: 675–680.

27 Kaltenbrunner, M., White, M.S., Głowacki, E.D. et al. (2012). Ultrathin and lightweight organic solar cells with high flexibility. *Nat. Commun.* 3: 770.

28 Park, S., Heo, S.W., Lee, W. et al. (2018). Self-powered ultra-flexible electronics via nano-grating-patterned organic photovoltaics. *Nature* 561: 516–521.

29 Espinosa, N., Hosel, M., Angmo, D., and Krebs, F.C. (2012). Solar cells with one-day energy payback for the factories of the future. *Energy Environ. Sci.* 5: 5117–5132.

30 Krebs, F.C., Espinosa, N., Hösel, M. et al. (2014). 25th anniversary article: rise to power – OPV-based solar parks. *Adv. Mater.* 26: 29–39.

31 Krebs, F.C., Fyenbo, J., Tanenbaum, D.M. et al. (2011). The OE-A OPV demonstrator anno domini 2011. *Energy Environ. Sci.* 4: 4116–4123.

32 Krebs, F.C., Tromholt, T., and Jorgensen, M. (2010). Upscaling of polymer solar cell fabrication using full roll-to-roll processing. *Nanoscale* 2: 873–886.

33 Espinosa, N., Garcia-Valverde, R., Urbina, A. et al. (2012). Life cycle assessment of ITO-free flexible polymer solar cells prepared by roll-to-roll coating and printing. *Sol. Energy Mater. Sol. Cells* 97: 3–13.

34 Jørgensen, M., Hagemann, O., Alstrup, J., and Krebs, F.C. (2009). Thermo-cleavable solvents for printing conjugated polymers: application in polymer solar cells. *Sol. Energy Mater. Sol. Cells* 93: 413–421.

35 Hu, X., Chen, L., Zhang, Y. et al. (2014). Large-scale flexible and highly conductive carbon transparent electrodes via roll-to-roll process and its high performance lab-scale indium tin oxide-free polymer solar cells. *Chem. Mater.* 26: 6293–6302.

36 Hu, X., Chen, L., Ji, T. et al. (2015). Roll-to-roll production of graphene hybrid electrodes for high-efficiency, flexible organic photoelectronics. *Adv. Mater. Interfaces* 2: 1500445.

37 Hu, X., Meng, X., Xiong, J. et al. (2017). Roll-to-roll fabrication of flexible orientated graphene transparent electrodes by shear force and one-step reducing post-treatment. *Adv. Mater. Technol.* 2: 1700138.

38 Meng, X., Hu, X., Yang, X. et al. (2018). Roll-to-roll printing of meter-scale composite transparent electrodes with optimized mechanical and optical properties for photoelectronics. *ACS Appl. Mater. Interfaces* 10: 8917–8925.

39 Jeon, N.J., Na, H., Jung, E.H. et al. (2018). A fluorene-terminated hole-transporting material for highly efficient and stable perovskite solar cells. *Nat. Energy* 3: 682–689.

40 Jiang, Q., Zhang, L., Wang, H. et al. (2016). Enhanced electron extraction using $SnO_2$ for high-efficiency planar-structure $HC(NH_2)_2PbI_3$-based perovskite solar cells. *Nat. Energy* 2: 16177.

41 Leblebici, S.Y., Leppert, L., Li, Y. et al. (2016). Facet-dependent photovoltaic efficiency variations in single grains of hybrid halide perovskite. *Nat. Energy* 1: 16093.

42 Saliba, M., Orlandi, S., Matsui, T. et al. (2016). A molecularly engineered hole-transporting material for efficient perovskite solar cells. *Nat. Energy* 1: 15017.

43 Snaith, H.J. (2018). Present status and future prospects of perovskite photovoltaics. *Nat. Mater.* 17: 372–376.

44 Brenner, T.M., Egger, D.A., Kronik, L. et al. (2016). Hybrid organic–inorganic perovskites: low-cost semiconductors with intriguing charge-transport properties. *Nat. Rev. Mater.* 1: 15007.

45 Bella, F., Griffini, G., Correa-Baena, J.-P. et al. (2016). Improving efficiency and stability of perovskite solar cells with photocurable fluoropolymers. *Science* 354: 203–206.

46 Chen, W., Wu, Y., Yue, Y. et al. (2015). Efficient and stable large-area perovskite solar cells with inorganic charge extraction layers. *Science* 350: 944–948.

47 Eperon, G.E., Leijtens, T., Bush, K.A. et al. (2016). Perovskite-perovskite tandem photovoltaics with optimized band gaps. *Science* 354: 861–865.

48 Li, X., Bi, D., Yi, C. et al. (2016). A vacuum flash–assisted solution process for high-efficiency large-area perovskite solar cells. *Science* 353: 58–62.

49 Rong, Y., Hu, Y., Mei, A. et al. (2018). Challenges for commercializing perovskite solar cells. *Science* 361: eaat8235.

50 Jiang, H., Feng, J., Zhao, H. et al. (2018). Low temperature fabrication for high performance flexible $CsPbI_2Br$ perovskite solar cells. *Adv. Sci.* 5: 1801117.

51 Kumar, M.H., Yantara, N., Dharani, S. et al. (2013). Flexible, low-temperature, solution processed ZnO-based perovskite solid state solar cells. *Chem. Commun.* 49: 11089–11091.

52 Liu, D. and Kelly, T.L. (2014). Perovskite solar cells with a planar heterojunction structure prepared using room-temperature solution processing techniques. *Nat. Photonics* advance online publication 8: 133–138.

53 You, J., Hong, Z., Yang, Y. et al. (2014). Low-temperature solution-processed perovskite solar cells with high efficiency and flexibility. *ACS Nano* 8: 1674–1680.

54 Docampo, P., Ball, J.M., Darwich, M. et al. (2013). Efficient organometal trihalide perovskite planar-heterojunction solar cells on flexible polymer substrates. *Nat. Commun.* 4: 2761.

55 Kaltenbrunner, M., Adam, G., Glowacki, E.D. et al. (2015). Flexible high power-per-weight perovskite solar cells with chromium oxide-metal contacts for improved stability in air. *Nat. Mater.* 14: 1032–1039.

56 Hwang, K., Jung, Y.-S., Heo, Y.-J. et al. (2015). Toward large scale roll-to-roll production of fully printed perovskite solar cells. *Adv. Mater.* 27: 1241–1247.

57 Heo, Y.-J., Kim, J.-E., Weerasinghe, H. et al. (2017). Printing-friendly sequential deposition via intra-additive approach for roll-to-roll process of perovskite solar cells. *Nano Energy* 41: 443–451.

58 Roldán-Carmona, C., Malinkiewicz, O., Soriano, A. et al. (2014). Flexible high efficiency perovskite solar cells. *Energy Environ. Sci.* 7: 994–997.

59 Huang, Z., Hu, X., Liu, C. et al. (2017). Nucleation and crystallization control via polyurethane to enhance the bendability of perovskite solar cells with excellent device performance. *Adv. Funct. Mater.* 27: 1703061.

60 Hu, X., Huang, Z., Zhou, X. et al. (2017). Wearable large-scale perovskite solar-power source via nanocellular scaffold. *Adv. Mater.* 29: 1703236.

61 Dou, B., Whitaker, J.B., Bruening, K. et al. (2018). Roll-to-roll printing of perovskite solar cells. *ACS Energy Lett.* 3: 2558–2565.

62 Qiu, L., Deng, J., Lu, X. et al. (2014). Integrating perovskite solar cells into a flexible fiber. *Angew. Chem. Int. Ed.* 53: 10425–10428.

63 Qiu, L., He, S., Yang, J. et al. (2016). Fiber-shaped perovskite solar cells with high power conversion efficiency. *Small* 12: 2419–2424.

64 Deng, J., Qiu, L., Lu, X. et al. (2015). Elastic perovskite solar cells. *J. Mater. Chem. A* 3: 21070–21076.

65 Li, R., Xiang, X., Tong, X. et al. (2015). Wearable double-twisted fibrous perovskite solar cell. *Adv. Mater.* 27: 3831–3835.

66 Hu, H., Yan, K., Peng, M. et al. (2016). Fiber-shaped perovskite solar cells with 5.3% efficiency. *J. Mater. Chem. A* 4: 3901–3906.

67 Li, L., Wu, Z., Yuan, S., and Zhang, X.-B. (2014). Advances and challenges for flexible energy storage and conversion devices and systems. *Energy Environ. Sci.* 7: 2101–2122.

68 Ding, J., Hu, W., Paek, E., and Mitlin, D. (2018). Review of hybrid ion capacitors: from aqueous to lithium to sodium. *Chem. Rev.* 118: 6457–6498.

69 Dubal, D.P., Chodankar, N.R., Kim, D.H., and Gomez-Romero, P. (2018). Towards flexible solid-state supercapacitors for smart and wearable electronics. *Chem. Soc. Rev.* 47: 2065–2129.

70 Simon, P., Gogotsi, Y., and Dunn, B. (2014). Where do batteries end and supercapacitors begin? *Science* 343: 1210–1211.

71 Miller, J.R. and Simon, P. (2008). Electrochemical capacitors for energy management. *Science* 321: 651–652.

72 Janek, J. and Zeier, W.G. (2016). A solid future for battery development. *Nat. Energy* 1: 16141.

73 Armand, M. and Tarascon, J.M. (2008). Building better batteries. *Nature* 451: 652–657.

74 Shao, Y., El-Kady, M.F., Sun, J. et al. (2018). Design and mechanisms of asymmetric supercapacitors. *Chem. Rev.* 118: 9233–9280.
75 Lukatskaya, M.R., Dunn, B., and Gogotsi, Y. (2016). Multidimensional materials and device architectures for future hybrid energy storage. *Nat. Commun.* 7: 12647.
76 Simon, P. and Gogotsi, Y. (2008). Materials for electrochemical capacitors. *Nat. Mater.* 7: 845–854.
77 Salanne, M., Rotenberg, B., Naoi, K. et al. (2016). Efficient storage mechanisms for building better supercapacitors. *Nat. Energy* 1: 16070.
78 Zhang, L.L. and Zhao, X.S. (2009). Carbon-based materials as supercapacitor electrodes. *Chem. Soc. Rev.* 38: 2520–2531.
79 Wang, Y., Song, Y., and Xia, Y. (2016). Electrochemical capacitors: mechanism, materials, systems, characterization and applications. *Chem. Soc. Rev.* 45: 5925–5950.
80 Wang, F., Wu, X., Yuan, X. et al. (2017). Latest advances in supercapacitors: from new electrode materials to novel device designs. *Chem. Soc. Rev.* 46: 6816–6854.
81 Simon, P. and Gogotsi, Y. (2013). Capacitive energy storage in nanostructured carbon–electrolyte systems. *Acc. Chem. Res.* 46: 1094–1103.
82 Xu, Y., Shi, G., and Duan, X. (2015). Self-assembled three-dimensional graphene macrostructures: synthesis and applications in supercapacitors. *Acc. Chem. Res.* 48: 1666–1675.
83 Zhai, Y., Dou, Y., Zhao, D. et al. (2011). Carbon materials for chemical capacitive energy storage. *Adv. Mater.* 23: 4828–4850.
84 Shao, Y., El-Kady, M.F., Wang, L.J. et al. (2015). Graphene-based materials for flexible supercapacitors. *Chem. Soc. Rev.* 44: 3639–3665.
85 Yang, X., Cheng, C., Wang, Y. et al. (2013). Liquid-mediated dense integration of graphene materials for compact capacitive energy storage. *Science* 341: 534–537.
86 Lin, T., Chen, I.-W., Liu, F. et al. (2015). Nitrogen-doped mesoporous carbon of extraordinary capacitance for electrochemical energy storage. *Science* 350: 1508–1513.
87 Lv, T., Liu, M., Zhu, D. et al. (2018). Nanocarbon-based materials for flexible all-solid-state supercapacitors. *Adv. Mater.* 30: e1705489.
88 Vonlanthen, D., Lazarev, P., See, K.A. et al. (2014). A stable polyaniline-benzoquinone-hydroquinone supercapacitor. *Adv. Mater.* 26: 5095–5100.
89 Wu, J.F., Zhang, Q.E., Wang, J.J. et al. (2018). A self-assembly route to porous polyaniline/reduced graphene oxide composite materials with molecular-level uniformity for high-performance supercapacitors. *Energy Environ. Sci.* 11: 1280–1286.
90 Zhao, F., Shi, Y., Pan, L., and Yu, G. (2017). Multifunctional nanostructured conductive polymer gels: synthesis, properties, and applications. *Acc. Chem. Res.* 50: 1734–1743.
91 Pan, L., Yu, G., Zhai, D. et al. (2012). Hierarchical nanostructured conducting polymer hydrogel with high electrochemical activity. *Proc. Natl. Acad. Sci. U.S.A.* 109: 9287–9292.

92 Zhao, Y., Liu, B., Pan, L., and Yu, G. (2013). 3D nanostructured conductive polymer hydrogels for high-performance electrochemical devices. *Energy Environ. Sci.* 6: 2856–2870.

93 Ibanez, J.G., Rincon, M.E., Gutierrez-Granados, S. et al. (2018). Conducting polymers in the fields of energy, environmental remediation, and chemical-chiral sensors. *Chem. Rev.* 118: 4731–4816.

94 Huang, J., Wei, J., Xiao, Y. et al. (2018). When Al-doped cobalt sulfide nanosheets meet nickel nanotube arrays: a highly efficient and stable cathode for asymmetric supercapacitors. *ACS Nano* 12: 3030–3041.

95 Liu, B.-T., Shi, X.-M., Lang, X.-Y. et al. (2018). Extraordinary pseudocapacitive energy storage triggered by phase transformation in hierarchical vanadium oxides. *Nat. Commun.* 9: 1375.

96 Le, Y., Lei, Z., Bin, W.H., and Wen, L.X. (2014). Formation of $Ni_xCo_3{-}_xS_4$ hollow nanoprisms with enhanced pseudocapacitive properties. *Angew. Chem. Ger. Ed.* 126: 3785–3788.

97 Xining, Z., Caiwei, S., Emmeline, K. et al. (2018). Titanium disulfide coated carbon nanotube hybrid electrodes enable high energy density symmetric pseudocapacitors. *Adv. Mater.* 30: 1704754.

98 Zhang, Q., Xu, W., Sun, J. et al. (2017). Constructing ultrahigh-capacity zinc–nickel–cobalt oxide@$Ni(OH)_2$ core–shell nanowire arrays for high-performance coaxial fiber-shaped asymmetric supercapacitors. *Nano Lett.* 17: 7552–7560.

99 Zhao, R., Wang, M., Zhao, D. et al. (2018). Molecular-level heterostructures assembled from titanium carbide MXene and Ni–co–Al layered double-hydroxide nanosheets for all-solid-state flexible asymmetric high-energy supercapacitors. *ACS Energy Lett.* 3: 132–140.

100 Yao, Y.X. and Wen, L.X. (2018). Mixed metal sulfides for electrochemical energy storage and conversion. *Adv. Energy Mater.* 8: 1701592.

101 Xiao, X., Song, H., Lin, S. et al. (2016). Scalable salt-templated synthesis of two-dimensional transition metal oxides. *Nat. Commun.* 7: 11296.

102 Geng, X., Zhang, Y., Han, Y. et al. (2017). Two-dimensional water-coupled metallic $MoS_2$ with nanochannels for ultrafast supercapacitors. *Nano Lett.* 17: 1825–1832.

103 Acerce, M., Voiry, D., and Chhowalla, M. (2015). Metallic 1T phase $MoS_2$ nanosheets as supercapacitor electrode materials. *Nat. Nanotechnol.* 10: 313–318.

104 Xia, Y., Mathis, T.S., Zhao, M.-Q. et al. (2018). Thickness-independent capacitance of vertically aligned liquid-crystalline MXenes. *Nature* 557: 409–412.

105 Wang, J., Tang, J., Ding, B. et al. (2017). Hierarchical porous carbons with layer-by-layer motif architectures from confined soft-template self-assembly in layered materials. *Nat. Commun.* 8: 15717.

106 Zhang, C.J., Anasori, B., Seral-Ascaso, A. et al. (2017). Transparent, flexible, and conductive 2D titanium carbide (MXene) films with high volumetric capacitance. *Adv. Mater.* 29: 1702678.

107 Yu, L., Hu, L., Anasori, B. et al. (2018). MXene-bonded activated carbon as a flexible electrode for high-performance supercapacitors. *ACS Energy Lett.* 3: 1597–1603.

108 Yang, S., Zhang, P., Wang, F. et al. (2018). Fluoride-free synthesis of two-dimensional titanium carbide (MXene) using a binary aqueous system. *Angew. Chem. Int. Ed.* 57: 15491–15495.

109 Ghidiu, M., Lukatskaya, M.R., Zhao, M.-Q. et al. (2014). Conductive two-dimensional titanium carbide 'clay' with high volumetric capacitance. *Nature* 516: 78–81.

110 Feng, D., Lei, T., Lukatskaya, M.R. et al. (2018). Robust and conductive two-dimensional metal–organic frameworks with exceptionally high volumetric and areal capacitance. *Nat. Energy* 3: 30–36.

111 Sheberla, D., Bachman, J.C., Elias, J.S. et al. (2017). Conductive MOF electrodes for stable supercapacitors with high areal capacitance. *Nat. Mater.* 16: 220–224.

112 Pachfule, P., Shinde, D., Majumder, M., and Xu, Q. (2016). Fabrication of carbon nanorods and graphene nanoribbons from a metal–organic framework. *Nat. Chem.* 8: 718–724.

113 Bendi, R., Kumar, V., Bhavanasi, V. et al. (2016). Metal organic framework-derived metal phosphates as electrode materials for supercapacitors. *Adv. Energy Mater.* 6: 1501833.

114 Wu, H.B. and Lou, X.W. (2017). Metal–organic frameworks and their derived materials for electrochemical energy storage and conversion: promises and challenges. *Sci. Adv.* 3: eaap9252.

115 Wang, L., Feng, X., Ren, L. et al. (2015). Flexible solid-state supercapacitor based on a metal–organic framework interwoven by electrochemically-deposited PANI. *J. Am. Chem. Soc.* 137: 4920–4923.

116 Xu, F., Xu, H., Chen, X. et al. (2015). Radical covalent organic frameworks: a general strategy to immobilize open-accessible polyradicals for high-performance capacitive energy storage. *Angew. Chem. Int. Ed.* 54: 6814–6818.

117 Li, P.F., Schon, T.B., and Seferos, D.S. (2015). Thiophene, selenophene, and tellurophene-based three-dimensional organic frameworks. *Angew. Chem. Int. Ed.* 54: 9361–9366.

118 Kou, Y., Xu, Y., Guo, Z., and Jiang, D. (2011). Supercapacitive energy storage and electric power supply using an Aza-fused $\pi$-conjugated microporous framework. *Angew. Chem. Int. Ed.* 50: 8753–8757.

119 Li, X.-C., Zhang, Y., Wang, C.-Y. et al. (2017). Redox-active triazatruxene-based conjugated microporous polymers for high-performance supercapacitors. *Chem. Sci.* 8: 2959–2965.

120 Liao, Y., Wang, H., Zhu, M., and Thomas, A. (2018). Efficient supercapacitor energy storage using conjugated microporous polymer networks synthesized from Buchwald–Hartwig coupling. *Adv. Mater.* 30: 1705710.

121 Halder, A., Ghosh, M., Khayum, M.A. et al. (2018). Interlayer hydrogen-bonded covalent organic frameworks as high-performance supercapacitors. *J. Am. Chem. Soc.* 140: 10941–10945.

122 Peurifoy, S.R., Russell, J.C., Sisto, T.J. et al. (2018). Designing three-dimensional architectures for high-performance electron accepting pseudocapacitors. *J. Am. Chem. Soc.* 140: 10960–10964.

123 Wei, H., Ning, J., Cao, X. et al. (2018). Benzotrithiophene-based covalent organic frameworks: construction and structure transformation under Ionothermal condition. *J. Am. Chem. Soc.* 140: 11618–11622.

124 Yang, M., Choi, B.G., Jung, S.C. et al. (2014). Polyoxometalate-coupled graphene via polymeric ionic liquid linker for supercapacitors. *Adv. Funct. Mater.* 24: 7301–7309.

125 Hao, C., Yang, B., Wen, F. et al. (2016). Flexible all-solid-state supercapacitors based on liquid-exfoliated black-phosphorus nanoflakes. *Adv. Mater.* 28: 3194–3201.

126 Liu, Z., Wu, Z.S., Yang, S. et al. (2016). Ultraflexible in-plane micro-supercapacitors by direct printing of solution-processable electrochemically exfoliated graphene. *Adv. Mater.* 28: 2217–2222.

127 El-Kady, M.F., Strong, V., Dubin, S., and Kaner, R.B. (2012). Laser scribing of high-performance and flexible graphene-based electrochemical capacitors. *Science* 335: 1326–1330.

128 Yu, C., Masarapu, C., Rong, J. et al. (2009). Stretchable supercapacitors based on buckled single-walled carbon-nanotube macrofilms. *Adv. Mater.* 21: 4793–4797.

129 Huang, R., Huang, M., Li, X. et al. (2018). Porous graphene films with unprecedented elastomeric scaffold-like folding behavior for foldable energy storage devices. *Adv. Mater.* 30: e1707025.

130 Peng, X., Liu, H., Yin, Q. et al. (2016). A zwitterionic gel electrolyte for efficient solid-state supercapacitors. *Nat. Commun.* 7: 11782. https://doi.org/10.1038/ncomms11782.

131 Lu, Z., Foroughi, J., Wang, C.Y. et al. (2018). Superelastic hybrid CNT/Graphene fibers for wearable energy storage. *Adv. Energy Mater.* 8: 1702047.

132 Wang, J., Li, X., Zi, Y. et al. (2015). A flexible fiber-based supercapacitor-triboelectric-nanogenerator power system for wearable electronics. *Adv. Mater.* 27: 4830–4836.

133 Wang, B., Fang, X., Sun, H. et al. (2015). Fabricating continuous supercapacitor fibers with high performances by integrating all building materials and steps into one process. *Adv. Mater.* 27: 7854–7860.

134 Qu, G., Cheng, J., Li, X. et al. (2016). A fiber supercapacitor with high energy density based on hollow graphene/conducting polymer fiber electrode. *Adv. Mater.* 28: 3646–3652.

135 Yang, Y., Huang, Q., Niu, L. et al. (2017). Waterproof, ultrahigh areal-capacitance, wearable supercapacitor fabrics. *Adv. Mater.* 29: 1606679.

136 Mai, L., Tian, X., Xu, X. et al. (2014). Nanowire electrodes for electrochemical energy storage devices. *Chem. Rev.* 114: 11828–11862.

137 Liu, L., Yu, Y., Yan, C. et al. (2015). Wearable energy-dense and power-dense supercapacitor yarns enabled by scalable graphene–metallic textile composite electrodes. *Nat. Commun.* 6: 7260. https://doi.org/10.1038/ncomms8260.

138 Bryan, A.M., Santino, L.M., Lu, Y. et al. (2016). Conducting polymers for pseudocapacitive energy storage. *Chem. Mater.* 28: 5989–5998.

139 Dubal, D.P., Ayyad, O., Ruiz, V., and Gomez-Romero, P. (2015). Hybrid energy storage: the merging of battery and supercapacitor chemistries. *Chem. Soc. Rev.* 44: 1777–1790.

140 Guo, Y., Li, W., Yu, H. et al. (2016). Flexible asymmetric supercapacitors via spray coating of a new electrochromic donor–acceptor polymer. *Adv. Energy Mater.*: 1601623.

141 Zhu, M.S., Huang, Y., Deng, Q.H. et al. (2016). Highly flexible, freestanding supercapacitor electrode with enhanced performance obtained by hybridizing polypyrrole chains with MXene. *Adv. Energy Mater.* 6: 1600969.

142 Ko, Y., Kwon, M., Bae, W.K. et al. (2017). Flexible supercapacitor electrodes based on real metal-like cellulose papers. *Nat. Commun.* 8: 536.

143 Li, Z., Ma, G., Ge, R. et al. (2016). Free-standing conducting polymer films for high-performance energy devices. *Angew. Chem. Int. Ed.* 55: 979–982.

144 Li, W., Gao, F., Wang, X. et al. (2016). Strong and robust polyaniline-based supramolecular hydrogels for flexible supercapacitors. *Angew. Chem. Int. Ed.* 55: 9196–9201.

145 Dong, L., Xu, C., Li, Y. et al. (2016). Breathable and wearable energy storage based on highly flexible paper electrodes. *Adv. Mater.* 28: 9313–9319.

146 Li, P., Jin, Z., Peng, L. et al. (2018). Stretchable all-gel-state fiber-shaped supercapacitors enabled by macromolecularly interconnected 3D graphene/nanostructured conductive polymer hydrogels. *Adv. Mater.* 30: e1800124.

147 Wang, L., Yang, H., Liu, X. et al. (2017). Constructing hierarchical tectorum-like $\alpha$-$Fe_2O_3$/PPy nanoarrays on carbon cloth for solid-state asymmetric supercapacitors. *Angew. Chem. Int. Ed.* 56: 1105–1110.

148 Sun, G., Zhang, X., Lin, R. et al. (2015). Hybrid fibers made of molybdenum disulfide, reduced graphene oxide, and multi-walled carbon nanotubes for solid-state, flexible, asymmetric supercapacitors. *Angew. Chem. Int. Ed.* 54: 4651–4656.

149 Xiao, Y., Huang, J., Xu, Y. et al. (2018). Hierarchical 1D nanofiber-2D nanosheet-shaped self-standing membranes for high-performance supercapacitors. *J. Mater. Chem. A* 6: 9161–9171.

150 Huang, Y., Zhong, M., Huang, Y. et al. (2015). A self-healable and highly stretchable supercapacitor based on a dual crosslinked polyelectrolyte. *Nat. Commun.* 6: 10310.

151 Huang, Y., Zhong, M., Shi, F. et al. (2017). An intrinsically stretchable and compressible supercapacitor containing a polyacrylamide hydrogel electrolyte. *Angew. Chem. Int. Ed.* 56: 9141–9145.

152 Miyasaka, T. and Murakami, T.N. (2004). The photocapacitor: an efficient self-charging capacitor for direct storage of solar energy. *Appl. Phys. Lett.* 85: 3932–3934.

153 Sun, Y. and Yan, X. (2017). Recent advances in dual-functional devices integrating solar cells and supercapacitors. *Solar RRL* 1: 1700002.

154 Liang, J., Zhu, G., Lu, Z. et al. (2018). Integrated perovskite solar capacitors with high energy conversion efficiency and fast photo-charging rate. *J. Mater. Chem. A* 6: 2047–2052.

155 Ng, C.H., Lim, H.N., Hayase, S. et al. (2018). Cesium lead halide inorganic-based perovskite-sensitized solar cell for photo-supercapacitor application under high humidity condition. *ACS Appl. Energy Mater.* 1: 692–699.
156 Fu, Y., Wu, H., Ye, S. et al. (2013). Integrated power fiber for energy conversion and storage. *Energy Environ. Sci.* 6: 805–812.
157 Chen, X., Sun, H., Yang, Z. et al. (2014). A novel "energy fiber" by coaxially integrating dye-sensitized solar cell and electrochemical capacitor. *J. Mater. Chem. A* 2: 1897–1902.
158 Zhang, Z., Chen, X., Chen, P. et al. (2014). Integrated polymer solar cell and electrochemical supercapacitor in a flexible and stable fiber format. *Adv. Mater.* 26: 466–470.
159 Xu, J., Ku, Z., Zhang, Y. et al. (2016). Integrated photo-supercapacitor based on PEDOT modified printable perovskite solar cell. *Adv. Mater. Technol.* 1: 1600074.
160 Xu, X., Li, S., Zhang, H. et al. (2015). A power pack based on organometallic perovskite solar cell and Supercapacitor. *ACS Nano* 9: 1782–1787.
161 Liu, Z., Zhong, Y., Sun, B. et al. (2017). Novel integration of perovskite solar cell and supercapacitor based on carbon electrode for hybridizing energy conversion and storage. *ACS Appl. Mater. Interfaces* 9: 22361–22368.
162 Li, C., Islam, M.M., Moore, J. et al. (2016). Wearable energy-smart ribbons for synchronous energy harvest and storage. *Nat. Commun.* 7: 13319.

# 6

# Flexible and Wearable Lithium-Ion Batteries

*Zhiwei Zhang, Peng Wang, Xianguang Miao, Peng Zhang, and Longwei Yin*

*Shandong University, School of Material Science and Engineering, Nr. 17923, Jingshi Street, Shandong 250061, China*

## 6.1 Introduction

Intelligent and wearable devices have been applied in the fields of fitness and health, medical and health care, industry and military, information and entertainment, and so on, and have not only attracted wide attention but also become an integral aspect of consumer electronics in recent years. The latest report from the market research institute IDTechEx shows that the market will triple by 2026, compared with what it was 10 years ago, and reach \$150 billion. When lightweight and miniaturization are the focal points of competition, the breakthroughs in flexible battery technology will be one of the most important to enter the intelligent and wearable high-speed device market.

Being lightweight and flexible is an important development trend in portable electronic products and folding or bendable flexible portable electronic products in the near future that is likely to greatly affect and even change the human way of life. With the development of science and technology, as an important energy storage device of flexible electronics, the flexible battery also has a bright future, with increasing prospects of its application and richer types, which will add value to the lives of people looking forward to working with change.

Portable electronics based on lightweight and high flexibility has attracted significant attention recently. One of the most important issues in such electronics is to develop high-performance lithium-ion batteries (LIBs) with good flexibility because the traditional LIBs are heavy and not bendable. Obviously, to prepare flexible LIBs, all its parts should be bendable and a flexible electrode is critically important.

## 6.2 Typical Lithium-Ion Batteries

Considering the rapid consumption of fossil fuels and the resulting environmental pollution problems, renewable energy such as wind, solar, and wave are expected to account for a much greater market share in the future. Naturally,

*Flexible and Wearable Electronics for Smart Clothing*, First Edition.
Edited by Gang Wang, Chengyi Hou, and Hongzhi Wang.

energy conversion and storage devices are urgently needed for large-scale application of renewable energy. As the most promising candidate for the energy storage field, LIBs have been widely used in many aspects such as electronic vehicles and portable electric devices [1–3].

Current commercial cells adopt three main cell types, as shown in Figure 6.1. Each cell usually consists of five main parts: cathode materials, anode materials, electrolytes, separator, and outer shells [4]. Oxides such as $LiFePO_4$, $LiCoO_2$, and hybrid ternary nickel-manganese-cobalt oxide (NMC) materials are the main types of commercial cathode materials that show a high redox potential in electrochemical cycling. The current commercial anode material is graphite, which possesses a layered structure that provides channels and sites for $Li^+$ transport and storage. However, the theoretical capacity of graphite is too slow to be used in high-energy LIBs, which has resulted in alloy-based tin, silicon, transition metal oxides, sulfides, phosphides, or even lithium metal being researched nowadays. Sometimes, cathode and anode materials are supported by current collectors to gather charges and facilitate the charge transfer process. The separator mainly consists of a polymer film subjected to special processing that acts as a barrier

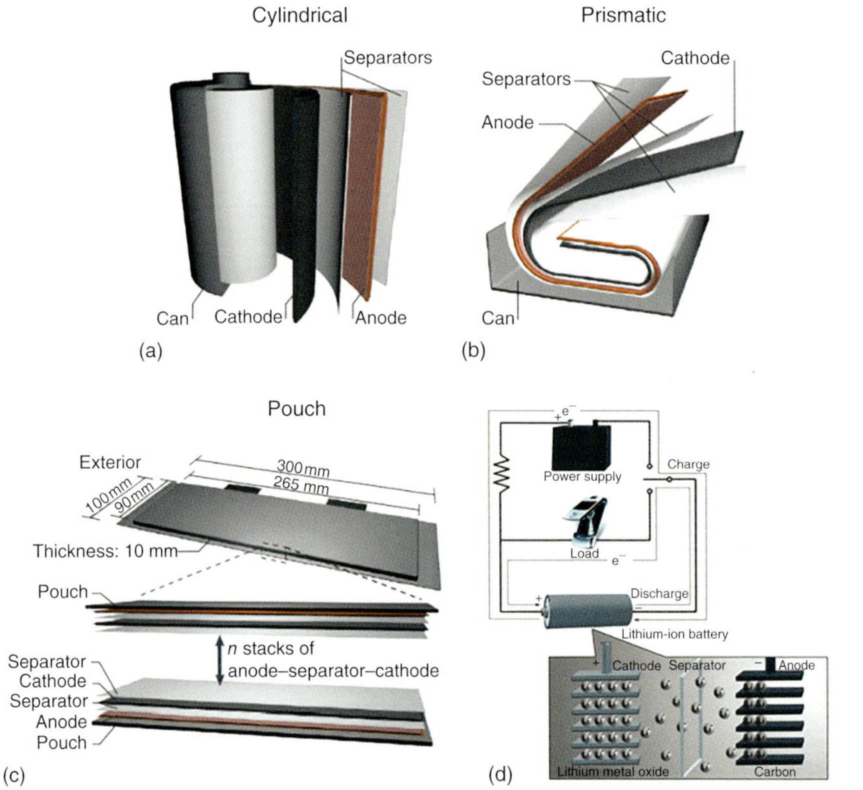


**Figure 6.1** (a–c) Three main commercial cell types and (d) a schematic diagram of the charge–discharge process. Source: Choi and Aurbach 2016 [4]. Copyright 2016, Reprinted with permission of Springer Nature.

to avoid direct contact between the cathode and anode materials. Electrolytes, whether they are of liquid type or solid state, allow the transfer of $Li^+$ and insulate the charge transfer occurring within internal batteries. The separator and the electrolyte together form a selective film that allows $Li^+$ transfer and avoids short circuit. The outer shell usually consists of a steel shell, an aluminum shell, or an aluminum-plastic film, which is used to encapsulate the electrode materials, electrolytes, and other components together.

LIBs typically follow the rocking-chair design mechanism. During the charging process, $Li^+$ transfers from the cathodes to the anodes. Conversely, during discharging, $Li^+$ intercalates into the cathodes via the electrolytes, thus realizing energy conversion and storage within the batteries. The chemical reactions of typical LIBs are described as follows [5].

$$\text{Cathode}: \ \text{LiCoO}_2 \leftrightarrow \text{Li}_{1\text{-}x}\text{CoO}_2 + x\text{Li}^+ + x\text{e}^-$$

$$\text{Anode}: \ 6\text{C} + x\text{Li}^+ + x\text{e}^- \leftrightarrow \text{Li}_x\text{C}_6$$

## 6.3 Electrode Materials for Flexible Lithium-Ion Batteries

### 6.3.1 Three-Dimensional (3D) Electrodes

Three-dimensional architecture materials based on carbon nanotubes (CNTs) and graphene commonly exhibit high aspect ratios and tunable porous structures. Their loosely compacted architectures and highly tunable internal voids render 3D materials with superior characteristics such as being lightweight and compressible, which are extremely important and necessary for flexible energy storage systems. For CNT materials, 3D architectures can be currently prepared by wet gel-initiated method and template-directed assembling method. The former largely depends on the self-gelation behavior of CNTs. More specifically, CNT gels or sols are homogeneously fabricated in aqueous conditions as precursors, and frameworks are obtained via the interactions among neighboring CNT bundles. Through the subsequent freeze drying or supercritical drying process, the solvents captured among the voids are wept, thus boosting the formation of CNT materials with 3D architectures. The steadiness and stabilization of the CNT skeleton frame are of vital importance for flexible energy storage devices. The template-directed assembling strategy (using ice or cube sugar as the template) is more facile and environmentally friendly. Highly orientated porous CNT materials can be synthesized through the procedure of matching cryogen to produce high specific capacities when they are implemented as electrodes in LIBs [6].

The 3D interconnected graphene network (such as foams, aerogels, and hydrogels) can also be inclined to loosely compact the 3D architectures. It is worth pointing out that the inner walls need to be made up of single or several layers of graphene sheets to form 3D architectures with high aspect and space ratios. Graphene oxide (GO) is diffusely employed as the precursor to fabricate 3D graphene materials. Apart from wet methods, which are the same as those employed for CNTs, 3D graphene materials can be also prepared by

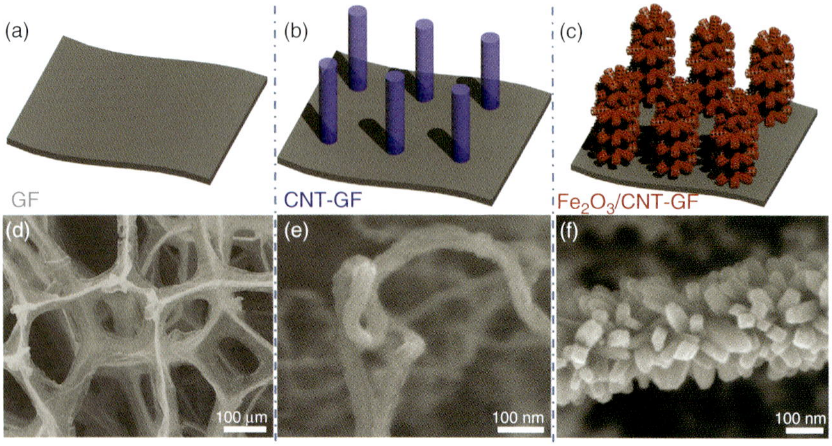


**Figure 6.2** (a–c) Schematics of the sample fabrication stage. (d–f) The corresponding SEM images: (d) graphene foam (GF), (e) CNT grown on GF, and (f) $Fe_2O_3$ nanorods grown on one CNT. Source: Chen et al. 2011 [11]. Copyright 2014, Reprinted with permission of Elsevier.

hydrothermal techniques, in which functional components or heterogeneous atoms can be incorporated or doped into graphene sheets before the hydrothermal reactions [6, 7]. The 3D interconnected porous graphene foams can also be synthesized by chemical vapor deposition, which is a unique method that adopts nickel foams as the template and methane ($CH_4$) as the carbon source. The fabricated graphene foams consist of interconnected networks of graphene sheets and exhibit excellent electrical conductivity, large surface area, and good flexibility. The incorporation of high-capacity electrode materials, such as alloys (of tin [8], silicon [9], and germanium [10]) and metal oxides ($Fe_2O_3$ [11], $SnO_2$ [8], and $Co_3O_4$ [12]), into the 3D interconnected graphene network matrix is one of the most effective and promising strategies to alleviating the huge volume expansion and achieving excellent cycling stability. Shen's group [11] successfully prepared 3D interconnected porous carbon nanotubes–graphene foam (CNT-GF) loading porous $\alpha$-$Fe_2O_3$ nanorods (Figure 6.2), which could offer plenty of internal void spaces that better accommodated the volume expansion during lithiation. The conductive 3D network played a critical role in accelerating reaction kinetics and electron transport.

### 6.3.2 Two-Dimensional (2D) Electrodes

#### 6.3.2.1 Conductive Substrate-Based Electrodes

The common strategy associated with constraining or ameliorating the volume expansion is to prepare in situ binder-free integrated electrodes that are based on transition metallic compounds linked with conductive and flexible matrices (such as metallic substrates, carbon paper, and carbon cloth). Directly growing active materials on conductive substrates is one of the interesting ways of fabricating highly integrated flexible electrodes. The proposed fabrication approach and the hierarchical integrated electrode design with intimate contact between the active particles and the graphite substrates are important for realizing electrochemical energy storage devices with the preferred properties, especially for next-generation flexible/stretchable LIBs. The novel construction philosophy can not only circumvent the "dead surface" in conventional slurry-derived electrodes and facilitate more facile electron transfer but also simplify the electrode manufacturing process. Additionally, the free space within the nanostructures can better tolerate the volume variation of the whole electrode during the $Li^+$ uptake/release process. In a traditional fabrication procedure, the slurry embracing active materials, binding agents, and the conductive additive is coated onto current collectors, which can further increase the interfacial resistance between the particles and the current collectors, impede the lithiation kinetics, and complicate the electrode design. The mechanical adhesion between the active materials and the collectors is also not robust enough to transport the strain arising from the volume expansion occurring during the intercalation/deintercalation cycles. Moreover, the water-soluble binder involves the use of toxic substances.

Typically, aluminum and copper foils are employed as the cathode and anode current collectors in LIBs, respectively, owing to their high electrical conductivities and the overwhelming advantages with regard to cost-effectiveness for large-scale commercial viability. On the other hand, the metal substrates employed in the design of flexible LIBs are too stiff to be bent or folded. Although flexibility can be ameliorated to a certain extent by reducing the foil thickness, the mechanical property deteriorates noticeably. The electrochemical performance of LIBs also reveals severe decay due to the continuous deterioration of the interface between the electrolyte and the electrodes during bending or folding [13]. To improve their mechanical flexibility, researchers usually employ carbon cloth as a superior current collector in flexible LIBs.

Carbon cloth has been demonstrated as an outstanding current collector in flexible batteries owing to its distinguished structural stability, prominent electrical conductivity, and considerable commercial availability. The bendable and soft matrix can markedly improve the electron diffusivity and electrical conductivity. The integrated hierarchical electrode presents all the advantageous characteristics of directly building high-capacity nanostructures on a carbon cloth conductive substrate, such as the short diffusion lengths of lithium ions, large material/electrolyte contact area, and good mechanical strength against volume variation in the active materials, which can directly enhance the rate capability and cycling performance during lithium insertion/extraction. For example, the integrated hierarchical electrode composed of electroactive metal oxides (i.e. $TiO_2$ [13], $V_6O_{13}$ [14], $Fe_2O_3$ [15], and NiO [16]) and carbon cloth have been extensively used to tune the electrochemical performance. Bimetal oxides, such as $Zn_2SnO_4$ [17], $Co_3V_2O_8$ [18], and $Li_4Ti_5O_{12}$ [19], are highly desirable as anode materials for LIBs because they can reveal several unique functionalities that are a benefit of the synergism existing between the diverse metal oxide constituents, such as a tunable working potential and increased electronic conductivity, which result from the mixed valence status [17]. Transition metal sulfides with outstanding electrical conductivities are also embedded into carbon cloth as electrodes for flexible LIBs and demonstrate excellent flexibility

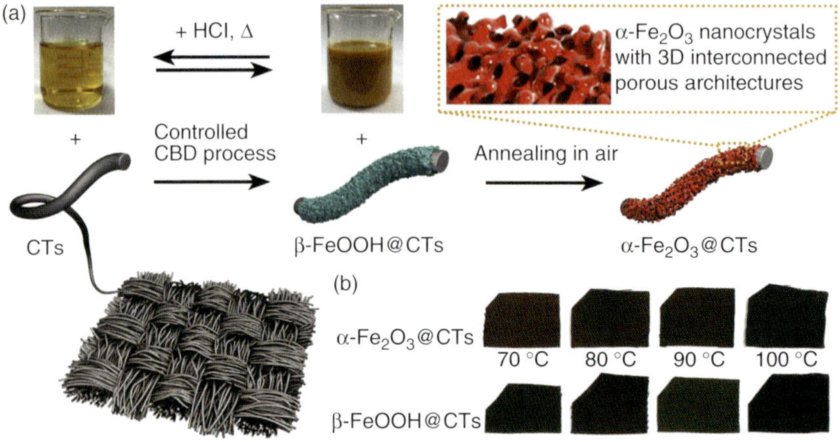


**Figure 6.3** (a) Schematic of the recyclable process for scalable controlled assembly of $\alpha$-$Fe_2O_3$ nanocrystals on flexible carbon textiles; (b) photographs of FeOOH@CTs and $Fe_2O_3$@CTs obtained under different chemical bath deposition (CBD) temperature. Source: Xiong et al. 2018 [20]. Copyright 2018, Reprinted with permission of Elsevier.

and impressive $Li^+$ storage capacity. For example, Zhao's group [20] reported a novel flexible high-performance LIB by designing $\alpha$-$Fe_2O_3$ nanocrystals with 3D interconnected porous architectures and in situ assembling them on carbon textiles (Figure 6.3). The resultant flexible electrode demonstrated superior mechanical strength and outstanding electrochemical performances.

#### 6.3.2.2 Freestanding Film-Based Electrodes

Briefly, two procedures are implemented to synthesize CNTs and graphene-based paper-like hybrid film electrodes. The former involves first converting graphene sheets or CNTs into a paper-like form, followed by deposition of the active materials onto them. This is a facile approach; however, the realization of high-performance electrodes is still overwhelmed by the inevitable stacking of graphene due to the $\pi$–$\pi$ interactions between the graphene sheets and by the severe aggregation of the active material nanoparticles. The latter is used to obtain paper-like electrodes directly by vacuum filtration, electrospinning, or vapor-phase polymerization of a homogeneous solution of the active materials and graphene or CNT mixture. This method can yield continuous paper-like freestanding films with active materials that are uniformly distributed in the carbon matrix.

#### 6.3.2.3 Graphene Papers

Graphene, a 2D monolayer of carbon atoms with packed honeycomb lattices, displays extraordinary characteristics such as large surface area, excellent thermal and chemical stability, outstanding conductivity, and mechanical flexibility. Furthermore, the unprecedented features of graphene make it possible to serve as building blocks for further development of macroscopic self-supporting graphene materials, which are satisfactory candidates for the electrodes of flexible LIBs [21]. Graphene paper with dual functions can act as both a conducting

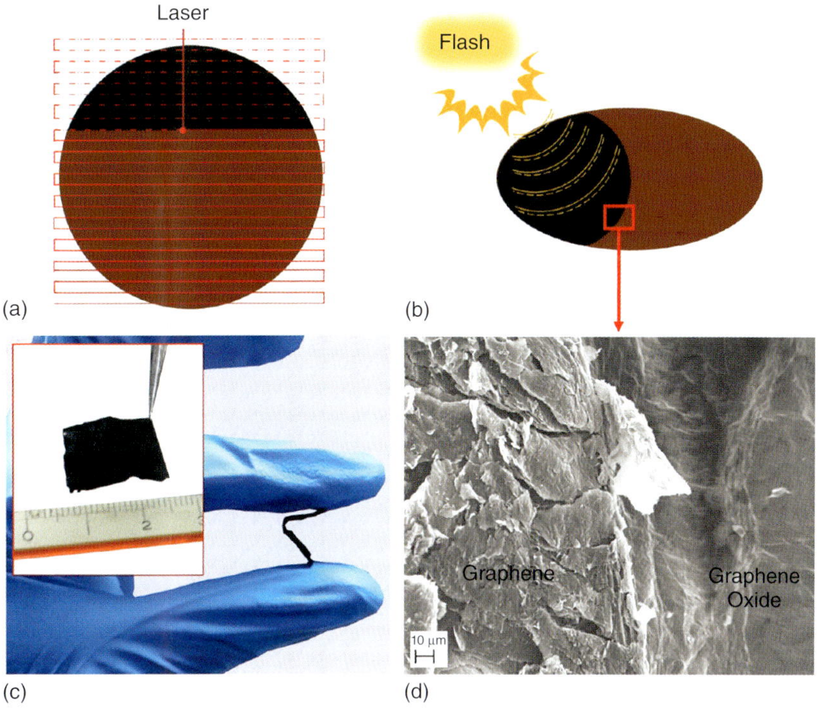


**Figure 6.4** Photothermal reduction of graphene oxide. (a–c) Photographs of flash-reduced graphene and (d) scanning electron microscopy image. Source: Mukherjee et al. 2012 [22]. Copyright 2012, Reprinted with permission of American Chemical Society.

agent and a current collector. Obviously, it is well known that graphite oxide can be fabricated from graphite powder by oxidation by using the typical Hummers method. Then, graphene paper can be commonly synthesized by filtration of graphene dispersions. Koratkar's group [22] introduced photothermal reduction of graphene oxide to obtain graphene paper anodes. The freestanding graphene paper displays considerable bending characteristics. Therefore, it is a promising candidate for employment in flexible LIBs, as it demonstrates outstanding cycling stability over 1000 discharge/charge cycles and good rate capability. Zhang's group [23] fabricated flexible graphene paper from graphene aerogel that was pre-synthesized by freeze-drying an aqueous GO dispersion followed by thermal reduction. The freestanding graphene paper obtained showed excellent flexibility (Figure 6.4).

#### 6.3.2.4 CNT Papers

Freestanding paper-like electrodes based on CNTs and graphene have received widespread attention for flexible energy storage devices. Much reduced overall electrode weight and the decrease in cost due to the avoiding of the binder and metallic foils are both advantages of using these carbon-based papers as freestanding electrodes. Although these paper-like electrodes have outstanding characteristics such as being lightweight and functioning as flexible LIB anodes, their unsatisfactory capacity is still a troublesome issue, which can be overcome by using these papers as current collectors, but not as the main active material. Among the different freestanding papers, the CNT paper has proved to be a competitive substrate for flexible high-performance LIBs owing to its merits of distinct electrical and thermal conductivities, large specific surface area, high aspect ratio, preferable chemical stability, and impressive mechanical properties [24]. With these characteristics, CNTs can be simply assembled into flexible membranes via a number of methods, such as vacuum filtration, self-assembly, dry-drawing, blade-coating, and so on. In this case, the CNT current collectors work as competitive mechanical supports, boost electron transport, and reduce the contact resistance at the electrode–CNT interface. The electrodes with CNT current collectors exhibit improvements in cycling stability, rate capability, and gravimetric energy density over those of the electrodes with metal current collectors. Additionally, the relatively smooth surface of the metal current collectors offers weak adhesion to the active materials, whereas CNTs are much rougher and thus can strengthen the interface contact. In this regard, the term CNT current collector suggests that CNTs are laminated with active material layers, which is different from the case where CNTs are mixed homogeneously with the active materials to form an integrated flexible electrode. Sanaee's group [24c] developed a $SnO_2$@a-Si/CNT paper as an effective binder-free anode for flexible LIBs. The freestanding CNT paper was prepared by the simple vacuum filtration method. Then, $SnO_2$ nanowires and a-Si thin film were deposited on the CNT paper. A similar idea was proposed by Fan's group [25], who used the CNT films drawn from superaligned carbon nanotube (SACNT) arrays as lightweight, thin, chemically stable, and flexible current collectors in LIBs. The CNT film served as a lightweight current collector to replace metal foils, and an electrode slurry was coated on the surface of the CNT film. This novel current collector showed better wetting, stronger adhesion, greater mechanical durability, and lower contact resistance at the electrode–current collector interface, compared with the copper current collector. The CNT current collector serves as a good matrix for maintaining the structural integrity of the electrode and transfers electrons efficiently to the active materials.

#### 6.3.2.5 Fabrication of Carbon Films by Vacuum Filtration Process

Wet methods such as vacuum-assisted filtration have been extensively investigated to fabricate 2D film-like electrodes using CNTs or/and graphene. The common procedures for fabricating CNT and graphene films involve the use of surface-functionalized CNTs and surfactants, which enhance the dispersion of CNTs and graphene in an aqueous solution that can then be vacuum filtered to obtain homogeneous films.

Pure CNT and graphene films as electrodes of LIBs usually output limited capacity. To solve this problem, electrochemically active materials with higher capacities are incorporated into self-supporting CNT and graphene films. During the vacuum-assisted filtration process, functional guest components can be incorporated into carbon sources to produce hybrid films. As a result, alloy-type silicon and several metal oxides such as $Co_3O_4$, $TiO_2$, $SnO_2$, $Mn_3O_4/MnO_2$, and $Fe_2O_3/Fe_3O_4$ are adopted as competitive electrode materials for flexible LIBs.

Regarding the graphene-based films, for example, Gao's group [26] proposed a lamellar graphene/nanocellulose/silicon (GN/NC/Si) film that was assembled in the form of interpenetrated GN nanosheets, which were synthesized via the facile vacuum-assisted filtration approach that was accompanied by the covalent cross-linking effect of glutaraldehyde. The hybrid film consists of the highly conductive GN matrix as an effective current collector, hydroxylated silicon nanoparticles (Si NPs) that are embedded uniformly within the GN interlayers, and NC as the adhesive to cross-link the GN and Si NPs. Regarding the CNT-based films, for example, Liu's group [27] (Figure 6.5) reported an in situ and scalable sol–gel method followed by vacuum filtration to fabricate freestanding $TiO_2$/SACNT hybrid films as flexible anodes for Li-ion batteries. The as-prepared $TiO_2$/SACNT anodes not only inherit the high flexibility of the SACNT framework but also enhance the rate capability and charge/discharge reversibility through its pseudocapacitive storage mechanism.

Composite films of CNTs and graphene have also been developed in recent years to take advantage of the synergistic effects between these two different carbon nanomaterials. Che's group [28] reported a freestanding flexible film featuring a 3D multilevel porous conductive architecture as the anode for high-performance LIBs. Specifically, the stable structure consists of CNT/reduced graphene oxide (rGO)@$MnO_2$ particles wrapped by rGO nanosheets. Micrometer-sized pores among continuously waved rGO layers and nanosized pores among $MnO_2$ nanosheets and CNT/rGO@$MnO_2$ particles exist in the sandwich-like film. Meanwhile, the continuously waved rGO layers with abundant pore channels can provide adequate tenacity as the current collector. Owing to the 3D multilevel porous architecture, the freestanding rGO-wrapped CNT/rGO@$MnO_2$ porous electrode exhibits ultrahigh capacity. Xie's group

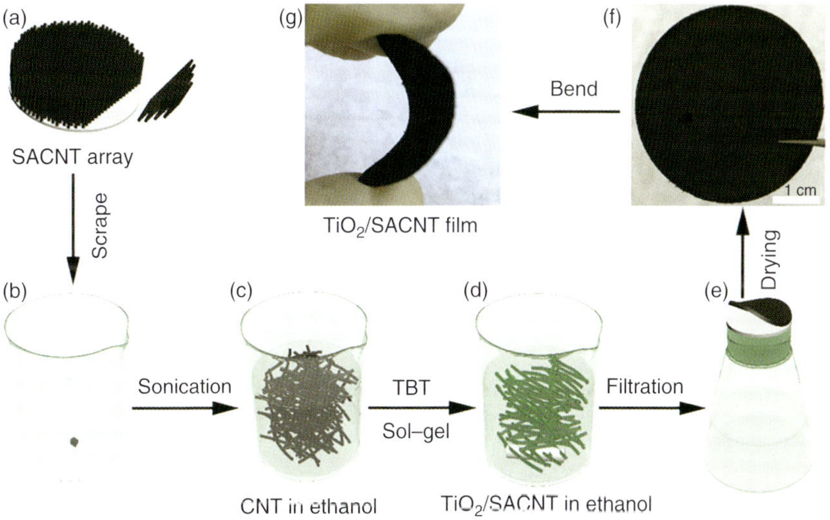


**Figure 6.5** Schematic image of the preparation of freestanding flexible $TiO_2$/CNT films. Source: Zhu et al. 2018 [27]. Copyright 2018, Reprinted with permission of American Chemical Society.

[29] presented a highly flexible membrane anode employing two-dimensional rGO sheets and a three-dimensional $MnO_2$–reduced graphene oxide–CNT nanocomposite that was fabricated by a vacuum filtration and thermal annealing approach. All the components in the 2D/3D thin film anode revealed a synergistic effect that improved the performance even at high current density. Liu's group [30] also successfully fabricated a flexible and lightweight $TiO_2$/graphene/CNTs electrode via the facile vacuum filtration approach. The long multiwalled carbon nanotubes (MWCNTs) were found to promote the formation of the highly conductive 3D network, prohibit the $TiO_2$/graphene from being heavily stacked, and facilitate the formation of a self-supported film with good mechanical integrity.

#### 6.3.2.6 Fabrication of Carbon Nanofiber Films by Electrospinning

Electrospinning technique is a competitive method for preparing carbon nanofiber (CNF) films and their composites. Unlike the other procedures for manufacturing film electrodes, the electrospinning method does not need intricate equipment or special templates. In addition, electrospinning can offer a relatively high production rate when using the continuous fiber-formation process and multiple syringes, which work simultaneously. In particular, transition metal carbides, sulfides, and oxides/carbon composite nanofibers can be facilely fabricated by electrospinning. It is crucial to homogeneously disperse the nanomaterials in CNFs. The electrospinning technology is also a high-efficiency method to synthesize flexible self-standing carbon-based films. Such freestanding films can be directly used as electrodes without adding an inactive binder, carbon black, or metal substrate; therefore, the electrodes exhibit lighter weight, higher conductivity, and better electrochemical performance.

For example, Liu's group [31] proposed an in situ strategy to fabricate electrospun robust freestanding silica/CNFs, which were constructed by encapsulating uniformly amorphous $SiO_2$ nanoparticles with nitrogen-doped CNFs. Tetraethyl orthosilicate and polyacrylonitrile (PAN) were used as the sources of $SiO_2$ and carbon, respectively. As shown in Figure 6.6, Huang's group [32] reported a facile strategy to fabricate flexible phosphorus-doped CNFs (PCNFs) as freestanding papers by using black phosphorus as the doping source that involved electrospinning and subsequent evaporation/deposition. The flexible and self-supported carbon papers exhibited superior lithium storage properties. Park's group [33] proposed molybdenum carbide (MoC) embedded in CNFs as an anode for LIBs.

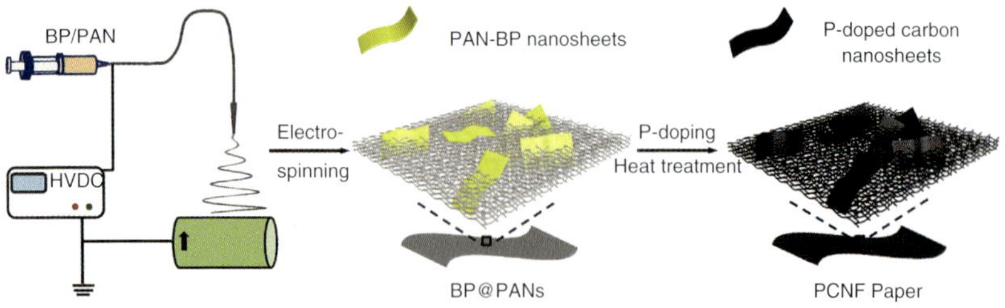


**Figure 6.6** Schematic image of the preparation of PCNF paper. Source: Li et al. 2018 [32]. Copyright 2018, Reprinted with permission of Elsevier.

The 3D flexible MoC/CNF was fabricated by using the electrospinning technique with an amorphous molybdenum precursor and PAN as the molybdenum and carbon sources, respectively, along with heating under $N_2$ atmosphere. Shao's group [34] reported successful fabrication of a nonwoven flexible $Li_4Ti_5O_{12}$/CNF composite film electrode for LIBs by coupling the electrospinning fabrication of a flexible nonwoven CNF as the substrate and the impregnation growth of the nanosized $Li_4Ti_5O_{12}$ active phase. The flexible $Li_4Ti_5O_{12}$/CNF composite displayed a remarkably high rate performance.

#### 6.3.2.7 Fabrication of Carbon Films by Vapor-Phase Polymerization

Till date, the vapor-phase polymerization approach has been successfully employed to deposit conducting polymers onto substrates. Polypyrrole (ppy) is universally accepted as one of the conducting polymers with the highest potential owing to its good conductivity, superior environmental friendliness, good biocompatibility, and high availability. Therefore, freestanding carbon films can be synthesized by a vapor-phase polymerization bottom-up self-assembly procedure that involves the pyrolysis process with ppy as the carbon source. Zheng's group [35] reported a modified route to nitrogen/sulfur dual-doped flexible carbon films as self-standing LIB anodes, which combines a novel vapor-phase polymerization film fabrication technology with a subsequent pyrolysis process under sulfur vapor atmosphere. The doped nitrogen and sulfur atoms can not only affect the equilibrium state of the electrons in carbon films but also stretch the interlayer distance of carbon via robust N—C and S—C chemical bonding for boosting $Li^+$ accommodation; thus, the fabricated NS—C hybrid film anodes demonstrate superior reversible capacities as well as outstanding rate capability and stable cycling durability for LIBs. They also employed the vapor-phase polymerization technology that was followed by a pyrolysis process to fabricate red phosphorus nanoparticles that are evenly embedded into cross-link-structural carbon films. The as-obtained phosphorus–carbon film anodes exhibit considerable reversible capacities, together with good rate and cycling properties.

### 6.3.3 One-Dimensional (1D) Electrodes

Recently, there has been tremendous scientific interest in developing 1D fiber-shaped LIBs. In contrast to planar LIBs, the construction of fibers offers some unique and promising merits. Developing high-efficiency fiber electrodes is quite vital for obtaining fiber-like batteries. In recent years, increasing interest has resulted in the preparation of aligned CNT fiber electrodes that can efficiently operate as matrices to hold up the active materials and current collectors to transport electrons simultaneously. Until now, several procedures have been adopted for fabricating CNT fibers that mainly embrace wet spinning of the fibers from a CNT solution and dry spinning of the fibers from either a drawable nanotube forest or a floating catalytic chemical-vapor-deposition reaction zone [36]. The wet spinning method can be simply scaled up for mass manufacturing. CNTs are first scattered in a solution, then extruded out of a spinneret, and finally coagulated into a solid fiber by extracting the dispersant. The wet spinning procedure commonly outputs a disordered structure of CNTs, which possesses unsatisfactory mechanical integrity and electrical conductivity. There has been increasing attention on fabricating highly orientated CNT fibers via the drying spinning procedure. CNTs can be easily assembled into fibers after being drawn out of the highly aligned CNT arrays. During the drawing procedure, reciprocities among the adjacent CNT bundles are beneficial for the transformation into a continuous fiber with the CNTs aligned along the spinning direction. The highly aligned fabrication of as-synthesized CNT fibers results in high mechanical integrity and outstanding electron transfer capability. Moreover, a series of active materials and functional constituents can be incorporated into the fiber by a co-spinning technique with high gravimetric specific capacity. These characteristics are of vital importance for the design of flexible LIBs.

Park's group [37] reported a cost-effective procedure to fabricate flexible and freestanding LIBs by modifying the CNT yarn (CNTY) synthesis process (Figure 6.7). The iron catalysts in CNTYs are transformed into electrochemically active iron trioxide nanoparticles in a straightforward manner, and combine various advantages, such as uniform dispersion of the active nanoparticles without any agglomeration, monodispersity, and intimate incorporation into the CNTs. This novel construction in the electrochemically activated CNTY electrode leads to a superior electrochemical performance (in terms of capacity, cycle performance, and rate capability) of the as-prepared CNTY when considered as an anode material for LIBs. Peng's group [38] reported 1D wire-like LIBs that were fabricated by incorporating $Li_4Ti_5O_{12}$ (LTO) and $LiMn_2O_4$ (LMO) nanoparticles into two aligned MWCNT yarns, which operated as the anode and the cathode, respectively. Owing to the aligned nanostructure and the high electrical conductivity, no current collector agents or binder additives are required during the manufacturing process. The as-synthesized MWCNT/LTO and MWCNT/LMO composite yarns can be paired as anode and cathode to fabricate a cell displaying conspicuous electrochemical performance.

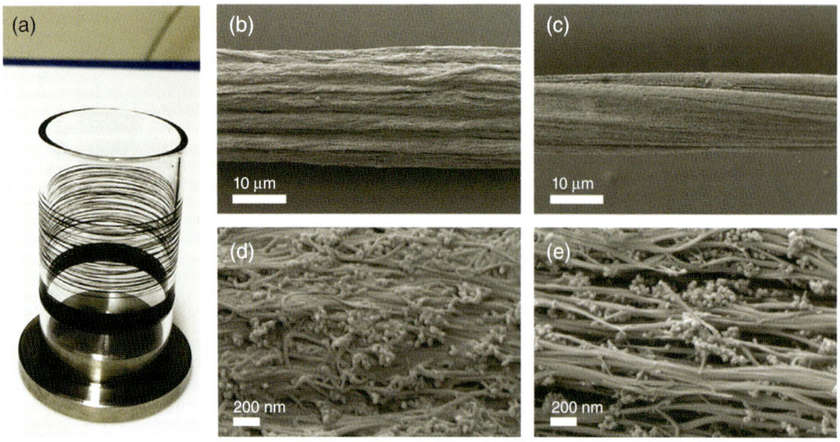


**Figure 6.7** (a) Photograph of as-produced R-CNTY wound on a roller, and low- and high-magnification SEM images of (b, d) R-CNTY and (c, e) A-CNTY. Source: Jung et al. 2016 [37]. Copyright 2016, Reprinted with permission of Elsevier.

## 6.4 Flexible Lithium-Ion Batteries Based on Electrolytes

Apart from the electrode materials mentioned above, the electrolyte also has an important impact on the mechanical and electrochemical performances of flexible LIBs, as it not only provides the mobile $Li^+$ access to the inside of the cell but also insulates the e-transfer process between the cathodes and anodes at the same time. In the past few decades, liquid-type electrolytes have been widely used in the configuration and production of flexible batteries owing to their high ionic conductivity and sufficient contact with the electrode materials. In addition, considering the thermal stability, mechanical stability, and cell safety of flexible batteries, readily deformable solid electrolytes are strongly desired and believed to be more suitable for prospective industrialization of flexible batteries [39]. In the 6.4.1 and 6.4.2 sections, a brief introduction of liquid-type electrolytes will be presented, along with a detailed description of the recommended flexible solid-state electrolytes.

### 6.4.1 Liquid-State Electrolytes

Electrolytes are supposed to be stable against both lithiated anode materials and high-voltage cathode materials during cycling. For this reason, aprotic organic solvents with lithium salts and performance-enhancing additives are used as the main type of electrolytes, instead of the aqueous electrolytes used in many other conventional batteries [40]. The fundamental states of the aprotic organic solvent, lithium salt, and additives will be expounded in the 6.4.1.1–6.4.1.3 sections.

#### 6.4.1.1 Aprotic Organic Solvent

Being able to dissolve and transport lithium salts and additives, the aprotic organic solvent plays a crucial role in the physical and electrochemical performance of electrolytes. In the past few years, many solvents have been revealed to possess suitable properties for practical application. For example, ethylene carbonate, usually abbreviated as EC, is widely used in first-generation flexible LIBs. It exhibits a large dipole moment, good salt solubility, and low vapor pressure, which can endow the electrolyte with satisfactory insulativity, outstanding ion saturability, and stable physical properties. Furthermore, it can form a stable solid electrolyte interface (SEI) on the surface of electrode materials, protecting them from further degeneration due to constant contact with the electrolyte. However, the high melting point (34–37 °C) and viscosity (1.90 cP at 40 °C) have limited the operating temperature range and ion mobility of EC. Similarly, propylene carbonate (PC), dimethyl carbonate (DMC), ethyl methyl carbonate, diethyl carbonate (DEC), fluoroethylene carbonate (FEC), and other single solvents also present internal paradoxes, possessing merits as well as demerits [40].

People choose the compound of the abovementioned single solvents perceptibly, attempting to centralize their advantages and avoid their weaknesses when employed in flexible LIBs. Tuan's group [41] used a mixture of solvents of DEC and FEC in a volumetric ratio of 7 : 3 to match the bilayer mesh anode composed of germanium and copper nanowires, which is flexible and shows good cycling and rate performance. Yang's group [10] considered the mixture of EC, DEC, and DMC in a volumetric ratio of 1 : 1 : 1 as the solvent of LIBs, and the 3D nitrogen-doped graphene foam shows good flexibility. Therefore, rational manufacturing of the existing solvents, such as fluorinated carbonates, sulfone-based electrolytes, ether-based electrolytes, and others, as well as further exploration of new types of solvents, is required for the practical application of liquid-type electrolytes in flexible LIBs.

#### 6.4.1.2 Lithium Salts

Lithium salts are crucial to the performance of electrolytes, which are supposed to display high ion conductivities, wide electrochemical potential windows, enhanced thermal and chemical stability against both extreme temperature and atmospheric environment, and the ability to form a stable SEI film on the surface of electrode materials. In the flexible LIBs synthesized with liquid-type electrolytes, $LiPF_6$ is the most widely used electrolyte owing to its excellent overall performance compared with others. Arias's group [42] used freestanding CNTs as the current collector to develop flexible LIBs and chose 1.0 M solution of $LiPF_6$ in EC/DEC (1 : 1) as the electrolyte, thereby realizing high areal capacity and flexibility performance simultaneously. However, poor thermal stability (decomposing at about 70 °C when dissolved in organic solvents) and moisture sensibility have hindered its range of application.

Not only $LiPF_6$ but also other salts have their shortcomings. For example, the ionic conductivity of $LiBF_4$ is relatively low, while $LiAsF_6$ is toxic and $LiClO_4$ can easily explode at high temperatures. Therefore, it is undeniable that great attention should be paid to the stability and kinetics of lithium salts in liquid-type electrolytes, especially in high- and low-temperature environments that arise due to the tremendous latitudinal discrepancies on the earth.

#### 6.4.1.3 Additives

Considering the unresolved intrinsic defects of the existing solvents and lithium salts, the addition of a small amount of additives is believed to be effective and economical for the improvement of the safety and electrochemical performance of LIBs. To be more specific, although the weight percentage of additives in the entire electrolyte is usually below 5%, it can help stabilize the SEI film on the electrode/electrolyte interface, enhance the ionic conductivity and thermal stability of electrolytes, and protect the cathode materials from dissolution and overcharging [43]. A variety of additives have been explored for improving the performances of the anode and cathode materials, as well as for the formation of a stable SEI film. Since many articles and books have introduced them specifically, we shall not repeat the information on electrolyte additives in this section.

### 6.4.2 Solid-State Electrolytes

In normal LIBs, the organic liquid-type electrolyte is believed to be the main blasting fuse responsible for the ignition and explosion of batteries. In addition, it is difficult for the energy density of the batteries with liquid-type electrolyte to meet the increasing requirements for both electronic devices and electric vehicles. Specifically, when flexible performance is further put forward, the possibility of electrolyte leakage, short circuit, or even burning of the whole battery will be infinitely increased, resulting in serious safety issues. Solid-state electrolytes, which have been researched since the 1960s, are the most promising candidates nowadays for replacing the liquid-type electrolyte. Furthermore, these electrolytes can help suppress the growth of lithium dendrites in lithium metal batteries, which are believed to have the ability to satisfy both the energy density and power density simultaneously. Therefore, the exploration of solid-state lithium-ion electrolytes is crucial and will certainly realize tremendous benefits in the near future.

However, there are still some drawbacks that hinder the large-scale application of solid-state electrolytes, such as low room temperature ionic conductivity and high interfacial resistance in contact with the electrode materials. Although the overall electrochemical performance of solid-state electrolytes is inferior to that of the commercial liquid-type electrolytes, researchers globally have made continuing efforts to boost the performance of flexible solid-state LIBs, which will be presented in the 6.4.2.1–6.4.2.3 sections.

#### 6.4.2.1 Inorganic Electrolytes

Most inorganic electrolytes are non-flammable and exhibit their own specific framework that is composed of coordination polyhedra. Their anisotropic conductivity is a result of the vacant and interstitial sites in the framework structure, which can dramatically lower the migration barrier energies between the adjacent available sites, thus accelerating the ion transport process.

In the past few decades, owing to their high ionic conductivity, oxide and sulfide solid-state electrolytes are the main types of inorganic electrolytes developed for solid-state batteries, for example, LISICON-type structure ($Li_{14}Zn(GeO_4)_4$), NASICON-type structure ($LiTi_2(PO_4)_3$), perovskite structure ($Li_{3.3}La_{0.56}TiO_3$), garnet structure ($Li_7La_3Zr_2O_{12}$), sulfide-based electrolytes ($Li_{10}GeP_2S_{12}$), and others. Generally, most of them are stable and suitable for use in all-solid LIBs. However, when the flexible property is taken into consideration, a few of them can still function effectively, owing to their brittle nature. Among the few cases reported in the literature, Ihlefeld's group [44] integrated a fast lithium-ion-conducting thin-film $La_{1/3-x}Li_{3x}TaO_3$ with a thin copper foil current collector to form flexible high-power all-solid-state batteries. Even though the copper substrates can help realize the flexibility demands, the electrochemical performance of the batteries was not presented. Another work using pure inorganic electrolytes was performed by Hussain and coworkers [45], in which a solid-state lithium phosphorus oxynitride (LIPON) electrolyte film was used to form flexible LIBs, showing an enhanced normalized capacity of 146 mAh/cm$^2$.

#### 6.4.2.2 Organic Electrolytes

Organic electrolytes possess a macromolecular or supramolecular architecture composed of a lithium salt that is immobilized in a polymer matrix, in which the lithium salt should exhibit a low lattice energy, and the polymer matrix is supposed to display a high dielectric constant, thus realizing sufficient $Li^+$

dissociation and fast $Li^+$ transportation. More importantly, flexible organic electrolytes reveal natural advantages when used in flexible LIBs [46].

However, the ionic conductivity of most polymer systems is relatively low at room temperature. Skorobogatiy's group [47] reported a solid-state full LIB constructed by coupling a polyethylene oxide (PEO) electrolyte with the strain-free $LiFePO_4$ cathode and $Li_4Ti_5O_{12}$ anode. Furthermore, the constructed battery film could endure a further cutting process, which produced several strips of width 1 cm and length 10 cm that could be integrated into a flexible textile LIB. Unfortunately, the electrochemical performance of the full battery was unsatisfactory, due to the low ionic conductivity of PEO.

To improve the ionic conductivity, much effort has been taken, such as cross-linking, using diblock copolymers, adding plasticizers, and other methods, to lower the glass transition temperatures of organic electrolytes. Wan's group [48] synthesized flexible double-network solid polymer electrolytes (DN-SPE) by one-pot polymerization of poly(ethylene glycol) diacrylate and poly(ethylene glycol) diglycidyl ether. The prepared DN-SPE showed excellent electrochemical performance, exhibiting enhanced ionic conductivity (from $10^{-5}$ to $10^{-4.5}$ S/cm at room temperature), high mechanical flexibility, wide electrochemical potential window (4.7 V vs. $Li^+/Li$), high thermal stability (up to 200 °C), and a good ability to suppress dendrites. Furthermore, when it was assembled into a solid-state lithium metal battery with lithium metal anode and $LiFePO_4$ cathode, impressive electrochemical properties could be realized even under stringent bending and truncated conditions.

#### 6.4.2.3 Organic/Inorganic Hybrid Electrolytes

Considering the relatively high ionic conductivity of inorganic electrolytes and the flexible performance of the organic counterparts, it is instinctive to couple them together to obtain synergistic effects in full solid-state LIBs. Moreover, composite electrolytes, developed by the integration of inorganic ceramic fillers (oxides, sulfides, hydrides, halides, LIPON, and others) into an organic polymer host (PEO, PAN, PMMA, PVC, and so on), could affect the ionic transport mechanism of each other and benefit both the electrochemical and physical performances of flexible LIBs [49].

Many performances have been improved by the use of composite electrolytes. First, in terms of ionic conductivity, Cui's group [50] incorporated 15 wt% perovskite-type $Li_{0.33}La_{0.557}TiO_3$ nanowires into the traditional PAN–$LiClO_4$ electrolyte and demonstrated that ceramic nanowire fillers can facilitate the formation of ionic conduction networks in a polymer-based solid electrolyte. The resulting electrolyte shows three orders of magnitude enhancement in the ionic conductivity, compared with a filler-free electrolyte, and its flexible performance also makes it possible for use in wearable devices. Secondly, it is widely accepted that garnet electrolytes face the problem of poor contact at the interface with electrodes. Fan's group [51] constructed a flexible solid polymer electrolyte (SPE) soft interface layer between garnet electrolyte (Ta-doped $Li_7La_3Zr_2O_{12}$ [LLZTO]) and electrodes to address the poor contact interface between them, as shown in Figure 6.8. The introduction of the inorganic electrolyte can help improve the ionic conductivity and enhance the mechanical

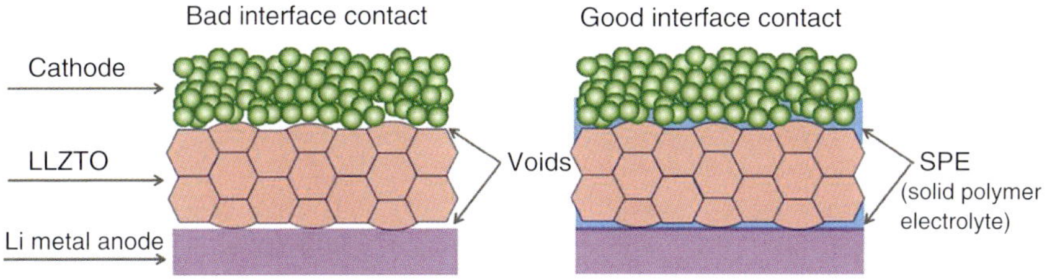


**Figure 6.8** Schematic of the flexible solid polymer electrolyte interface layer for solid-state Li metal batteries. For the conventional solid-state battery, it has a bad interface contact because of the voids in the garnet LLZTO and electrodes. With the SPE soft interface layer, the contact between electrodes and LLZTO is improved. Source: Chi et al. 2018 [51]. Copyright 2018, Reprinted with permission of Elsevier.

properties, while the organic layer endows the composite electrolyte with better wettability to the electrode. Furthermore, the flexible performance of the device can also be conspicuously enhanced by virtue of the organic layer. Thirdly, for the case of electrolyte stability over wide temperature and voltage ranges, Cui's group [52] reported a type of freestanding poly(propylene carbonate)/$Li_{6.75}La_3Zr_{1.75}Ta_{0.25}O_{12}$ composite solid electrolyte for ambient temperature flexible solid-state lithium batteries, which deliver high ionic conductivities over a large temperature range and a wide electrochemical potential window and satisfactory mechanical strengths. Moreover, the flexible $LiFePO_4/Li_4Ti_5O_{12}$ LIB developed shows excellent rate capability and cycling stability. Therefore, the strategy of coupling organic/inorganic electrolytes is one of the most effective in solving electrolyte problems in flexible LIBs.

Besides inorganic solid-state electrolytes, some types of simple inorganic substances are also integrated into organic electrolyte substrates. Ardebili's group [53] reported a high-performance flexible thin-film LIB with solid composite electrolytes consisting of PEO and 1 wt% GO nanosheets. This battery exhibits high maximum operating voltage, high energy density, and as many as 6000 bending cycles, with a capacity retention of 93%. Tsai's group [54] synthesized a silicon-doped flexible comb-like polyethylene glycol copolymer (Si-PEG) film as the electrolyte for the all-solid-state LIB, where silicon is introduced into the PEG chains by chemical bonding. The resulting Si-PEG is more flexible than PEG, which can be explained by the fact that the Si–O–C bonds exhibit an even lower glass transition temperature than the C–O–C bonds in the PEG molecular chains. The ionic conductivity, operating temperature range, tensile strength, and electrochemical performance of Si-PEG-based all-solid-state batteries (with $LiFePO_4$ and lithium as the electrode materials) are all superior to those of the electrolytes without silicon doping, which presents a new strategy in the development of solid-state flexible electrolytes.

It must be mentioned that unlike the above examples in which polymer electrolytes act as the substrates while their inorganic counterparts act as the additives, the relative contents of the electrolytes can be adjusted. Goodenough and coworkers [55] (Figure 6.9) reported two types of PEO/garnet composite electrolytes for flexible solid-state LIBs: "ceramic-in-polymer"

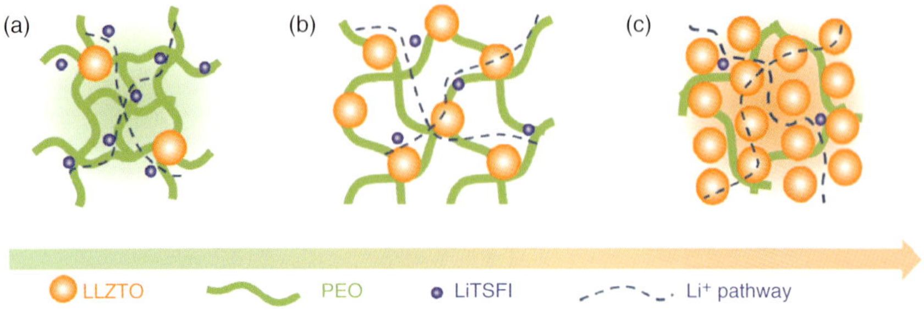


**Figure 6.9** Schematic illustration of PEO-LLZTO ceramic/polymer solid electrolyte: (a) "ceramic-in-polymer"; (b) "intermediate"; (c) "polymer-in-ceramic". Source: Chen et al. 2018 [55]. Copyright 2018, Reprinted with permission of Elsevier.

and "polymer-in-ceramic," both with flexible and mechanically robust performances. When they are integrated into $LiFePO_4$/Li batteries, both the types of all-solid-state batteries exhibit excellent cycling stability and can endure bending or twisting tests.

## 6.5 Inactive Materials and Components of Flexible LIBs

### 6.5.1 Separators

The separator is a critical component in flexible and traditional LIBs because it prevents the cathode and the anode from contacting each other to avoid short circuit of the batteries. However, separators will increase the ion resistance and weight of the batteries. Therefore, the structure and the physical and chemical properties of separators play important roles in the performance of batteries. The separator must be completely microporous to facilitate ionic transport, be chemically and electrochemically stable, and have sufficient mechanical strength to ensure the safety of LIBs.

#### 6.5.1.1 Types of Separators

Separators can be categorized into four types on the basis of their compositions and structures: microporous membranes, composite membranes, nonwoven mats, and electrolyte membranes.

Microporous membrane separators possess pores of micrometer scale, and can be further divided into monolayer membranes and multilayer membranes, according to the number of layers. Polyolefins are the most common material for preparing monolayer membranes. Many researchers have investigated the structural characteristics and properties of polypropylene (PP) [56], polyethylene (PE) [57], and their blends [58] because of their good mechanical strength and chemical stability. However, poor thermal stability and weak electrolyte wettability limit their application in LIBs. In order to overcome these shortcomings, additional materials such as poly(vinylidene fluoride) (PVDF) [59], PAN [60], and poly(methylmethacrylate) (PMMA) [61, 62] have been studied owing to their excellent ionic conductivity, chemical stability, mechanical strength, and wettability. Multilayer membranes, such as PP/PE bilayer separators and PP/PE/PP trilayer separators [63, 64], have been developed to improve the mechanical strength, thermal resistance, and electrochemical performance of monolayer separators.

Incorporated with nanosized inorganic particles, composite membranes and nonwoven mats have better physical and chemical properties. Some ceramic compounds such as $Al_2O_3$, $SiO_2$ [65], and $TiO_2$ [66] are hard and hydrophilic, which can significantly improve the mechanical strength, thermal stability, and wettability of microporous membranes and nonwoven mats. However, the addition of inorganic particles increases the weights of the batteries and decreases the tensile strengths of the original polymeric membranes. Consequently, it is of great significance to rationally design the components and structure of composite membranes.

Nonwoven mat separators can be prepared from both natural materials, including celluloses and their chemically modified derivatives, and synthetic materials, which consist of a single polyolefin or a combination of polyolefins, through chemical and mechanical methods. PVDF, PAN [67], PMMA [68], and their modified derivatives are used most commonly in nonwoven mat separators. In addition, nonwoven mat separators are usually used as a support layer for separators.

The last type of separators is electrolyte membranes, which can act as both an electrolyte and a separator between two electrodes. Unlike the separators mentioned above, which are filled with a liquid electrolyte to facilitate the lithium ion transport, electrolyte membranes are formed by swelling the polymer matrix with the electrolyte. Compared with the other types of separators, electrolyte membranes exhibit wider electrochemical potential windows, better compatibility, and higher ionic conductivities. Recently, four types of electrolyte membranes have been investigated: solid ceramic electrolyte, SPE, gel polymer electrolyte, and composite electrolyte.

#### 6.5.1.2 Physical and Chemical Properties of Separators

Although the separator is not involved in the electrochemical reaction during the charge/discharge process, it still plays a critical role in ensuring the safety and good performance of batteries. Physical and chemical properties such as structure, mechanical strength, wettability, permeability, thermal stability, and chemical stability are essential to achieve excellent battery performance.

The separators are mainly characterized by thickness, porosity, pore size, and pore distribution. Thick separators will increase the resistance, whereas thin separators will decrease the mechanical strength. Normally, a thickness of 20–25 μm is considered to be standard [69]. Typically, the separators require a porosity of 40% [70]. Too low a porosity can result in not enough of the electrolytes being absorbed, whereas too high a porosity suggests that the electrolyte cannot be effectively enclosed. The pore size needs to be appropriate to facilitate ionic transport and prevent the formation of lithium dendrites. Besides, the uniformity of the pores can also affect the performance.

The separators of flexible LIBs must have sufficient tensile strength attributable to the bending and puncture strengths in order to withstand the penetration of lithium dendrites. In addition, the separators should not curl up, shrink, and wrinkle when immersed in a liquid electrolyte.

Wettability is the ability of a separator to absorb a liquid electrolyte quickly during cell assembly and retain a certain amount of the electrolyte during cell operation. Excellent wettability can facilitate ionic transport, and thus minimize the cell resistance. It is noteworthy that wettability is related to the types of separators, porosity, and pore size.

Permeability is critical to achieving good performances of batteries, since nonuniform permeability can lead to a nonuniform current and the formation of lithium dendrites. MacMullin number is the ratio of the resistance of a separator filled with a liquid electrolyte to that of the original separator, which can be used to describe the permeability [71]. Air permeability, which is usually characterized by the Gurley value [72], is used to estimate MacMullin number. If the Gurley value is high, the separator has a low porosity and a high tortuosity. Typically, the Gurley values for flexible LIB separators are less than 0.025 s/μm.

In order to maintain the integrity during cell operation, separators must be both chemically and electrochemically stable. In other words, the separator should not react with or dissolve in the electrolyte or be oxidized or reduced during the charge/discharge process. In addition, the separator should not shrink and wrinkle obviously at high temperatures. Typically, the thermal shrinkage is less than 5% after 60 minutes at 90 °C [73].

#### 6.5.1.3 Manufacture of Separators

As mentioned above, separators can be divided into four types. Therefore, different manufacturing processes are applied to cater to the different requirements of separators. All in all, there are two main methods to produce microporous separators: dry process and wet process. In addition, the phase inversion method is also used to produce microporous separators. Nonwoven mats are prepared by four methods: dry-laid, wet-laid, melt-blown, and electrospinning processes.

The wet process and the dry process both include an extrusion step to produce polymer thin films and a stretching step to obtain porous structures. Figure 6.10 shows the scanning electron microscopy (SEM) images of microporous membrane separators prepared by both the dry process and the wet process. It is obvious that the microporous membranes prepared by the dry process have slit-like pores, whereas those fabricated by the wet process show interconnected and elliptical pores [70]. The dry process is typically a batch process and, since no solvent needs to be used, is a relatively easier method than the wet process. The separators prepared by the wet process reveal significantly improved durability since the interconnected pores and tortuous structure can effectively prevent the formation of lithium dendrites during the charge/discharge process. The structure and properties of the separators depend on the solution composition, temperature, stretching rate, and solvent evaporation for both the dry and wet processes. Phase inversion is a traditional method to obtain porous structures. Compared to the dry/wet process, it does not require the structure to be stretched. As a consequence, this method can minimize the thermal shrinkage of the membrane and the amount of defects produced during the stretching step.

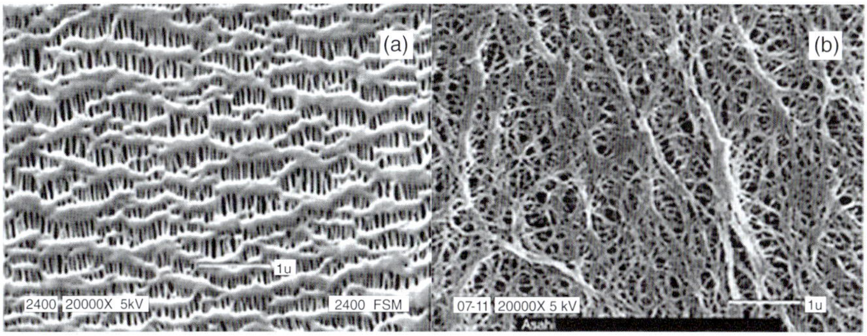


**Figure 6.10** SEM images of microporous membrane separators prepared by (a) dry process and (b) wet process. Source: Zhang 2004 [70]. Copyright 2004, Reprinted with permission of American Chemical Society.

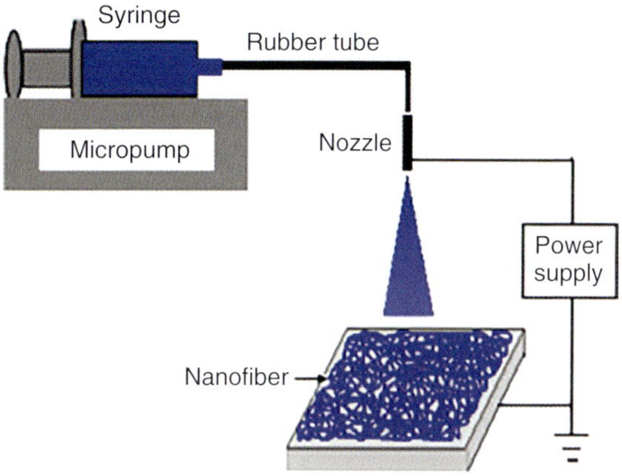


**Figure 6.11** Schematic illustration of the preparation process for polyethylene terephthalate (PET) fibrous membrane. Source: Hao et al. 2013 [74]. Copyright 2013, Reprinted with permission of Elsevier.

Nonwoven mat separators are commonly prepared by bonding together irregularly directed fibers through chemical or physical methods. Dry-laid and wet-laid methods are conventional methods to produce nonwoven mat separators owing to the good flexibility and strength that can be realized. However, the relatively low conductivity usually causes a loss of battery energy density. In order to overcome these drawbacks, the electrospinning method, which can be used to produce thin nanofiber separators at the scale of nanometers, is applied to flexible LIBs. Figure 6.11 shows a schematic illustration of the preparation process [74]. Besides, the melt-blown method is also used to produce nonwoven separators.

### 6.5.2 Casing/Packaging

All the components of batteries need to be packed in a sealed shell in order to avoid the leakage of the liquid electrolyte and undesirable parasitic reactions. Traditional LIBs, such as cylindrical, prismatic, pouch, or coin, which use aluminum or copper current collector and stainless steel shells, are not suitable for flexible LIBs. Owing to the requirements of flexibility, the packing components and structures in flexible LIBs are apparently different from those of conventional LIBs.

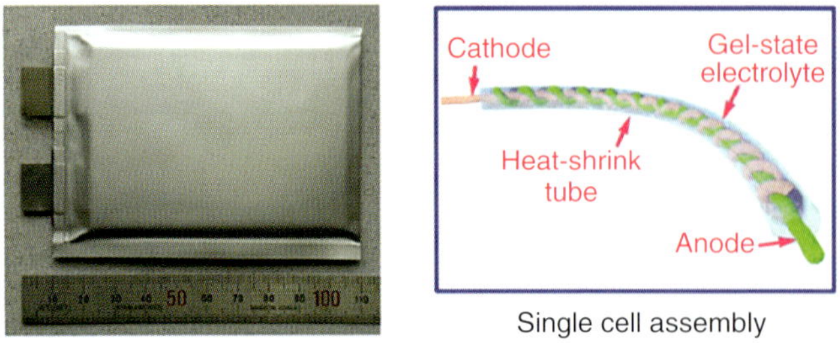


**Figure 6.12** Different types of casing components for flexible lithium-ion batteries. (a) Polymer-coated aluminum pouch and (b) plastic tube. Source: (a) Source: Yeo et al. 2016 [75]. Copyright 2016, Reproduced with permission of Elsevier; (b) Jin et al. 2018 [76]. Copyright 2018, Reproduced with permission Elsevier.

#### 6.5.2.1 Casing/Package Components

Aluminum and stainless steel are the traditional materials used to seal LIBs in a rigid/semi-rigid container. However, the rigid/semi-rigid shell is restricted in flexible LIBs owing to the low flexibility and high weight. Therefore, different flexible casing components have been investigated in recent years, as Figure 6.12 shows [75, 76]. The flexible package includes a base film that seals the battery components and an additional layer on top that covers the base film to encapsulate the battery.

#### 6.5.2.2 Casing/Packaging Structure

Compared with planar papers or film structures, cable-structured devices can be easily knitted into fabrics or transformed to all kinds of shapes for different uses. Recently, Kwon et al. [77] used a hollow spiral multihelix structure of nickel–tin as the anode and the conventional $LiCoO_2$ as the cathode, which were separated by using a modified polyethylene terephthalate nonwoven separator, to accomplish a high capacity of 1 mAh/cm between 2.5 and 4.2 V.

Some special electronic devices such as conformable skin sensors [78] need to be not only bent but also stretched and deformed to complicated shapes. As a consequence, stretchable-type batteries, which can remain functional under the conditions of bending, twisting, stretching, and compressing, have attracted great attention recently. Polydimethylsiloxane (PDMS) is the most commonly used elastomeric substrate because of its high mechanical flexibility and chemical/thermal stability. However, the electrical conductivity of PDMS is very low. In order to solve this problem, Lee et al. [79] designed porous CNT-embedded PDMS nanocomposites as the anodes of flexible LIBs, which are capable of good electrochemical performances. Figure 6.13 shows the bendable LIBs integrated into a flexible LED that were investigated by Koo et al. [80].

### 6.5.3 Current Collectors

Current collectors are conductive substrates that touch the electrodes and gather the current from the active materials to produce an external current during the charge/discharge process. Conventional current collectors are copper and

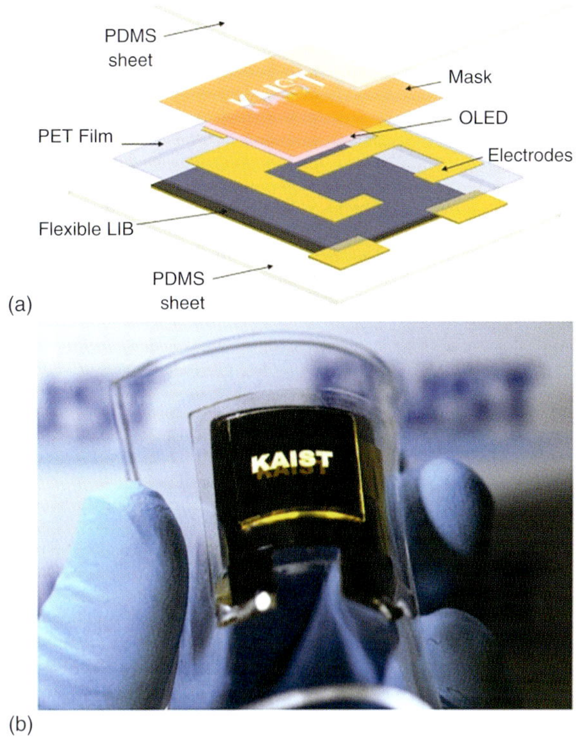


**Figure 6.13** (a) Schematic diagram of an all-flexible LED system. (b) Picture of an all-in-one flexible LED system integrated with a bendable LIB. Source: Koo et al. 2012 [80]. Copyright 2012, Reprinted with permission American Chemical Society.

aluminum foils for the anode and the cathode, respectively. However, copper and aluminum are relatively heavy and exhibit low flexibility. Therefore, Choi et al. [81] prepared a thin copper layer on a porous polymeric membrane by radio frequency magnetron sputtering as a current collector for flexible LIBs. However, current collectors are always integrated with the electrodes. In other words, the current collectors are not separate, and sometimes contribute to the capacity as electrodes, as mentioned above. Figure 6.14 shows three types of flexible current collectors [82, 83].

### 6.5.4 Electrode Additive Materials

Electrode additive materials such as binders and conductive additives are mixed with the active materials in order to promote the mechanical strength and conductivity of the electrodes. The addition of electrode additive materials is a double-edged sword in flexible LIBs: they can enhance the mechanical durability, strength, and uniformity, but can also decrease the conductivity (binders) and capacity of the batteries.

#### 6.5.4.1 Binders

Binders are polymers such as Polyvinylidene Fluoride (PVDF), polytetrafluoroethylene (PTFE), and carboxymethylcellulose (CMC) that produce adhesion between the electrodes and the current collectors. Binders can improve the

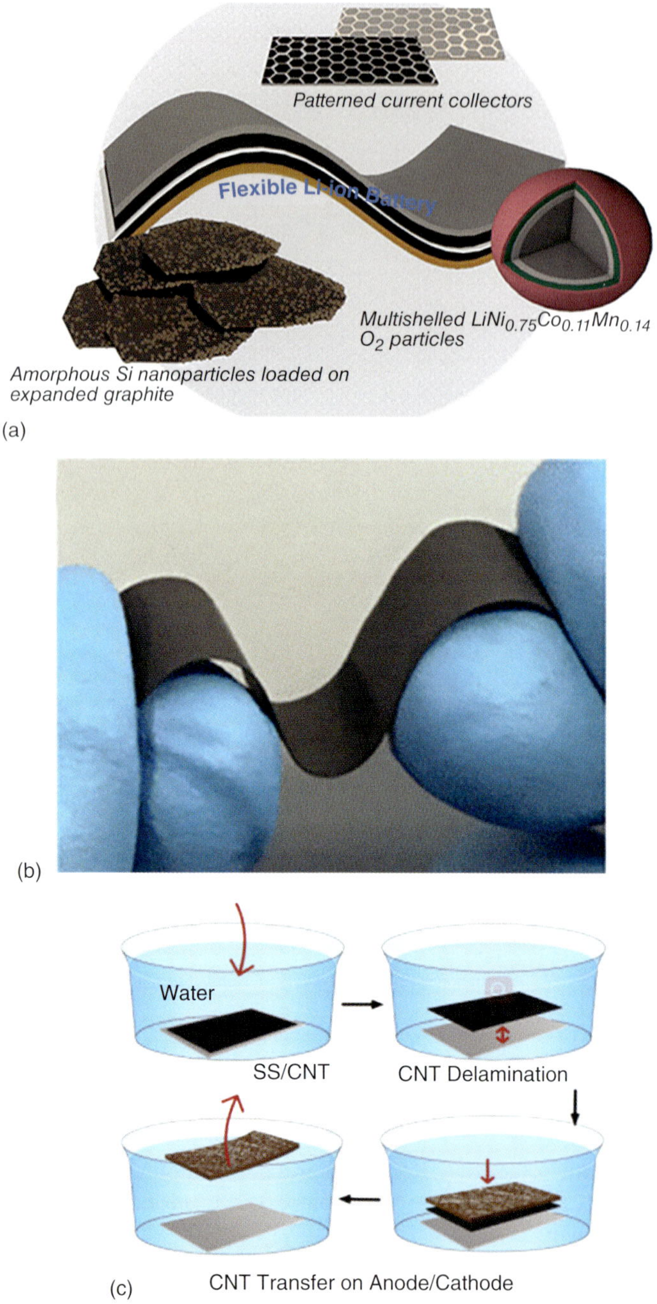


**Figure 6.14** Different materials used in flexible current collectors. (a) Metal-based; (b) paper-based; (c) carbon-based. (a) Source: Park et al. 2014 [82]. Copyright 2014, Reprinted with permission of American Chemical Society; (b) Lu et al. 2016 [83]. Copyright 2016, Reprinted with permission of American Chemical Society; (c) Gaikwad et al. 2014 [42]. Copyright 2014, Reprinted with permission of John Wiley & Sons.

mechanical strength and maintain the integrity of the electrodes, but can also have some undesirable effects. For example, binders can increase the resistance and polarization of electrodes. As a result, it is critical to use binders having high conductivities and relatively low densities. For example, PVDF–hexafluoropropylene (HFP) is widely used in flexible LIBs owing to its highly amorphous nature, high chemical stability, and low glass transition temperature, all of which can improve the cell performance significantly. It is noteworthy that self-healing polymer binders, which have the ability to rectify the mechanical damage caused to the electrodes and thus improve the durability of rechargeable batteries, have attracted the attention of scientists [84].

#### 6.5.4.2 Conductive Additives

Conductive additives such as carbon black, graphite, and ethylene black are commonly used to improve the conductivity of the active material and thus improve the cycle and rate performances of batteries. Similar to the binders, an excess of conductive additives can result in parasitic reactions with the electrolyte during cell operation. In flexible LIBs, graphene and CNTs are the conductive additives most used. Ban et al. [85] demonstrated that 5% addition of single-walled CNTs can improve the conductivity of $Fe_3O_4$, leading to its excellent performance in flexible LIBs.

## 6.6 Conclusions and Prospects

Considering the trend toward lightweight and wearable portable electronic products, flexible LIBs have already been employed in a wide range of practical applications, and will unquestionably attract more widespread attention in the next several years. The configuration and application of flexible LIBs, the research status of electrode materials, electrolytes, and inactive materials have been highlighted in this review. On one hand, similar to the typical LIBs, the components are supposed to follow the integrated design and charge/discharge mechanism of LIBs. On the other hand, some specific performances are required to meet the flexibility demands in order to be used in wearable and stretchable electronic products. First, fabrication of extraordinary electrode materials, investigation of innovative structures, and optimization of synthesis procedures have been demonstrated as three critical strategies that guarantee the flexibility of LIBs. It is urgently required to implant an elaborate structural design on the active materials of flexible LIBs in order to replace the traditional electrode components, thus improving the mechanical performance and ameliorating the electrical and electrochemical properties. As for the electrolytes, although the majority employed nowadays are still liquid-type electrolytes, solid-state electrolytes will undisputedly account for a lion's share of the market in the foreseeable future owing to their safety and other specific performance attributes for flexible LIBs such as organic/inorganic hybrid electrolytes, which are the most promising. Moreover, inactive materials, including separators, current collectors, binders, and conductive additives, play an important role in realizing high energy/power density, rate capacity, and long cycle life of batteries, although they have always been ignored, in comparison to the active materials.

In the past few years, many achievements have been recorded through the unremitting efforts of scientists all over the world, some of which even show the potential for practical application in wearable electronic products. However, there are still significantly challenging problems that need to be solved before large-scale production and commercial operation are possible. First, the selectivity of innovative electrode materials and the battery structural design call for an unprecedented material system that can replace the currently prevailing carbonaceous materials focused on CNTs and graphene, as well as the preferable processes for fabricating flexible electrodes, such as lithography and 3D printing. Another problem restricting the development of flexible LIBs is the lack of matching electrolytes. Finding suitable solid-state electrolytes having comparable ionic conductivities as the liquid-type electrolytes, as well as sufferable impedance at the interface with the electrodes, is the urgent challenge that needs to be addressed to guarantee the minimum flexibility requirement. Research hotpots of inactive materials include the components and the casing structure. Finding flexible current collectors and shells, choosing suitable separators, binders, and conductive additives, and designing rational casing structures are the crucial issues that need to be investigated to improve the overall performance.

Integration of the overall configuration together with detailed design of each challenge is the research principle in flexible LIBs. Considering the unstoppable progress in intelligent lifestyle and the urgent need for portable electronic products nowadays, we believe that flexible LIBs will enter the public domain in the near future through the unremitting efforts of researchers, thereby changing the lifestyle of the present world.

## References


1 Peters, J.F., Baumann, M., Zimmermann, B. et al. (2017). The environmental impact of Li-ion batteries and the role of key parameters – a review. *Renewable Sustainable Energy Rev.* 67: 491–506.

2 Manthiram, A. (2017). An outlook on lithium ion battery technology. *ACS Cent. Sci.* 3 (10): 1063–1069.

3 Nitta, N., Wu, F., Lee, J.T., and Yushin, G. (2015). Li-ion battery materials: present and future. *Mater. Today* 18 (5): 252–264.

4 Choi, J.W. and Aurbach, D. (2016). Promise and reality of post-lithium-ion batteries with high energy densities. *Nat. Rev. Mater.* 1 (4): 16013.

5 Ji, L., Lin, Z., Alcoutlabi, M., and Zhang, X. (2011). Recent developments in nanostructured anode materials for rechargeable lithium-ion batteries. *Energy Environ. Sci.* 4 (8): 2682.

6 Zhang, Y., Jiao, Y., Liao, M. et al. (2017). Carbon nanomaterials for flexible lithium ion batteries. *Carbon* 124: 79–88.

7 Li, N., Chen, Z., Ren, W. et al. (2012). Flexible graphene-based lithium ion batteries with ultrafast charge and discharge rates. *Proc. Natl. Acad. Sci. U.S.A.* 109 (43): 17360–17365.

8 Botas, C., Carriazo, D., Singh, G., and Rojo, T. (2015). Sn- and $SnO_2$-graphene flexible foams suitable as binder-free anodes for lithium ion batteries. *J. Mater. Chem. A* 3 (25): 13402–13410.

9 Wang, X., Li, G., Hassan, F.M. et al. (2015). Building sponge-like robust architectures of CNT–graphene–Si composites with enhanced rate and cycling performance for lithium-ion batteries. *J. Mater. Chem. A* 3 (7): 3962–3967.

10 Mo, R., Rooney, D., Sun, K., and Yang, H.Y. (2017). 3D nitrogen-doped graphene foam with encapsulated germanium/nitrogen-doped graphene yolk-shell nanoarchitecture for high-performance flexible Li-ion battery. *Nat. Commun.* 8: 13949.

11 Chen, M., Liu, J., Chao, D. et al. (2014). Porous $\alpha$-$Fe_2O_3$ nanorods supported on carbon nanotubes-graphene foam as superior anode for lithium ion batteries. *Nano Energy* 9: 364–372.

12 Yao, Y., Zhu, Y., Zhao, S. et al. (2018). Halide ion intercalated electrodeposition synthesis of $Co_3O_4$ nanosheets with tunable pores on graphene foams as free-standing and flexible Li-ion battery anodes. *ACS Appl. Energy Mater.* 1 (3): 1239–1251.

13 Lui, G., Li, G., Wang, X. et al. (2016). Flexible, three-dimensional ordered macroporous $TiO_2$ electrode with enhanced electrode–electrolyte interaction in high-power Li-ion batteries. *Nano Energy* 24: 72–77.

14 Xu, S., Cen, D., Gao, P. et al. (2018). 3D interconnected $V_6O_{13}$ nanosheets grown on carbonized textile via a seed-assisted hydrothermal process as high-performance flexible cathodes for lithium-ion batteries. *Nanoscale Res. Lett.* 13 (1): 65.

15 Balogun, M.-S., Wu, Z., Luo, Y. et al. (2016). High power density nitridated hematite ($\alpha$-$Fe_2O_3$) nanorods as anode for high-performance flexible lithium ion batteries. *J. Power Sources* 308: 7–17.

16 Long, H., Shi, T., Hu, H. et al. (2014). Growth of hierarchal mesoporous NiO nanosheets on carbon cloth as binder-free anodes for high-performance flexible lithium-ion batteries. *Sci. Rep.* 4: 7413.

17 Xia, J., Tian, R., Guo, Y. et al. (2018). $Zn_2SnO_4$-carbon cloth freestanding flexible anodes for high-performance lithium-ion batteries. *Mater. Des.* 156: 272–277.

18 Zhou, C., Lu, J., Hu, M. et al. (2018). High areal capacity Li-ion storage of binder-free metal vanadate/carbon hybrid anode by ion-exchange reaction. *Small* 14 (35): e1801832.

19 Yang, L.Y., Li, H.Z., Cheng, L.Z. et al. (2017). A three-dimensional surface modified carbon cloth designed as flexible current collector for high-performance lithium and sodium batteries. *J. Alloys Compd.* 726: 837–845.

20 Xiong, W.-S., Jiang, Y., Xia, Y. et al. (2018). A sustainable approach for scalable production of $\alpha$-$Fe_2O_3$ nanocrystals with 3D interconnected porous architectures on flexible carbon textiles as integrated electrodes for lithium-ion batteries. *J. Power Sources* 401: 65–72.

21 Chen, K., Wang, Q., Niu, Z., and Chen, J. (2018). Graphene-based materials for flexible energy storage devices. *J. Energy Chem.* 27 (1): 12–24.

22 Mukherjee, R., Thomas, A.V., Krishnamurthy, A., and Koratkar, N. (2012). Photothermally reduced graphene as high-power anodes for lithium-ion batteries. *ACS Nano* 6 (9): 7867–7878.

23 Liu, F., Song, S., Xue, D., and Zhang, H. (2012). Folded structured graphene paper for high performance electrode materials. *Adv. Mater.* 24 (8): 1089–1094.

24 (a) Li, L., Wu, Z., Yuan, S., and Zhang, X.-B. (2014). Advances and challenges for flexible energy storage and conversion devices and systems. *Energy Environ. Sci.* 7 (7): 2101. (b) Zhou, G., Li, F., and Cheng, H.-M. (2014). Progress in flexible lithium batteries and future prospects. *Energy Environ. Sci.* 7 (4): 1307–1338. (c) Abnavi, A., Sadati Faramarzi, M., Abdollahi, A. et al. (2017). $SnO_2$@a-Si core-shell nanowires on free-standing CNT paper as a thin and flexible Li-ion battery anode with high areal capacity. *Nanotechnology* 28 (25): 255404.

25 Wang, K., Luo, S., Wu, Y. et al. (2013). Super-aligned carbon nanotube films as current collectors for lightweight and flexible lithium ion batteries. *Adv. Funct. Mater.* 23 (7): 846–853.

26 Zhou, X., Liu, Y., Du, C. et al. (2018). Free-standing sandwich-type graphene/nanocellulose/silicon laminar anode for flexible rechargeable lithium ion batteries. *ACS Appl. Mater. Interfaces* 10 (35): 29638–29646.

27 Zhu, K., Luo, Y., Zhao, F. et al. (2018). Free-standing, binder-free titania/super-aligned carbon nanotube anodes for flexible and fast-charging Li-ion batteries. *ACS Sustainable Chem. Eng.* 6 (3): 3426–3433.

28 Li, S., Zhao, Y., Liu, Z. et al. (2018). Flexible graphene-wrapped carbon nanotube/graphene@$MnO_2$ 3D multilevel porous film for high-performance lithium-ion batteries. *Small* 14 (32): e1801007.

29 Li, Y., Ye, D., Shi, B. et al. (2017). Free-standing reduced graphene oxide/$MnO_2$-reduced graphene oxide-carbon nanotube nanocomposite flexible membrane as an anode for improving lithium-ion batteries. *Phys. Chem. Chem. Phys.* 19 (11): 7498–7505.

30 Luo, J., Li, F., Zhou, Y. et al. (2018). Paper-like $TiO_2$/graphene-carbon nanotube hybrid electrode with high mass loading: toward high-performance lithium ion battery. *J. Alloys Compd.* 749: 697–704.

31 Li, L., Liu, P., Zhu, K. et al. (2017). Flexible and robust N-doped carbon nanofiber film encapsulating uniformly silica nanoparticles: free-standing long-life and low-cost electrodes for Li- and Na-ion batteries. *Electrochim. Acta* 235: 79–87.

32 Li, D., Wang, D., Rui, K. et al. (2018). Flexible phosphorus doped carbon nanosheets/nanofibers: electrospun preparation and enhanced Li-storage properties as free-standing anodes for lithium ion batteries. *J. Power Sources* 384: 27–33.

33 Lee, G.-H., Moon, S.-H., Kim, M.-C. et al. (2018). Molybdenum carbide embedded in carbon nanofiber as a 3D flexible anode with superior stability and high-rate performance for Li-ion batteries. *Ceram. Int.* 44 (7): 7972–7977.

34 Zhang, Z., Deng, X., Sunarso, J. et al. (2017). Two-step fabrication of $Li_4Ti_5O_{12}$-coated carbon nanofibers as a flexible film electrode for high-power lithium-ion batteries. *ChemElectroChem* 4 (9): 2286–2292.

35 (a) Ruan, J., Yuan, T., Pang, Y. et al. (2018). Nitrogen and sulfur dual-doped carbon films as flexible free-standing anodes for Li-ion and Na-ion batteries. *Carbon* 126: 9–16. (b) Ruan, J., Yuan, T., Pang, Y. et al. (2017). Red phosphorus-embedded cross-link-structural carbon films as flexible anodes for highly reversible Li-ion storage. *ACS Appl. Mater. Interfaces* 9 (41): 36261–36268.
36 Di, J., Zhang, X., Yong, Z. et al. (2016). Carbon-nanotube fibers for wearable devices and smart textiles. *Adv. Mater.* 28 (47): 10529–10538.
37 Jung, Y., Jeong, Y.C., Kim, J.H. et al. (2016). One step preparation and excellent performance of CNT yarn based flexible micro lithium ion batteries. *Energy Storage Mater.* 5: 1–7.
38 Ren, J., Zhang, Y., Bai, W. et al. (2014). Elastic and wearable wire-shaped lithium-ion battery with high electrochemical performance. *Angew. Chem.* 53 (30): 7864–7869.
39 Wang, X., Lu, X., Liu, B. et al. (2014). Flexible energy-storage devices: design consideration and recent progress. *Adv. Mater.* 26 (28): 4763–4782.
40 Hu, L., Zhang, S.S., and Zhang, Z. (2012). Rechargeable batteries: materials, technologies and new trends. *Green Energy Technol.*
41 Chang, W.-C., Kao, T.-L., Lin, Y., and Tuan, H.-Y. (2017). A flexible all inorganic nanowire bilayer mesh as a high-performance lithium-ion battery anode. *J. Mater. Chem. A* 5 (43): 22662–22671.
42 Gaikwad, A.M., Khau, B.V., Davies, G. et al. (2015). A high areal capacity flexible lithium-ion battery with a strain-compliant design. *Adv. Energy Mater.* 5 (3): 1401389.
43 Zhang, S.S. (2006). A review on electrolyte additives for lithium-ion batteries. *J. Power Sources* 162 (2): 1379–1394.
44 Ihlefeld, J.F., Clem, P.G., Doyle, B.L. et al. (2011). Fast lithium-ion conducting thin-film electrolytes integrated directly on flexible substrates for high-power solid-state batteries. *Adv. Mater.* 23 (47): 5663–5667.
45 Kutbee, A. T., Ghoneim, M. T., Hussain, M. M. (2015). Flexible lithium-ion planer thin-film battery. *IEEE International Conference on Nanotechnology.*
46 Meyer, W.H. (1998). Polymer electrolytes for lithium ion batteries. *Adv. Mater.* 10 (6): 439–448.
47 Liu, Y., Gorgutsa, S., Santato, C., and Skorobogatiy, M. (2012). Flexible, solid electrolyte-based lithium battery composed of $LiFePO_4$ cathode and $Li_4Ti_5O_{12}$ anode for applications in smart textiles. *J. Electrochem. Soc.* 159 (4): A349–A356.
48 Duan, H., Yin, Y.-X., Zeng, X.-X. et al. (2018). In-situ plasticized polymer electrolyte with double-network for flexible solid-state lithium-metal batteries. *Energy Storage Mater.* 10: 85–91.
49 Chen, R., Qu, W., Guo, X. et al. (2016). The pursuit of solid-state electrolytes for lithium batteries: from comprehensive insight to emerging horizons. *Mater. Horiz.* 3 (6): 487–516.
50 Liu, W., Liu, N., Sun, J. et al. (2015). Ionic conductivity enhancement of polymer electrolytes with ceramic nanowire fillers. *Nano Lett.* 15 (4): 2740–2745.

51 Chi, S.-S., Liu, Y., Zhao, N. et al. (2018). Solid polymer electrolyte soft interface layer with 3D lithium anode for all-solid-state lithium batteries. *Energy Storage Mater.* 17: 309–316.

52 Zhang, J., Zang, X., Wen, H. et al. (2017). High-voltage and free-standing poly(propylene carbonate)/$Li_{6.75}La_3Zr_{1.75}Ta_{0.25}O_{12}$ composite solid electrolyte for wide temperature range and flexible solid lithium ion battery. *J. Mater. Chem. A* 5 (10): 4940–4948.

53 Kammoun, M., Berg, S., and Ardebili, H. (2015). Flexible thin-film battery based on graphene-oxide embedded in solid polymer electrolyte. *Nanoscale* 7 (41): 17516–17522.

54 Ji, X., Zeng, H., Gong, X. et al. (2017). A Si-doped flexible self-supporting comb-like polyethylene glycol copolymer (Si-PEG) film as a polymer electrolyte for an all solid-state lithium-ion battery. *J. Mater. Chem. A* 5 (46): 24444–24452.

55 Chen, L., Li, Y., Li, S.-P. et al. (2018). PEO/garnet composite electrolytes for solid-state lithium batteries: from "ceramic-in-polymer" to "polymer-in-ceramic". *Nano Energy* 46: 176–184.

56 Venugopal, G., Moore, J., Howard, J., and Pendalwar, S. (1999). Characterization of microporous separators for lithium-ion batteries. *J. Power Sources* 77: 34–41.

57 Love, C.T. (2011). Thermomechanical analysis and durability of commercial micro-porous polymer Li-ion battery separators. *J. Power Sources* 196 (5): 2905–2912.

58 (a) Djian, D., Alloin, F., Martinet, S. et al. (2007). Lithium-ion batteries with high charge rate capacity: influence of the porous separator. *J. Power Sources* 172 (1): 416–421. (b) Wu, M.-S., Chiang, P.-C.J., Lin, J.-C., and Jan, Y.-S. (2004). Correlation between electrochemical characteristics and thermal stability of advanced lithium-ion batteries in abuse tests – short-circuit tests. *Electrochim. Acta* 49 (11): 1803–1812.

59 (a) Djian, D., Alloin, F., Martinet, S., and Lignier, H. (2009). Macroporous poly(vinylidene fluoride) membrane as a separator for lithium-ion batteries with high charge rate capacity. *J. Power Sources* 187 (2): 575–580. (b) Saunier, J., Alloin, F., Sanchez, J.Y., and Maniguet, L. (2004). Plasticized microporous poly(vinylidene fluoride) separators for lithium-ion batteries. III. Gel properties and irreversible modifications of poly(vinylidene fluoride) membranes under swelling in liquid electrolytes. *J. Polym. Sci., Part B: Polym. Phys.* 42 (12): 2308–2317.

60 Jung, B., Yoon, J.K., Kim, B., and Rhee, H.-W. (2005). Effect of crystallization and annealing on polyacrylonitrile membranes for ultrafiltration. *J. Membr. Sci.* 246 (1): 67–76.

61 Bohnke, O., Frand, G., Rezrazi, M. et al. (1993). Fast ion transport in new lithium electrolytes gelled with PMMA. 1. Influence of polymer concentration. *Solid State Ionics* 66: 97–104.

62 Lee, M.-H., Kim, H.J., Kim, E. et al. (1996). Effect of phase separation on ionic conductivity of poly(methy1 methacrylate)-based solid polymer electrolyte. *Solid State Ionics* 85: 91–98.

63 Cao, C., Tan, L., Liu, W. et al. (2014). Polydopamine coated electrospun poly(vinyldiene fluoride) nanofibrous membrane as separator for lithium-ion batteries. *J. Power Sources* 248: 224–229.
64 Liang, Y., Cheng, S., Zhao, J. et al. (2013). Heat treatment of electrospun polyvinylidene fluoride fibrous membrane separators for rechargeable lithium-ion batteries. *J. Power Sources* 240: 204–211.
65 Du Pasquier, A., Warren, P.C., Culver, D. et al. (2000). Plastic PVDF-HFP electrolyte laminates prepared by a phase-inversion process. *Solid State Ionics* 135: 249–257.
66 Subramania, A., Kalyana Sundaram, N.T., Sathiya Priya, A.R., and Vijaya Kumar, G. (2007). Preparation of a novel composite micro-porous polymer electrolyte membrane for high performance Li-ion battery. *J. Membr. Sci.* 294 (1–2): 8–15.
67 Raghavan, P., Manuel, J., Zhao, X. et al. (2011). Preparation and electrochemical characterization of gel polymer electrolyte based on electrospun polyacrylonitrile nonwoven membranes for lithium batteries. *J. Power Sources* 196 (16): 6742–6749.
68 Rao, M., Geng, X., Liao, Y. et al. (2012). Preparation and performance of gel polymer electrolyte based on electrospun polymer membrane and ionic liquid for lithium ion battery. *J. Membr. Sci.* 399–400: 37–42.
69 Lin, D.-J., Chang, C.-L., Chang, C.-L. et al. (2004). Fine structure of poly(vinylidene fluoride) membranes prepared by phase inversion from a water/*N*-methyl-2-pyrollidone/poly(vinylidene fluoride) system. *J. Polym. Sci., Part B: Polym. Phys.* 42 (5): 830–842.
70 Arora, P. and Zhang, Z. (2004). Battery separators. *Chem. Rev.* 104: 4419–4462.
71 Patel, K.K., Paulsen, J.M., and Desilvestro, J. (2003). Numerical simulation of porous networks in relation to battery electrodes and separators. *J. Power Sources* 122 (2): 144–152.
72 Abraham, K.M. (1993). Directions in secondary lithium battery research and development. *Electrochim. Acta* 38 (9): 1233–1248.
73 Yang, M. and Hou, J. (2012). Membranes in lithium ion batteries. *Membranes* 2 (3): 367–383.
74 Hao, J., Lei, G., Li, Z. et al. (2013). A novel polyethylene terephthalate nonwoven separator based on electrospinning technique for lithium ion battery. *J. Membr. Sci.* 428: 11–16.
75 Yeo, J.-S., Yoo, E.-J., Ha, S.-H. et al. (2016). Electrochemical properties of large-sized pouch-type lithium ion batteries with bio-inspired organic cathode materials. *J. Power Sources* 313: 91–95.
76 Jin, Z., Li, P., Jin, Y., and Xiao, D. (2018). Superficial-defect engineered nickel/iron oxide nanocrystals enable high-efficient flexible fiber battery. *Energy Storage Mater.* 13: 160–167.
77 Kwon, Y.H., Woo, S.-W., Jung, H.-R. et al. (2012). Cable-type flexible lithium ion battery based on hollow multi-helix electrodes. *Adv. Mater.* 24: 5192–5197.
78 Rogers, J.A., Someya, T., and Huang, Y. (2010). Materials and mechanics for stretchable electronics. *Science* 327: 1603–1607.

79 Lee, H., Yoo, J.-K., Park, J.-H. et al. (2012). A stretchable polymer-carbon nanotube composite electrode for flexible lithium-ion batteries: porosity engineering by controlled phase separation. *Adv. Energy Mater.* 2 (8): 976–982.
80 Koo, M., Park, K.I., Lee, S.H. et al. (2012). Bendable inorganic thin-film battery for fully flexible electronic systems. *Nano Lett.* 12 (9): 4810–4816.
81 Choi, J.-Y., Lee, D.J., Lee, Y.M. et al. (2013). Silicon nanofibrils on a flexible current collector for bendable lithium-ion battery anodes. *Adv. Funct. Mater.* 23 (17): 2108–2114.
82 Park, M.H., Noh, M., Lee, S. et al. (2014). Flexible high-energy Li-ion batteries with fast-charging capability. *Nano Lett.* 14 (7): 4083–4089.
83 Lu, H., Behm, M., Leijonmarck, S. et al. (2016). Flexible paper electrodes for Li-ion batteries using low amount of TEMPO-oxidized cellulose nanofibrils as binder. *ACS Appl. Mater. Interfaces* 8 (28): 18097–18106.
84 Wang, C., Wu, H., Chen, Z. et al. (2013). Self-healing chemistry enables the stable operation of silicon microparticle anodes for high-energy lithium-ion batteries. *Nat. Chem.* 5 (12): 1042–1048.
85 Ban, C., Wu, Z., Gillaspie, D.T. et al. (2010). Nanostructured $Fe_3O_4$/SWNT electrode: binder-free and high-rate li-ion anode. *Adv. Mater.* 22 (20): 145–149.

# Part III

# Interacting

# 7

# Thermal and Humidity Management for Next-Generation Textiles

*Junxing Meng*[1], *Chengyi Hou*[1], *Chenhong Zhang*[1], *Qinghong Zhang*[2], *Yaogang Li*[2], *and Hongzhi Wang*[1]

[1] *Donghua University, State Key Laboratory for Modification of Chemical Fibres and Polymer Materials, College of Material Science and Engineering, 2999 North Renmin Road, Songjiang, Shanghai 201620, China*
[2] *Donghua University, Engineering Research Center of Advanced Glasses Manufacturing Technology, Ministry of Education, 2999 North Renmin Road, Songjiang, Shanghai 201620, China*

## 7.1 Introduction

Clothing's original function was to protect the body. For a long time, clothes were made directly from fur or plants. The development of technology allowed people to process natural materials and manufacture artificial fabrics. With these advances, clothing gained functions beyond mere protection. Modern textiles must be aesthetically pleasing, durable, and suitable for specific applications such as military use, health care, sports, and space exploration.

Progress in new materials, nanotechnology, and electronics has led to a new generation of clothing called smart clothing [1]. Smart clothing covers a wide range of newly developed fabrics and textiles that are integrated with intelligence, meaning they are able to interact with and respond to the body or the external environment. Comfort is still an important property of clothing. The two main factors related to comfort are temperature and humidity. The thermal energy of the body is an ideal power source for portable devices. The following Sections 7.2–7.4 will review thermal and humidity management systems based on innovative fabrics and textiles.

Smart materials can be divided into the following three categories according to their mechanism.

A. Passive smart materials (PSMs) are materials or systems that can sense and respond to environmental stimuli without input power. Their physical or chemical structure can change in response to changes in the environmental temperature or humidity.
B. Energy-harvesting materials (EHMs) are materials or systems that can transform body heat into electricity. The human body produces heat all the time. EHMs are used to harvest thermal energy and use it as an electrical power source.

*Flexible and Wearable Electronics for Smart Clothing*, First Edition.
Edited by Gang Wang, Chengyi Hou, and Hongzhi Wang.

C. Active smart materials (ASMs) are materials or systems that use external electric power to drive smart components. Their shape and temperature are controlled by conscious operations or pre-programmed commands.

## 7.2 Passive Smart Materials

Thermal comfort, which is related to the body's perception of temperature, is commonly affected by two main factors, namely heat and humidity exchange with the environment. PSMs, whose response results from an intrinsic physical or chemical structural change, can be effective in a changing environment. Phase change materials (PCMs), which can absorb or release heat in response to an environmental change, are candidates for thermal-regulating textiles [2]. The optimal phase change point is around the apparent body temperature.

Stucki et al. prepared a smart functional outdoor membrane based on thermoresponsive poly(*N*-isopropylacrylamide) (PNIPAM) microspheres embedded into a polyurethane matrix [3]. Owing to the selective dissolution of two-particle composite precursors, membranes produced using the template removal method can form a porous structure with PNIPAM microspheres located within the pore walls (Figure 7.1). The mean size of the thermoresponsive microplugs was adjusted to change the membrane breathability according to the outdoor temperature. This self-venting effect enhances water vapor transport above the critical temperature of the PNIPAM while keeping a barrier against liquid penetration. Stucki et al. also evaluated the practical effect of energy loss due to water evaporation. A test was conducted with an average adult fully covered with the thermoresponsive membrane (i.e. wearing a jacket and pants), with the status either passive or active. In warm conditions, the energy loss due to water evaporation was calculated to be about 80 W. In cold conditions, the energy loss for the passive status was only about 22 W, a reduction of 73% (Figure 7.2). In cold conditions, the textile fabric closes off its pores to reduce heat loss and thus keeps a person warm. In hot conditions, the PNIPAM microplugs open, allowing for high breathability and heat transfer.

Zhong et al. designed two kinds of humidity-induced, bendable smart clothing that can reversibly adjust its thermal insulation [4]. In the first design, Nafion

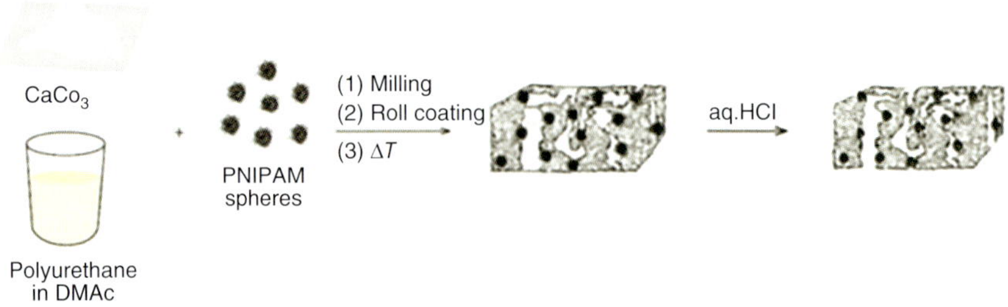


**Figure 7.1** Fabrication process of the thermoresponsive membrane proposed by Stucki et al. Source: Stucki et al. 2018 [3]. Copyright 2018, Reproduced with permission of John Wiley & Sons.

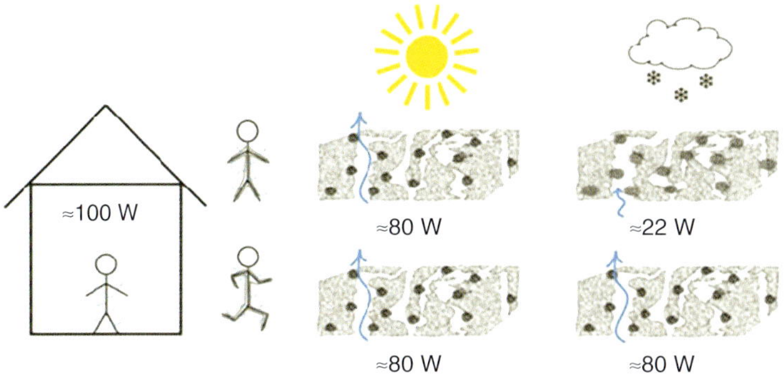


**Figure 7.2** Evaluation of thermal energy loss conducted by Stucki et al. Source: Stucki et al. 2018 [3]. Copyright 2018, Reproduced with permission of John Wiley & Sons.

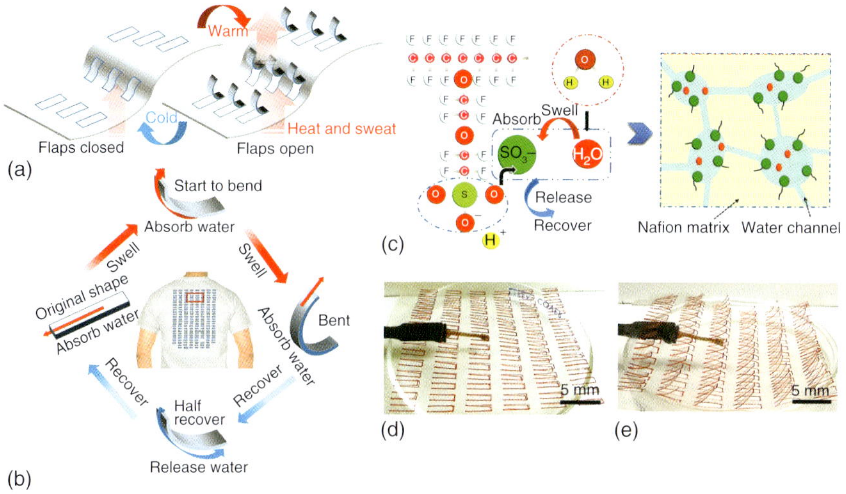


**Figure 7.3** Illustration of Zhong et al.'s first design. (a) Nafion sheet schematics; (b) reversible swelling, bending, release, and recovery behavior; (c) swelling and water transport mechanism of Nafion; (d–e) Nafion sheet with opened flaps when placed upon water vapor. Source: Zhong et al. 2017 [4]. Copyright 2017, Reproduced with permission of Springer Nature. https://www.nature.com/articles/srep44208#rightslink. http://creativecommons.org/licenses/by/4.0/. Licensed under CCBY 4.0.

films were pre-cut to form a continuous series of flaps, as shown in Figure 7.3. Like the pores in human skin, which expand when the body is warm to evaporate moisture, the flaps open once the humidity on the inner side of the Nafion membrane is higher than that on the outer side, which facilitates air flow between the human body and the environment. As a result, the humidity and temperature remain at comfortable levels even during sweating. The flaps go back to the closed state after humidity equilibrium is reached to keep the body warm. In the second design, a Nafion membrane with an arched shape was put between two fabrics and anchored to the outer layer, which acts as an insulation structure. During sweating, the inner side of the Nafion membrane absorbs moisture and expands,

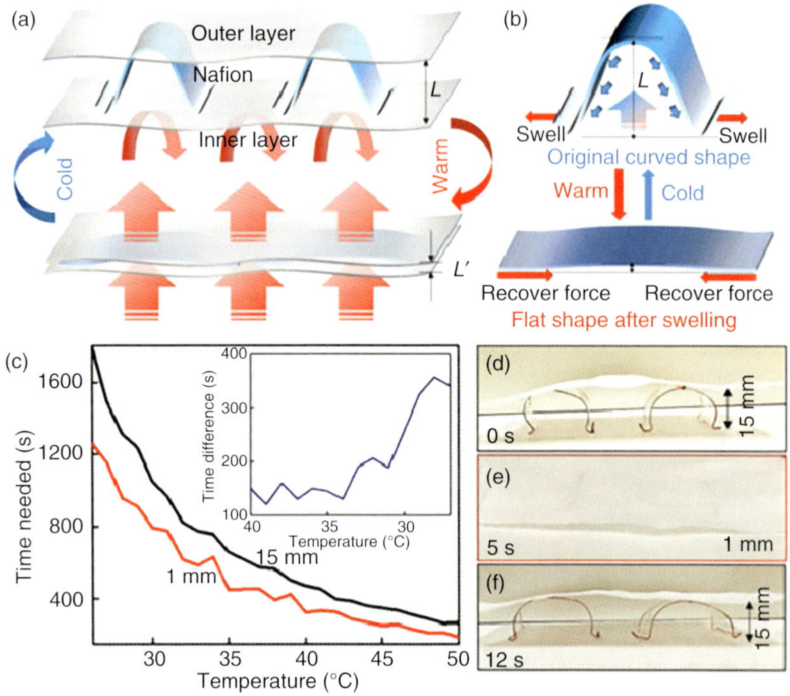


**Figure 7.4** Illustration of Zhong et al.'s second design. (a) Schematic of the Nafion film as a thermally adaptive interlayer. (b) Shape change of the Nafion film between cold state and warm state. (c) The time needed to cool by 1 °C with different insulation thicknesses. (d–f) Photographs of the arched Nafion ribbon for one cycle. Source: Zhong et al. 2017 [4]. Copyright 2017, Reproduced with permission of Springer Nature. https://www.nature.com/articles/srep44208#rightslink. http://creativecommons.org/licenses/by/4.0/. Licensed under CCBY 4.0.

which flattens the membrane. As shown in Figure 7.4, due to the shape change of Nafion films, the gap between the two fabrics decreases, reducing the thermal insulation between the wearer and the surrounding environment. In addition, the insulation structure can go back to its initial state to restore its warmth preservation function [4].

Mu et al. developed a molecular-channel-driven actuator for personal humidity and heat management. The actuation was based on the humidity sensitivity of Nafion-based materials. The laser-cut method was employed in Nafion film fabrication. The Nafion-based actuator has actuation arrays with a semilunar shape [5]. A macroscale geometric design was used to create a bilayer stimuli-responsive actuator with color-switching capability. An illustration of the system is shown in Figure 7.5. To determine the humidity adjustment performance of the Nafion-based actuator, water vapor transmission rate (WVTR) tests were conducted on a sports shirt that integrated the semilunar-patterned Nafion films. A standard terylene-based sports shirt and a sports shirt that integrated unpatterned Nafion films were also tested for comparison. At a relative humidity of 90%, the sports shirt that integrated the semilunar-patterned Nafion

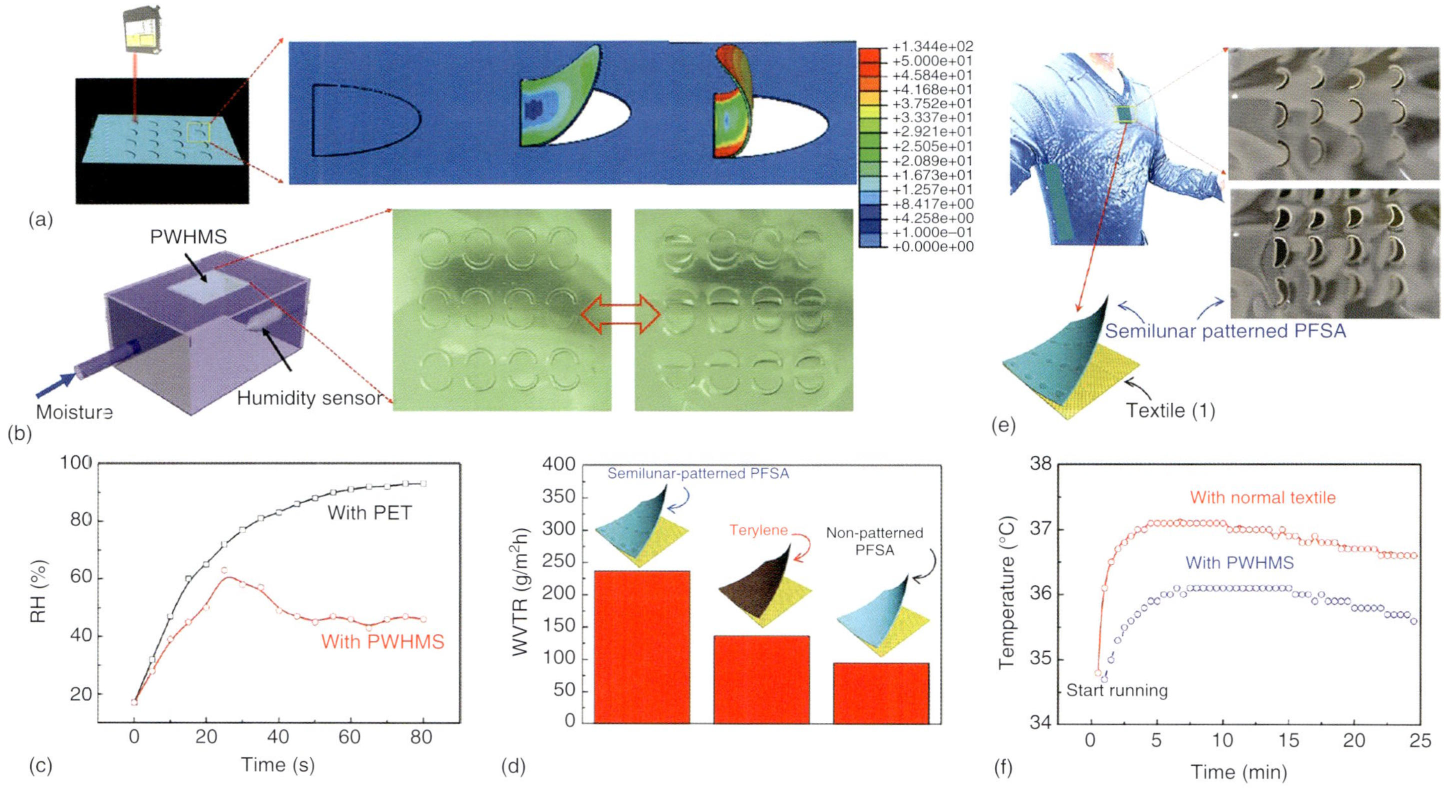


**Figure 7.5** Personal humidity and heat management system proposed by Mu et al. (a) Schematic illustration of the fabrication process using laser-cutting method. (b) Self-made humidity chamber used for the quantitative measurement of personal humidity and heat management system. (Scale bar is 1 cm). (c) Humidity change within the chamber affixed with normal PET film and personal humidity and heat management system. (d) Water vapor transmission rate of three different films (tested on shirts): semilunar patterned PFSA, terylene, and non-patterned PFSA. (e, f) Time-dependent skin temperature when the tester wears normal textiles and modifier cloth with personal humidity and heat management system (Scale bar is 1.5 cm). Source: Mu et al. 2018 [5]. Copyright 2018, Reproduced with permission of Springer Nature. https://www.nature.com/articles/s41467-018-03032-2#rightslink. http://creativecommons.org/licenses/by/4.0/. Licensed under CCBY 4.0.

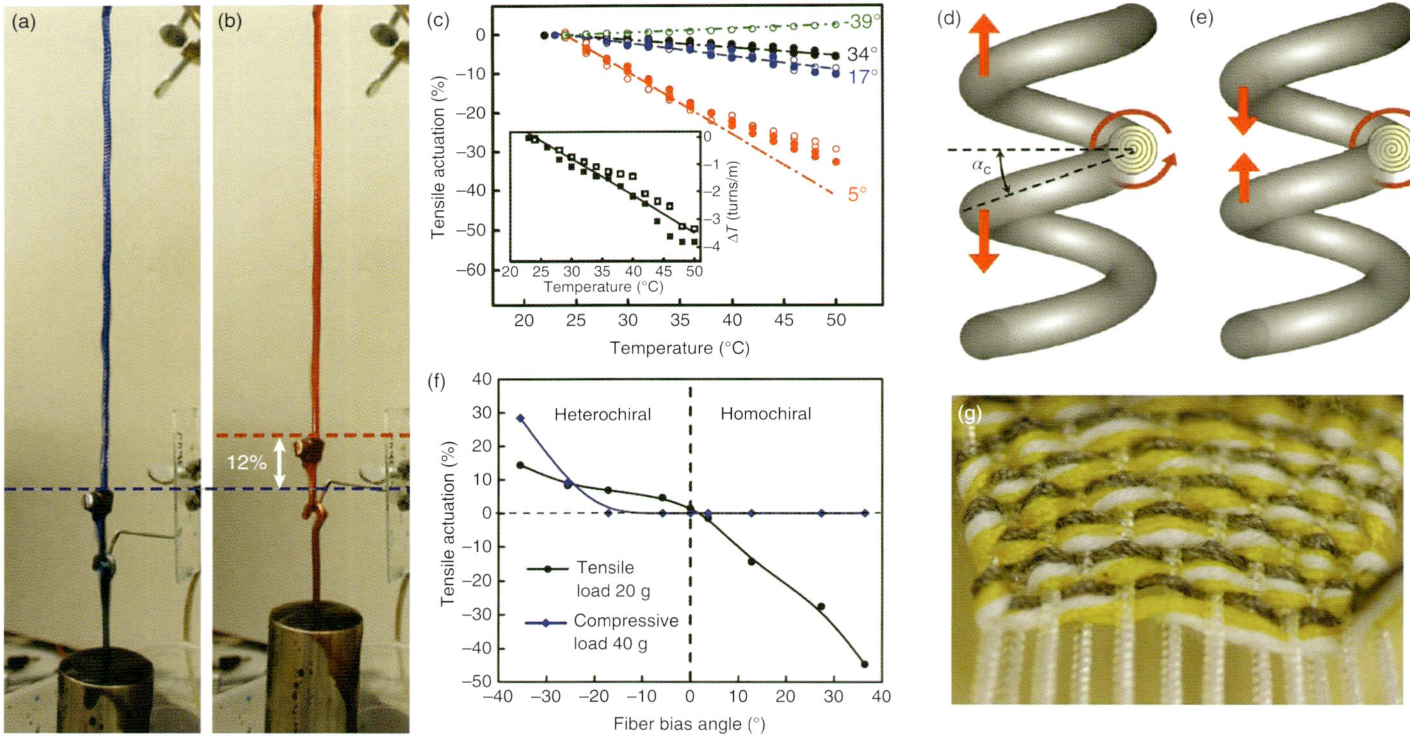


**Figure 7.6** Mechanism and applications of coiled nylon fibers proposed by Haines et al. (a, b) Hydrothermal actuation of a coiled 860 mm diameter nylon six fishing line lifting a 500 g load by 12% when switched at 0.2 Hz between ~25 °C water (dyed blue) and 95 °C water (dyed red). (c) Calculated temperature dependence of tensile actuation (dashed lines) compared to experimental results. (d, e) Schematic illustration of the mechanism by which torsional fiber actuation drives large-stroke tensile actuation for heterochiral (left) and homochiral (right) coiled fibers. (f) Measured tensile actuation versus fiber bias angle actuated between 25 °C and 95 °C. (g) An actuating textile woven from conventional polyester, cotton, and silver-plated nylon (to drive electrothermal actuation) yarn.Source: Haines et al. 2014[6]. Copyright 2014, Reproduced with permission of Science.

films showed a WVTR (up to 237 g/m$^2$ per hour) that was much higher than those of the other two shirts. The test participant who wore the former shirt had a skin temperature that was 1.3 °C lower than that of the participant who wore a normal commercial sports shirt after running 3 km.

Haines et al. developed torsional yarn made of extremely twisted nylon fibers that can contract by 49% and generate 5.3 kW of mechanical work per kilogram of yarn in response to a temperature change [6]. A textile woven from the coiled nylon fibers can change porosity in response to environmental temperature. This feature can increase wearer comfort or protect emergency responders from intense heat. Haines et al. demonstrated a braid with a coiled nylon muscle inserted in the center (Figure 7.6). Coiled nylon fibers can be woven from commercial conducting silver-plated yarn for electrothermal heating. Manual control of electrical heating can be introduced by connecting the conducting yarn to a circuit to change the porosity of the coiled nylon fibers. By changing the braid bias angle and muscle chirality, the open and closed state can be adjusted during electrothermal heating. Thermal contractions of up to 1.2%/°C can be achieved with these textile weaving designs. These results show the potential of comfort-adjusting clothing that uses nylon muscles.

Wang et al. used *Escherichia coli* cells to fabricate a biohybrid film that exhibits multifunctional responsiveness to human sweat [7]. The biohybrid film can reversibly change shape and biofluorescence intensity within a few seconds, in response to environmental humidity changes owing to the dehydration/hydration of the living cells. Wang et al. developed a sandwich structure in which cell layers are coated on both sides of a moisture-inert material. Under a moisture gradient, this biohybrid fabric bends toward the side with lower humidity, and reversibly recovers its flatness when contractile forces on the two sides balance. The bending degree can be modulated by changing the thickness and elasticity of the supporting layer. To adjust the moisture transfer and heat resistance of the fabric, a flap structure was adopted to control the skin exposure percentage. With this technology, a running suit and shoes were developed. For the running suit, during exercise, regions near the skin that produce a lot of heat and sweat need a large unit size and a high percentage of ventilating flaps to be open. In an experiment, five minutes after participants began to feel humid, the flaps opened and sweat efficiently evaporated through the pores, reducing humidity and temperature near the body. The designed shoes had multifunctional fluorescent flaps. When humidity changes, the flaps change shape and fluorescence intensity, which may reduce the discomfort and hygiene problems associated with the accumulation of foot sweat.

## 7.3 Energy-Harvesting Materials

Materials that can harvest thermal energy have been widely investigated. Thermoelectric (TE) materials based on the Seebeck effect have shown great potential as smart fabrics.

The Seebeck effect was discovered in 1821 by the German scientist Thomas Johann Seebeck. TE materials utilize this effect to directly covert high-entropy

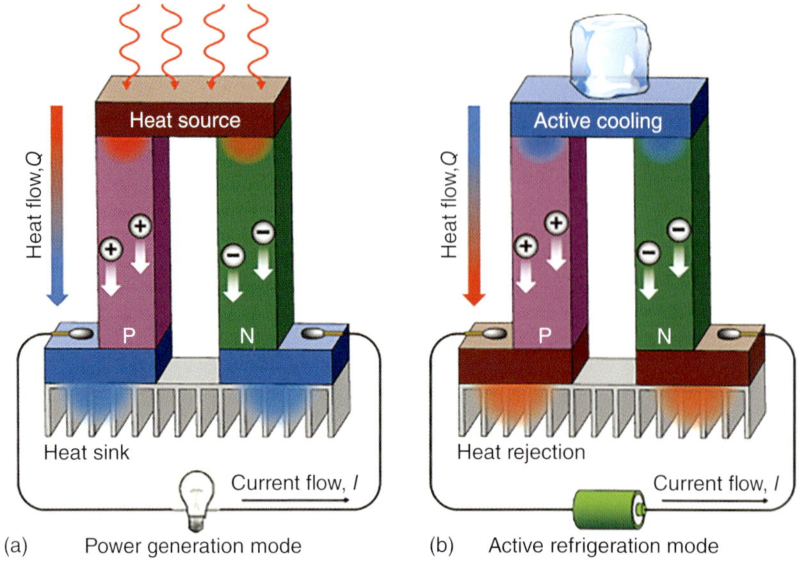


Figure 7.7 Schematic illustrations of a TE module for power generation (a. Seebeck effect) and active refrigeration (b. Peltier effect). Source: Li et al. 2010[9]. Copyright 2010, Reproduced with permission of Springer Nature.

heat energy into electricity [8]. The basic structures of TE devices are shown in Figure 7.7. In a typical TE energy-harvesting unit, two TE materials with different charge carriers (electrons for n-type materials or holes for p-type materials) are connected to each other. When a temperature gradient ($\Delta T$) is applied to the unit, the charge carriers diffuse from the hot side to the cold side, generating an electrostatic potential ($\Delta V$). If the ends of the two TE materials are connected, an electrical current will flow [9]. Generally, practically usable TE devices are made of dozens, or even hundreds, of TE couples to achieve high output voltage and power because the electrostatic potential generated by a single TE unit is extremely low (usually from several microvolts to millivolts). TE devices can also convert electrical energy into thermal energy (for cooling or heating) based on the Peltier effect, which was discovered in 1834 by the French scientist Jean Charles Athanase Peltier. TE devices lack moving parts and fluids, make no noise, and are highly reliable, and thus have great potential in the field of wearable technology [11].

Lu et al. synthesized nanostructured $Bi_2Te_3$ and $Sb_2Te_3$ as n- and p-type legs, respectively, and deposited them on the two sides of a silk fabric to produce TE columns, as shown in Figure 7.8 [10]. To ensure good contact between the TE legs and the silk substrate, a large number of pores were pricked at designated locations. This silk fabric had 12 thermocouples and used silver foil as the connector. One side of the fabric was in close contact with the heating stage, and the other side was in close contact with a cooling pad. The open-circuit voltage linearly increased from 5 to 10 mV and the power output linearly increased from 4 to 15 nW when $\Delta T$ increased from 5 to 35 K. To harvest thermal energy from the human body, the TE fabric was attached to a participant's arm. The voltage

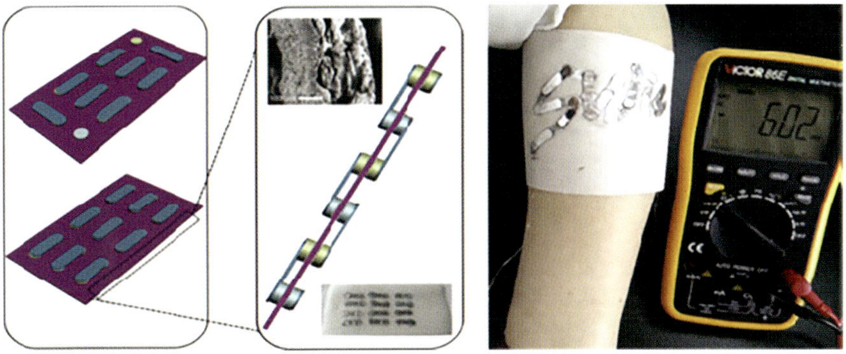


(a) (b)

**Figure 7.8** Silk-based TE fabric proposed by Lu et al. (a) Schematic illustration of the silk-based TE power generator. (b) A specially designed silk-based TE generator and its voltage output when attached to the arm at 20 °C. Source: Lu et al. 2016[10]. Copyright 2016, Reproduced with permission of Elsevier.

**Figure 7.9** T-shirts with integrated TEG devices developed by Hyland et al. Source: Hyland et al. 2016[12]. Copyright 2016, Reproduced with permission of Elsevier.

outputs were monitored before and after 30 minutes of walking. The output was higher after exercise owing to increased body temperature. A significant increase in output occurred during the first 30 seconds in both the stationary state and the exercise state. Thereafter, there was a temperature equilibrium across the cold and hot junctions of the TE fabric, resulting in an output reduction.

Hyland et al. reported an optimal thermoelectric generator (TEG) design based on n- and p-type bismuth telluride materials. The device employs polydimethylsiloxane as a soft substrate to improve insulation and reduce heat loss [12]. TEG devices that integrated a total of 50 thermocouples were fabricated and tested on a temperature-controlled hot plate and at various body locations, including the wrist, upper arm, and chest. The TEG devices were also integrated into a T-shirt, shown in Figure 7.9, and the output power was recorded during various human activities. In the tests using the human body, the highest output power was obtained for the upper arm (20 $\mu W/cm^2$). Of note, airflow significantly increased the output power when the wearer was moving.

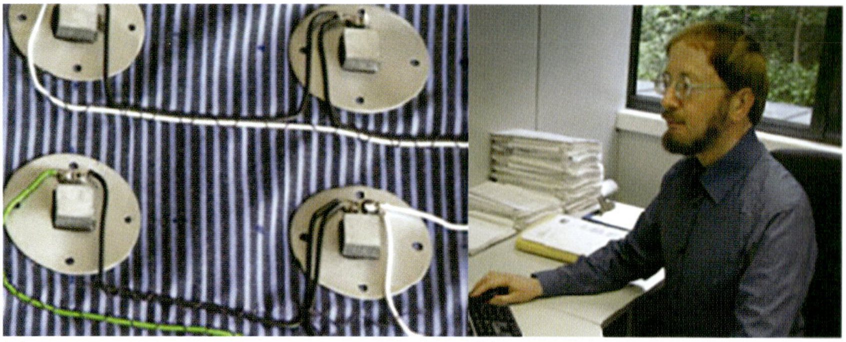

(a) (b)

**Figure 7.10** Shirt with hidden integrated TEGs developed by Leonov et al. (a) Four of eight thermopiles assembled on an additional piece of textile. (b) Shirt with hidden integrated TEH. Source: Leonov 2013 [13]. Copyright 2013, Reproduced with permission of IEEE.

Leonov et al. developed TE devices that can be directly glued or sewn to commercial shirts without requiring any modifications to existing garments. Because the textile covering only influences thermopile performance, hidden TE devices can be integrated into an office shirt, as shown in Figure 7.10. Such a shirt can generate power in the range of 5–0.5 mW at ambient temperatures of 15–27 °C. To improve comfort while maintaining electrical performance, the TE devices are attached to a cotton layer on the skin side, and the radiator is made of carbon fabric and cotton. Compared with alkaline batteries with the same thickness and weight, such a TE shirt can produce electric energy during nine months of use (if worn 10 h/d) [13].

Fiber-based TEG s (FTEGs) can accommodate three-dimensional deformation owing to their excellent tensile, bending, and in-plane shear properties, making them suitable for clothing applications. A schematic diagram of a typical FTEG unit is shown in Figure 7.11 [14]. Although FTEGs are more flexible and deformable than film-based TEGs, they have some problems that need to be addressed. The most crucial one is their high contact resistance. Generally, n-type TE materials are coated on the surface of a fiber/filament as TE elements. The interconnection layer is fabricated using a single-layer multifilament yarn. Thus, a porous region with air insulation pockets exists between the TE material and the connecting fabric. Furthermore, the interconnection layer is usually composed of multifilament yarn with several layers. A large number of contact points may be removed during the deformation of FTEGs, increasing contact resistance. For TE devices, the electrical and thermal conductivities have a large effect on electrical performance.

To achieve better integration of TE fabrics, a knitted fabric panel was fabricated using alumel and chromel wires around glass-epoxy substrates [15]. These metal wires were interconnected to form thermocouples, as shown in Figure 7.12. The maximum output power was 0.166 μW per couple at a temperature difference of 26 K. This design makes the fabric very flexible and comfortable.

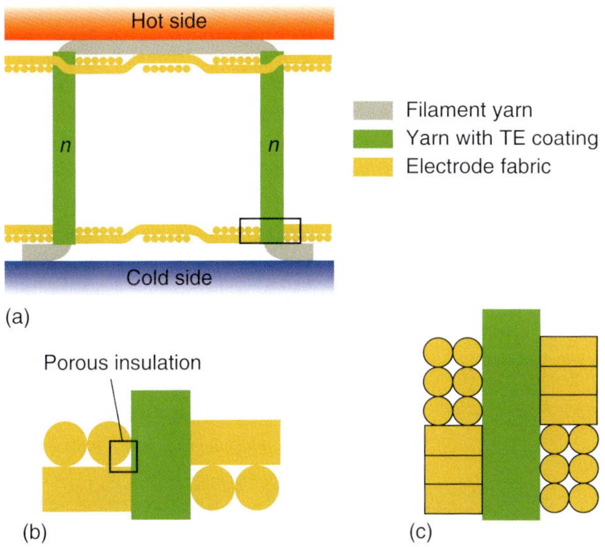


**Figure 7.11** (a) Schematic diagram of a typical FTEG unit. (b, c) The enlarged contact regions between the electrode fabric and TE coated yarns. Source: Zhang et al. 2018[14]. Copyright 2018, Reproduced with permission John Wiley & Sons.

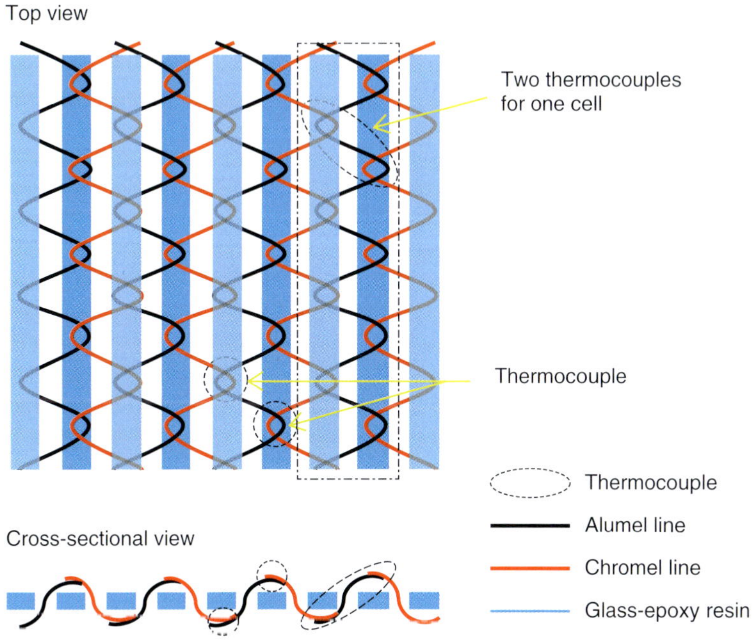


**Figure 7.12** Schematic diagrams of knitted panel. Source: Yamamoto and Takai 2002 [15]. Copyright 2002, Reproduced with permission of John Wiley & Sons.

Zhang et al. demonstrated the continuous fabrication of crystalline TE micro/nanowires by thermally drawing hermetically sealed high-quality inorganic TE materials in a flexible fiber-like substrate [16]. The resulting TE fibers were highly flexible, ultralong, and mechanically stable, and maintained the TE properties of their bulk counterparts. Two types of TEG covered on different curved surfaces were constructed and knitted into a two-dimensional textile to realize body temperature regulation, as shown in Figure 7.13. Owing to its flexibility, this TE device can efficiently connect hot and cold sources on arbitrarily curved surfaces. The output power density was on the $mW/cm^2$ level, which is suitable for a wearable energy generator. Furthermore, the thermal drawing method can be applied to other TE materials, such as SnSe and SiGe, to broaden flexible TE device design and applications.

Kim et al. fabricated a wearable TEG using a dispenser printing method. Twelve thermocouples consisting of $Bi_{0.5}Sb_{1.5}Te_3$ and $Bi_2Se_{0.3}Te_{2.7}$ were integrated into

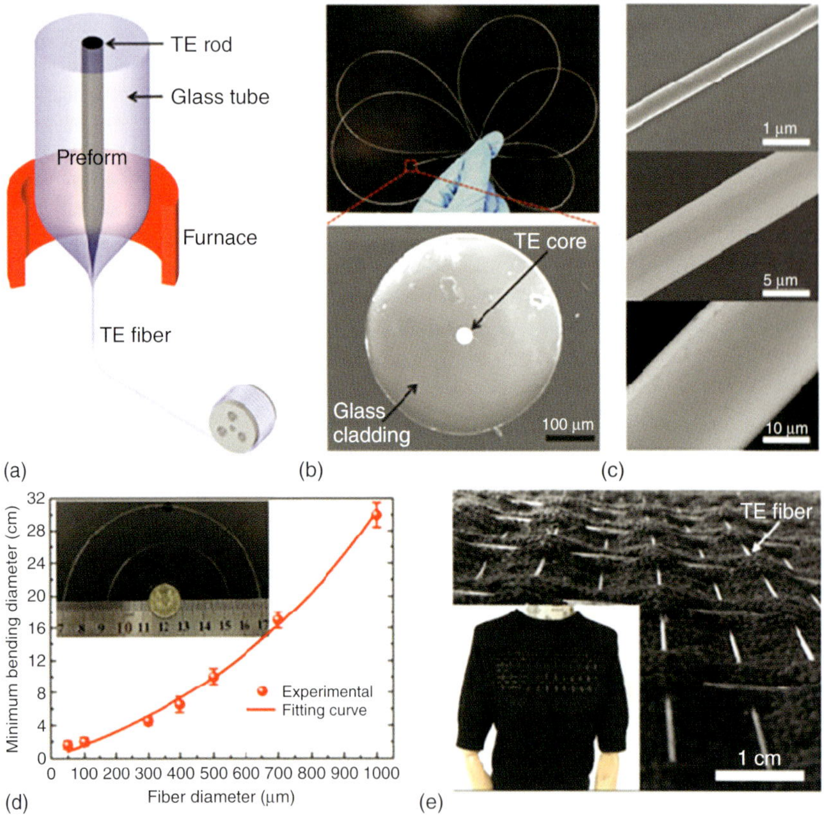


**Figure 7.13** (a) Schematic diagram of the thermal drawing process. (b) Single TE fiber showing a good flexibility and its cross-sectional SEM image. (c) SEM images of drawn TE fiber cores. (d) Minimum bending diameters of TE fibers with different fiber diameters. (e) Large-area fabric woven with TE fibers. Source: Zhang et al. 2017[16]. Copyright 2017, Reproduced with permission of Elsevier.

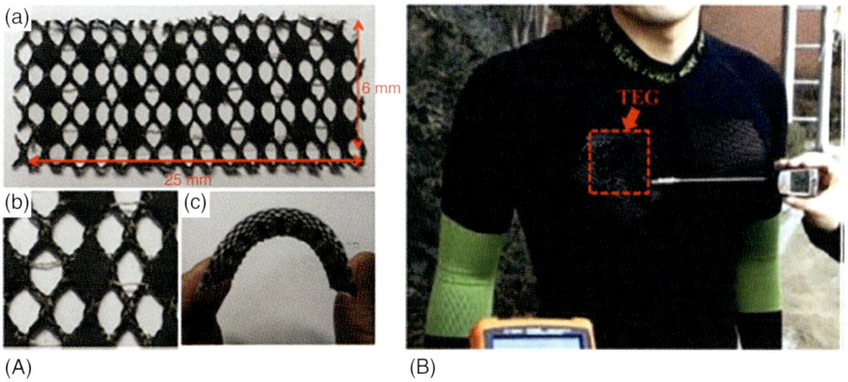


**Figure 7.14** Photographs of a shirt with an embedded TEG proposed by Kim et al. (A) Image of the printed 12-couple TEG and the bendable TEG unit. (B) Photographs of the shirt with the TEG embedded for clothing applications. Source: Kim et al. 2014[17]. Copyright 2014, Reproduced with permission of Institute of Physics.

a flexible and lightweight polymer-based fabric with an area of 6 mm × 25 mm [17]. A conducting thread containing silver was also integrated into the fabric as the connector, as shown in Figure 7.14. In a bending test, the TEG demonstrated stable electrical performance. The device generated 224 nW under a temperature difference of 15 K. Tests showed that in cold ambient air (5 °C), the TEG can generate 292.4 nW/cm$^2$ when worn by an average adult. The power generated by the TEG was enhanced by wearer movement, which led to air flow across the device and thus maintained the temperature gradient. The experimental results showed that a higher voltage was maintained during walking compared with that in a stationary state.

A large number of TEGs connected in parallel or series are needed to generate a large temperature gradient. There is also a scarcity of high-performance and flexible TE materials. Although a lot of effort has been devoted to integrating TEGs into clothing, more studies are required to increase the power output and other properties for practical use.

## 7.4 Active Smart Materials

ASMs usually have a wider range of thermal comfort adjustment compared with that for PSMs. For outdoor activities in winter or mountain climbing, preventing body heat loss is critical for safety. Electric heating is an alternative to thicker clothing for generating thermal comfort.

Wang et al. reported the preparation of multifunctional reduced graphene oxide (rGO)/polyethylene terephthalate (PET) fabrics. Suction filtration and reduction were used to fabricate the fabrics. The fabrication method is facile, low cost, and easy to control [18]. The rGO/PET fabrics, which operate based on the Joule heating effect, have a square resistance as low as 24.7 Ω/sq, making them

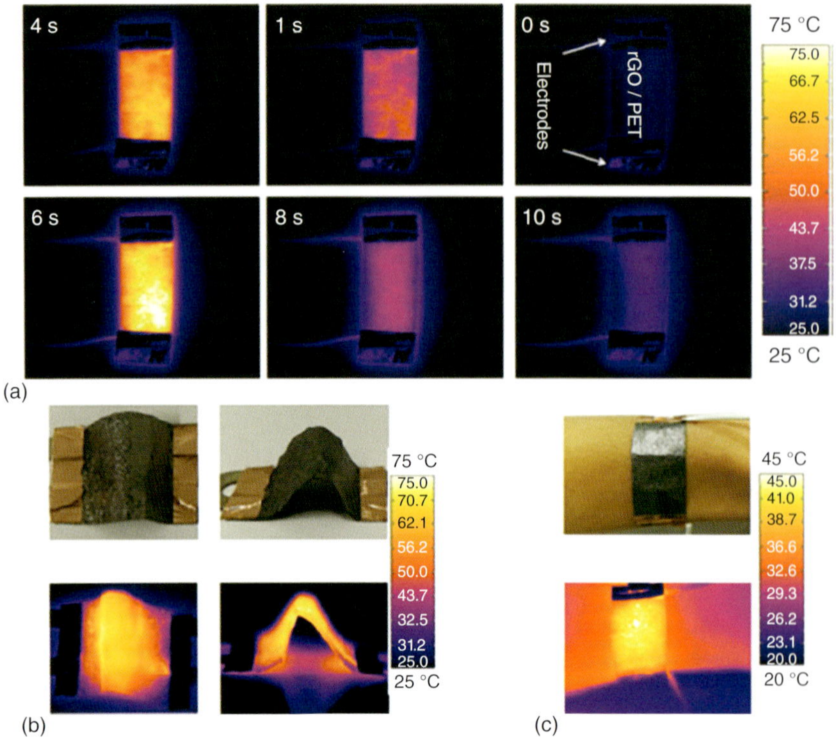


**Figure 7.15** Infrared images of rGO/PET fabric proposed by Wang et al. (a) Different moments with an applied voltage of 10 V. (b) Under the bending state with an applied voltage of 10 V. (c) Stuck on the arm with an applied voltage of 6 V Source: Wang et al. 2018 [18]. Copyright 2018, Reproduced with permission of Elsevier.

suitable for electric heating applications. The variation of electrical resistance was insignificant after 100 bending–releasing operations and 10-folding–releasing cycles. At a low voltage of ~6 V, the fabrics can rapidly reach an equilibrium temperature of about 50 °C within only three seconds. The fabric temperature can quickly drop back to ambient temperature after the Joule heating is turned off. The infrared images of the rGO/PET fabric shown in Figure 7.15 show this temperature change process. The rGO/PET fabric has potential for wearable smart devices. For example, a wearable heater could induce vasodilation and quicken blood flow around bruises or joints, reducing pain and enhancing rehabilitation.

Guo et al. reported freestanding, flexible/foldable, and wearable ultrathin graphene paper (GP) that has good thermal conductivity and a sensitive electrothermal response for personal thermal management (PTM). PTM devices are shown in Figure 7.16 [19]. PTM has two functions, namely heating and cooling. For a typical GP with a thickness of 4 μm, the electrical conductivity remained

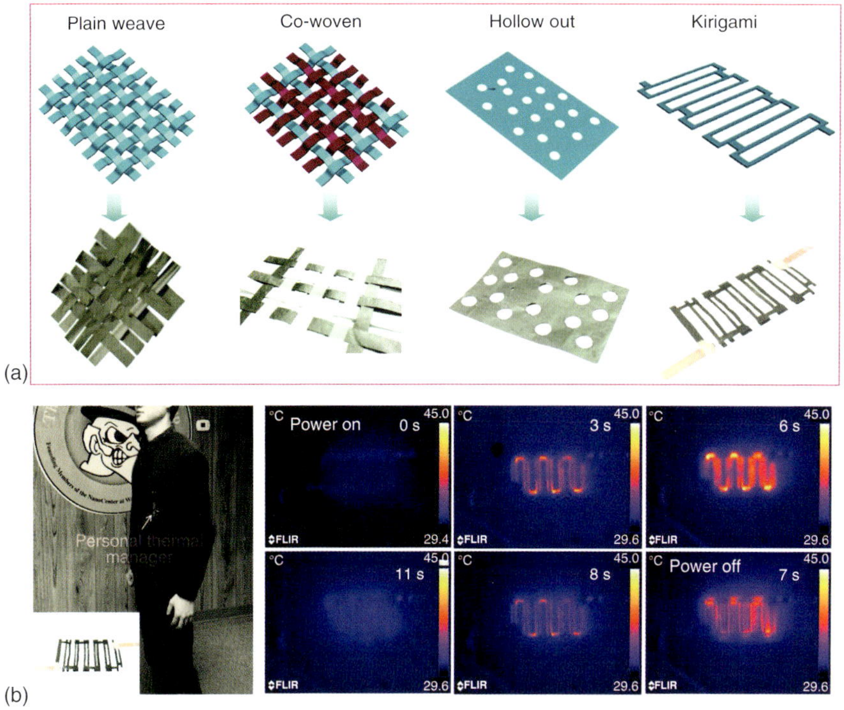


**Figure 7.16** Wearable PTM devices proposed by Guo et al. (a) Multiple device designs for PTM. (b) Demonstration of a wearable PTM device and the operational process of a kirigami structure PTM device in applied voltage of 3.2 V. Source: Guo et al. 2017 [19]. Copyright 2017, Reproduced with permission of John Wiley & Sons.

constant at about 110 S/cm after 1500 minutes of washing or 500 bending cycles. A voltage of 3.2 V can heat the GP from room temperature to body temperature within seconds. Owing to its high infrared emissivity, the GP also has a good cooling effect. The personal cooling response of the GP is only seven seconds, which is much lower than that for cotton fabric (150 seconds). In addition, the GP has high yield strength and can thus be shaped and woven into various structures to meet the necessary requirements.

To effectively control the heating process, Lee et al. developed user-interactive textile interfaces and mobile apps for temperature-regulating systems with consideration of the user's propensity for smart clothing [20]. The design of a temperature-controlled smart jacket with an all-fabric integration system is shown in Figure 7.17. The back and hood contain heating units. A textile switch and a mobile app are used to control the heating system. The textile switch provides feedback on the current state of the system and the specific areas being heated. For the smart clothing market to grow, user-centered design is required for the smart clothing interface.

**Figure 7.17** Design of temperature-controlled smart jacket with an all-fabric integration system proposed by Lee et al. Source: Lee et al. 2018 [20]. Copyright 2018, Reproduced with permission of Springer Nature.

## 7.5 Conclusion

For practical applications using smart materials, some critical challenges remain [21]. The efficient fabrication of textile-shaped devices on a large scale is a big challenge. Device performance usually decays when device size is increased from the micro to macro scale. Interface issues between electrodes also greatly influence contact resistance and reduce the electrical performance of TE devices. In addition, methods for sealing functional components are vital to device stability. Another crucial factor is washability, considering the negative effects of liquids on electronics. Overall, more focus should remain on the specific area, while further investigation should be devoted to the above-mentioned crucial factors.

## References


1 Cho, G., Lee, S., and Cho, J. (2009). *Int. J. Hum.-Comput. Interact.* 25: 582.
2 Mondal, S. (2008). *Appl. Therm. Eng.* 28: 1536.
3 Stucki, M., Stockli, M., and Stark, W.J. (2018). *Macromol. Mater. Eng.* 303: 1700562.
4 Zhong, Y., Zhang, F., Wang, M. et al. (2017). *Sci. Rep.* 7: 44208.
5 Mu, J., Wang, G., Yan, H. et al. (2018). *Nat. Commun.* 9: 590.
6 Haines, C.S., Lima, M.D., Li, N. et al. (2014). *Science* 343: 868.

7 Wang, W., Yao, L., Cheng, C.-Y. et al. (2017). *Sci. Adv.* 3: e1601984.
8 Seebeck, T.J. (1822). *Abh. Akad. Wiss. Berlin* 1820–1821: 289.
9 Li, J.F., Liu, W.S., Zhao, L.D., and Zhou, M. (2010). *NPG Asia Mater.* 2: 152.
10 Lu, Z., Zhang, H., Mao, C., and Li, C.M. (2016). *Appl. Energy* 164: 57.
11 Du, Y., Xu, J., Paul, B., and Eklund, P. (2018). *Appl. Mater. Today* 12: 366.
12 Hyland, M., Hunter, H., Liu, J. et al. (2016). *Appl. Energy* 182: 518.
13 Leonov, V. (2013). *IEEE Sens. J.* 13: 1.
14 Zhang, L., Lin, S., Hua, T. et al. (2018). *Adv. Energy Mater.* 8: 1700524.
15 Yamamoto, N. and Takai, H. (2002). *Electr. Eng. Jpn.* 140: 16.
16 Zhang, T., Li, K., Zhang, J. et al. (2017). *Nano Energy* 41: 35.
17 Kim, M.K., Kim, M.-S., Lee, S. et al. (2014). *Smart Mater. Struct.* 23: 105002.
18 Wang, D., Li, D., Zhao, M. et al. (2018). *Appl. Surf. Sci.* 454: 218.
19 Guo, Y., Dun, C., Xu, J. et al. (2017). *Small* 13: 1702645.
20 Lee, E., Roh, J.S., and Kim, S. (2018). *Fibers Polym.* 19: 238.
21 Weng, W., Chen, P., He, S. et al. (2016). *Angew. Chem. Int. Ed.* 55: 6140.

# 8

# Functionalization of Fiber Materials for Washable Smart Wearable Textiles

*Yunjie Yin, Yan Xu, and Chaoxia Wang*

*Jiangnan University, Key Laboratory of Eco-Textile, Ministry of Education, School of Textiles and Clothing, 1800 Lihu Road, Wuxi 214122, China*

## 8.1 Introduction

There have been many attempts during the last decades to design and develop a series of wearable devices that serve many purposes [1]. The new devices that emerged are more sophisticated than the previous ones with a potential to be used to perform a variety of tasks and enhance the operations of some fields, such as military, public safety, health care, and sports health. Smart wearable is a recently adopted term in the field of wearable electronic devices and smart conductive textiles. It can be worn directly on the body or integrated into clothes and accessories. Smart bracelets, smart watches, and smart glasses are the most common wearable electronic devices, and they can monitor the vital parameters, such as the electrocardiogram (ECG), pulse and oxygen saturation, blood pressure, at home as well in outdoor environments, keep the user informed about his health conditions, and/or alert medical doctors in case of emergencies [2]. Smart conductive textiles are not simple bracelets, watches, or glasses; they could be woven like typical fabrics and integrate advanced functions such as energy storage and conversion, sensors, color change, and drug release, or can deliver responses to external stimuli, among others [3].

### 8.1.1 Conductive Textiles

Conductive textiles are a part of smart textiles with widely differing specific electrical conductivity; they plays an important role in smart wearables. Conductive textiles include conductive fibers, yarns, fabrics, and final products made from them. Conductive textiles are mainly used to prepare supercapacitors and sensors for energy storage and health monitoring, respectively. The original manufacturing technology for conductive textiles was to attach various kinds of electronic devices to garments [4]. Now, the preparation of conductive textiles can be divided into four categories, that is, polymer conductive textiles, metal conductive textiles, carbon material conductive textiles, and composite material conductive textiles. These conductive textile materials will be subjected

*Flexible and Wearable Electronics for Smart Clothing*, First Edition.
Edited by Gang Wang, Chengyi Hou, and Hongzhi Wang.

to various environments in practical applications, including water and humid environments. Therefore, a major challenge in the applications of conductive textiles is washing durability, especially for health monitoring clothing. Washability is significant for conductive textile materials and flexible wearable electronic devices based on textile materials [5].

### 8.1.2 Waterproof Conductive Textiles

Conductive textiles can be fabricated by the incorporation of functional conductive components into textile substrates. For smart wearable devices, we not only value their conductivity, but also have to face the problem of washing durability during wearing process, which is closely related to their service life [6]. Conductive textiles will face a variety of harsh environments in application, as the basis for smart textiles or electronic textiles. On contact with water, the common conductive fabric will short-circuit, seriously affecting the normal functioning of smart textiles and electronic textiles, and even cause personal injury to the user. Therefore, if conductive textiles can take into account the waterproof or hydrophobic function, the practical application value of the conductive textile will be greatly improved, the scope of use of the textile can be greatly expanded, and the normal functioning of the device can be maintained even in rainy or underwater conditions. When the surface of the conductive fabric is exposed to water and humid environments, if it has a hydrophobic function, it can effectively resist the influence and loss of these water and humid environments, such as the "short circuit," which can not only resist the damage due to the liquid contaminant itself, but even make the device safe. The fabric surface cannot be spread by liquid, and only the spherical sliding can be formed, and the dust and dirt on the surface of the fabric can be self-cleaning, thereby improving the use strength of the conductive fabric and increasing the service life of the conductive fabric.

### 8.1.3 Washable Conductive Textiles

Washable conductive textiles are special conductive textile materials with the features of softness, general textile appearance, and stability in water and humid environments. They can be cut and sewn like a normal fabric, and more importantly, can be washed directly and dried. Because of these advantages, it is expected to combine electrical devices with clothing and garments for wearable textiles. The main factors of the water washability of the conductive textile material include the binding force of the conductive material to the textile substrate, its internal bonding force, and its own hydrophobic properties. The hydrophobicity of conductive textiles can improve the water washability and durability.

### 8.1.4 Evaluation of Washable Conductive Textiles

For washable conductive textiles, it is of great significance to evaluate the washability of conductive textiles. Washing methods or corresponding standards are necessary. We can use these methods or standards to evaluate whether conductive textiles can meet the wearing requirements. The evaluation of washable conductive textiles mainly focuses on the change in conductivity before and after washing, which is significant for wearing service life. At the same time, washing durability is also evaluated.

## 8.2 Fiber Materials Functionalization for Conductivity

Conductive textiles are obtained by combining conductive functional components with conventional textiles. Owing to the difference in conductive components, they can be divided into four types: polymer conductive textiles, metal conductive textiles, carbon material conductive textiles, and composite material conductive textiles. The preparation methods of conductive textiles are varied, such as coating, printing, graft modification, in situ chemical polymerization, vacuum filtration, and dip coating.

### 8.2.1 Conductive Fiber Substrates Based on Polymer Materials

Conducting polymers are a new class of organic materials that have extensive delocalization of $\pi$-electrons in a conjugated structure with wide applications, such as energy storage, molecular recognition, electromagnetic interference shielding, opto-electronic devices, corrosion protection, microwave absorption, gas separation, sensors, and heat generation. The $\pi$-electrons are delocalized over large segments of the polymer chain, which are responsible for the electronic properties of the conductive polymer. Among conducting polymers, polyaniline (PANI), polypyrrole (PPy), poly(3,4-ethylenedioxythiophene) (PEDOT), polyion complex, and polythiophene (PTh) have received widespread attention. PANI possesses thermal and chemical stability, and low production cost. It can be easily doped with inorganic and organic acids to prepare the conductive film. PPy is of great significance for its high conductivity, good environmental stability, good adhesion, and nontoxicity. PEDOT is characterized by simple molecular structure, small energy gap, and high conductivity. The intrinsic insolubility limits the application of PEDOT, which can be solved with the doping of water soluble polyelectrolyte poly(sodium-*p*-styrenesulfonate) (PSS). PSS is used as a charge balancing counterion during oxidative polymerization of the 3,4-ethylenedioxythiophene (EDOT) monomer. PEDOT:PSS is provided with high conductivity and stability. The polyion complex can be easily dispersed in water as colloidal gel particles with diameters of several tens of nanometers and it has high stability. It can be coated on hard surfaces of microelectronics as well as fibers and fabrics and other stretchable substrates. Conductive textiles based on conducting polymers can be prepared using dip coating, graft modification, and polymerization (such as in situ chemical, in situ electrochemical, in situ vapor phase polymerization) processes, etc.

#### 8.2.1.1 Dip Coating

Conducting polymers are insoluble in most of the solvents available, but through some treatment, the polymer can be made soluble in some selected solvents.

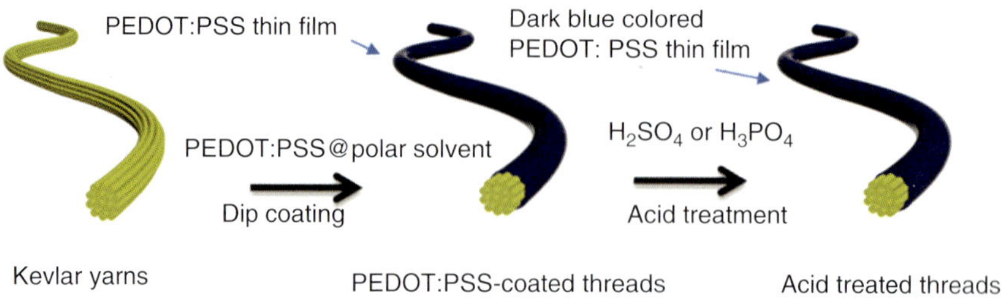


**Figure 8.1** Schematic illustration of preparation procedure of the conductive PEDOT:PSS coated Kevlar thread.

The main advantage of soluble conducting polymers is that they can be directly applied to any substrate. For example, the conductive poly-paraphenylene terephthalamide (Kevlar) threads were obtained by coating the conductive PEDOT:PSS on the surfaces via dip coating [7]. The preparation procedure of conductive PEDOT:PSS coated Kevlar threads is illustrated in Figure 8.1. The electrically conductive threads should be mechanically flexible with no significant increase in resistance when used in smart textiles. The conductive durability, flexibility, and washability of the conductive threads were shown in Figure 8.2. The surface resistance of the fibers slightly increased from 13 to 15 Ω after prolonged exposure to air, and the surface resistance gradually increased with repeated bending; the surface resistance increased by 13% (from 13.5 to 15.3 Ω) as a result of repeated bending of 300 times. For conductive textiles, it is valuable to improve the conductive durability of washing cycles. The surface resistance of the PEDOT:PSS coated Kevlar thread after four repeated washings in deionized water and detergent solution was increased from 13.6 to 17.6 Ω and from 15.1 to 41.0 Ω, respectively. It could be seen that the PEDOT:PSS coated Kevlar thread possessed excellent durability, flexibility, and washability.

Fully wearable electrodes also were fabricated by using dip-coating technology of conducting polymer PEDOT:PSS on a commercial knitted fabric [8]. Knitted fabrics were chosen as substrates because of their high level of stretchability, allowing a good contact with the skin. Fabrics coated with PEDOT:PSS were assembled into textiles structures, such as bras, by sewing with conductive yarn; then textile electrodes were connected to an ECG measurement device and high-quality ECG signals were successfully recorded. The surface resistivity of textile electrodes with substrate of cotton, polyamide, and polyester (PET) was 21.02, 34.26, and 36.31 kΩ, respectively. Washability tests of connected underwear were carried out up to 50 washing cycles, as shown in Figure 8.3, and ECG data, recorded from a healthy volunteer, were found to be stable. The research work paves the way for the fabrication of low-cost electrodes for cutaneous electrophysiology.

#### 8.2.1.2 Graft Modification

Conductive textiles can be fabricated by graft modification to improve the binding force between conductive polymers and textiles. The modification of polypropylene (PP) and viscose (VS) textiles with conductive polymer by chemical oxidative polymerization of pyrrole (Py) under static conditions using

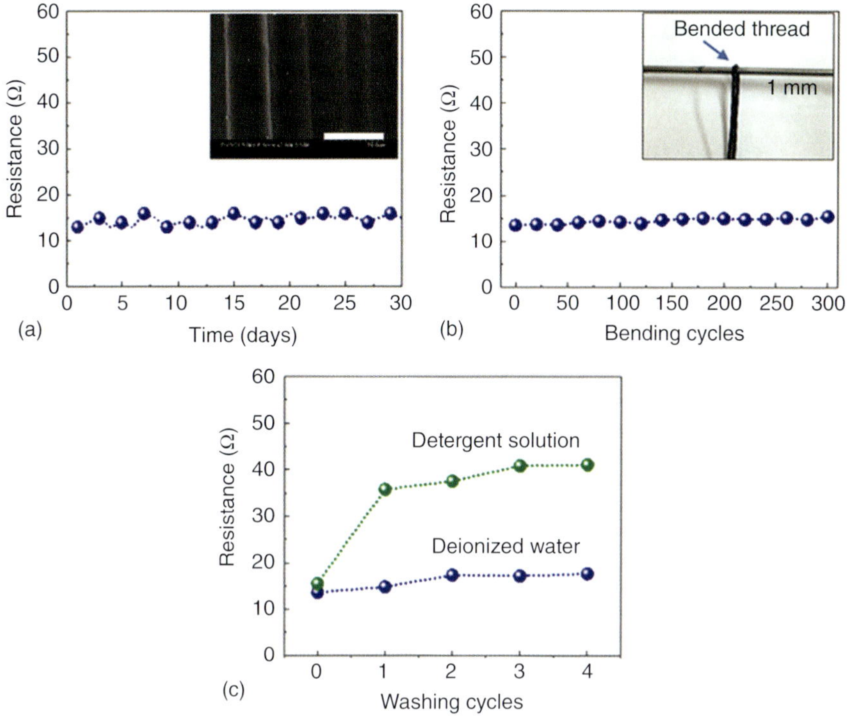


**Figure 8.2** Changes in surface resistance of PEDOTPSS coated Kevlar threads: (a) after prolonged exposure to air, (b) repeated bending, and (c) washing. The inset exhibits the scanning electron microscope (SEM) images with a 20 mm scale bar and a schematic diagram of bending test.

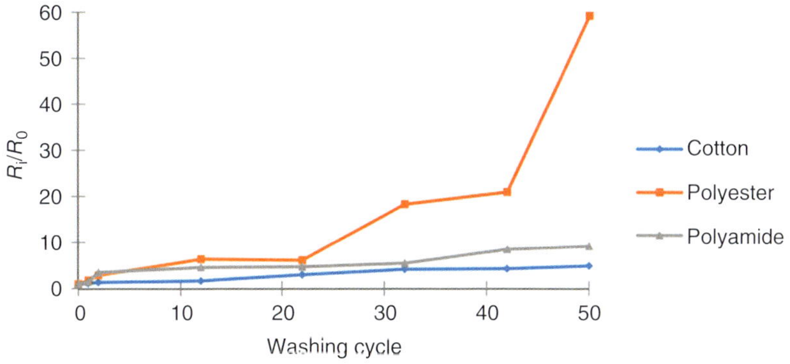


**Figure 8.3** The evolution of $R_i/R_0$ vs. washing cycle.

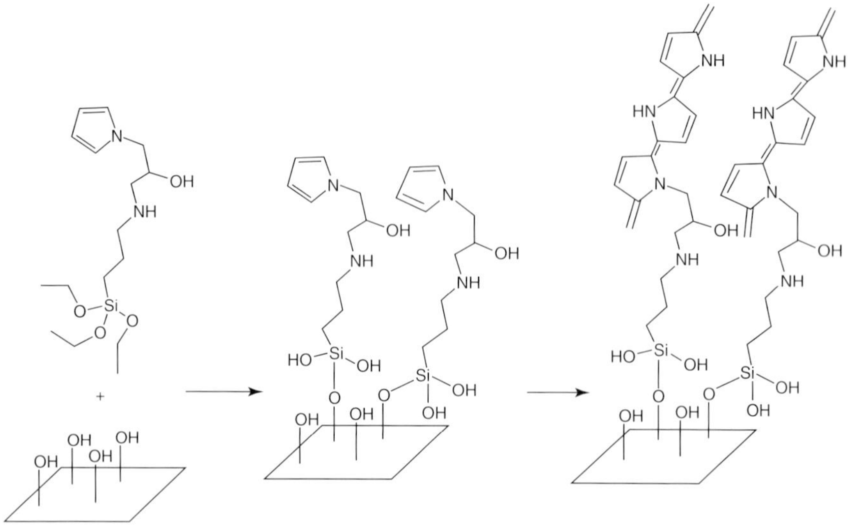

**Figure 8.4** Mechanism of PPy binding through SP to the textile surface.

ferric chloride as oxidant was reported [9]. To improve the resistance of textiles against scouring and abrasion of the outer PPy layers, a Py-functionalized silane, 1-(3-(triethoxysilyl)propylamino)-3-(1*H*-Py-1-yl)propan-2-ol (SP), was used. SP was bonded to the surface of VS by the reaction between hydroxyl groups of hydrolyzed SP and those of the VS. To ensure covalent bonding of SP onto PP fabrics, the fabrics were first grafted by vinyltrimethoxysilane (VTMS) using a radiofrequency plasma discharge. The binding of SP onto PP was ensured by reaction between the hydroxyl groups of the hydrolyzed SP and VTMS. The mechanism was shown in Figure 8.4. Electrical conductivity and its stability after washing the abovementioned fabrics were investigated, as shown in Figures 8.5 and 8.6. In the case of VS, PPy was found to penetrate into the substrate. A compromise was found between the influence of SP and penetration phenomenon. The best conductivity after washing was $3 \times 10^{-5}$ S/sq. In the case of PP, the effect of pretreatment with SP was much better than that for VS, and a higher concentration of SP led to improved fastness of the conductive layer (the best conductivity after washing was $8 \times 10^{-5}$ S/sq), which indicated that the coating promoted by means of SP was more favored than that for VS. In both cases, for VS and for PP textiles the presence of SP improved the modification with PPy and consequently also the fastness of PPy layers.

#### 8.2.1.3 In Situ Chemical Polymerization

For in situ chemical polymerization, solutions of monomer and a suitable oxidant (e.g. $FeCl_3$) are mixed together and subjected to constant stirring for prolonged duration. Oxidative polymerization occurs and polymers form in bulk. The coating of different materials with conducting polymers, i.e. PANI, PPy, PTh, and their derivatives, is possible by means of in situ chemical polymerization. The in situ chemical polymerization can be performed in a single- or double-bath process. For single-bath process, monomer and oxidant solutions are mixed in

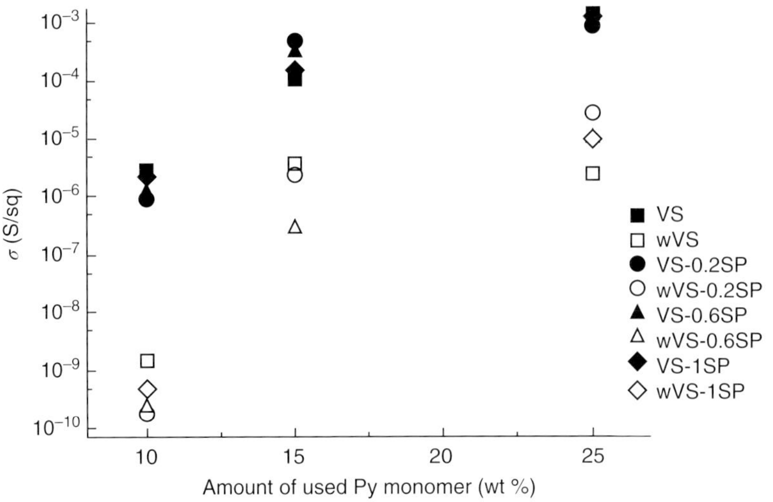


**Figure 8.5** Conductivity dependence of modified VS textile on amounts of Py and Py-functionalized triethoxysilane used, before and after washing.

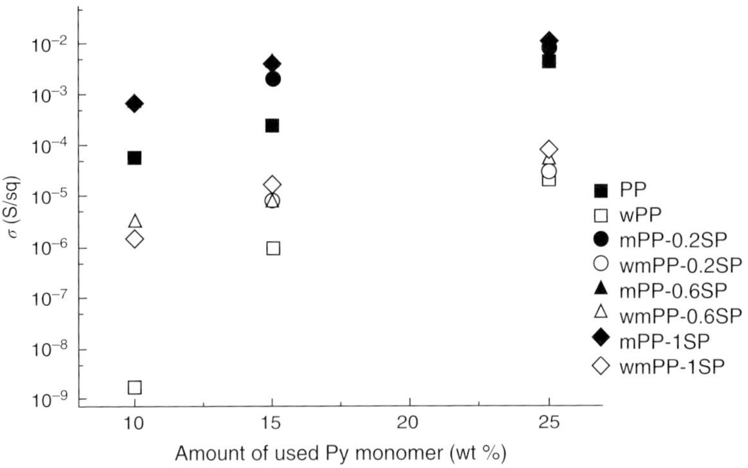


**Figure 8.6** Conductivity dependence of polypropylene textile on amounts of Py and Py-functionalized triethoxysilane used, before and after washing.

a single beaker and simultaneously the textile substrate is immersed into it. For double-bath process, the textile substrate is treated with monomer solution first and then the monomer-enriched substrate is immersed into oxidant solution or vice versa. As polymerization begins, some polymers deposit on the textile substrate due to adsorption and some present in the solution in bulk. Using PPy as an example, the color of the substrate as well as the solution changed to greenish black, which was the color of the polymer after polymerization [10]. It is suitable for laboratory preparation and mass production of conductive textiles. In order

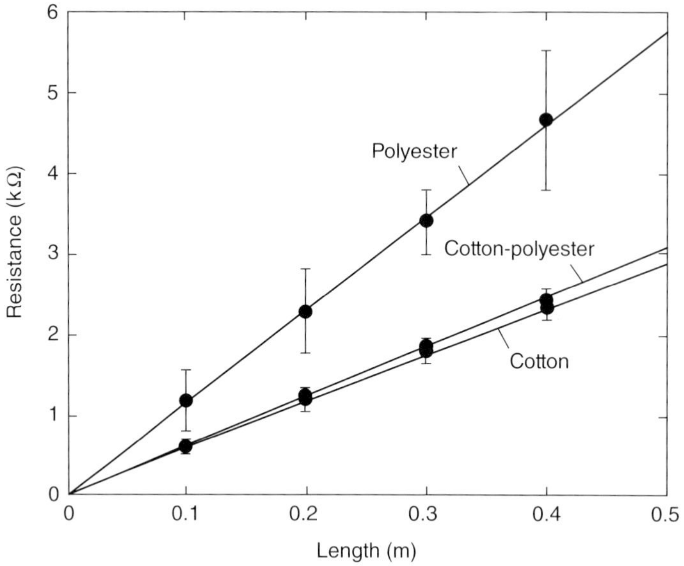


**Figure 8.7** Plot of electrical resistance vs. length of yarns.

to improve the washing durability, the requirement is that PPy should have some affinity to the textile substrate.

#### 8.2.1.4 Electrochemical Polymerization

Electrochemical polymerization is usually performed in a one-compartment cell where two electrodes such as anode and cathode are connected with an external power supply. The cell is provided with monomer solution with suitable electrolytes and dopant. In most of the cases, the electrolytes act as dopants as well. Electrochemical oxidation of the monomer results in polymer films deposited on anode surfaces. If the anode surface is covered with a textile substrate, the polymers will deposit on it. The polymerization rate and yield depend upon the material of the electrode, types of solvent, electrolyte, supply voltage, temperature, time, etc. The electroconductive yarns were prepared by combined in situ chemical and electrochemical polymerization of Py onto cotton, PET, and cotton-PET blended yarns [11]. The chemical polymerization was carried out prior to the electrochemical polymerization to impart some amount of electrical conductivity to the yarns, which was found to result in more uniform coating of PPy on the surface of the yarns during electrochemical polymerization. The electrical resistance of the electroconductive yarns was measured at different lengths along the yarn. The experimental results were presented in Figure 8.7. The resistivity of the cotton, PET, and cotton-PET blended yarns was obtained as 5.7750, 11.4900, and 6.0870 kΩ/m, respectively.

#### 8.2.1.5 In Situ Vapor Phase Polymerization

In situ vapor phase polymerization of conducting polymers is a suitable process for producing electroconductive textile in two steps as follows. Impregnation of textile is done in an aqueous solution of oxidant and dopant and followed by drying. After that, the fabric is exposed to monomer vapor for in situ polymerization. PPy coated wool, cotton, and nylon 66 fabrics were prepared using the continuous vapor-phase polymerization method [12]. Comparing the electrical conductivity obtained for wool, cotton, and nylon 66 fabrics with those obtained in other works, a higher conductivity was achieved in this work. It is concluded that the continuous vapor-phase polymerization of Py is an efficient coating method by which the higher conductivity can be obtained for natural as well as synthetic textile. Conductive fabric also can be prepared by vapor phase polymerization of EDOT on fabric in the presence of fluorinated decyl polyhedral oligomeric silsesquioxane (FD-POSS) and a fluorinated alkyl silane (FAS) [13]. When PEDOT was vapor-phase polymerized onto fabrics, the incorporation of FD-POSS and FAS into the coating layer during the polymerization process endowed the conductive coating with a durable superamphiphobic surface. The coated fabric had contact angles of 169° and 156° to water and hexadecane respectively, and a surface resistance in the range of 0.8–1.2 kΩ/sq. The incorporation of FD-POSS and FAS into the PEDOT layer showed a very small influence on the conductivity but improved the washing and abrasion stability considerably. The coated fabric could withstand at least 500 cycles of standard laundry and 10 000 cycles of abrasion without apparently changing the superamphiphobicity, while the conductivity only had a small reduction after the washing and abrasion. More interestingly, the coating had a self-healing ability to auto-repair from chemical damages to restore the liquid repellency. The fabrics prepared by vapor phase polymerization show a high uniform polymer coating on the fiber surface, which is provided with minimized surface resistivity changes, improved light fastness, and washing fastness. However, controlling of the add-on is difficult and equipment set-up is complicated.

### 8.2.2 Conductive Fiber Substrates Based on Metal Materials

Metal is the first material used to make conductive textiles, and the world's first conductive fiber was prepared by stretching steel wire. The use of copper, silver, and iron is more common. Metal conductive textiles can be fabricated by wire spinning, weaving, and the deposition of metal nanoparticles on textile substrates. The metal deposited textiles not only show electrical conductivity but also have excellent mechanical stability, flexibility, and strong attraction toward almost all types of textile fabrics. However, the bonding strength between the metal and the ordinary fabric is very weak, so it is generally required to be combined with the fabric by means of electroless plating or other deposition methods.

#### 8.2.2.1 Electroless Plating

Electroless plating is a method of coating in which metal ions in the plating solution are reduced to metal and deposited on the surface of the substrate with the help of an appropriate reducing agent without an external current. Electroless plating possesses the features of high density, uniform thickness, and low cost. Common solutions for electroless plating include electroless silver, nickel, copper,

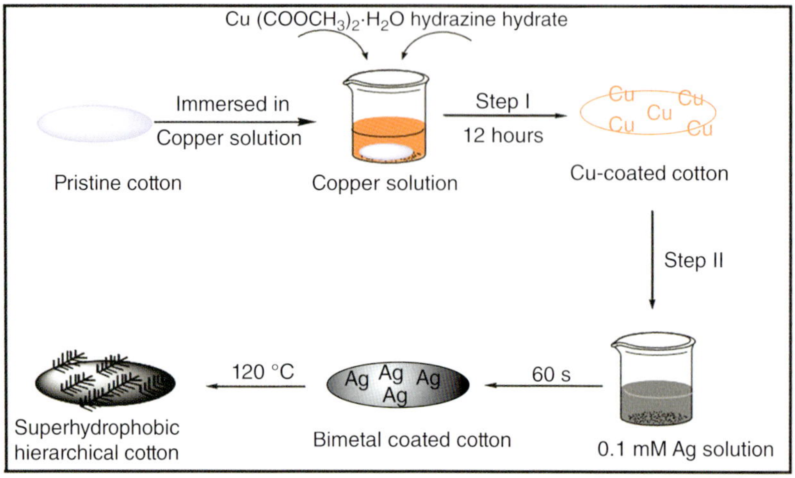


**Figure 8.8** Schematic representation of the fabrication of superhydrophobic cotton.

cobalt, nickel phosphorus, nickel phosphorus boron, etc. Moreover, electroless plating of two metals is available.

A superhydrophobic textile with electrical conductivity and antibacterial activity was developed through bimetallic deposition of copper and silver over cotton fabric [14]. As shown in Figure 8.8 copper and silver bimetallic coatings over cotton fabric were made using in situ deposition and electroless deposition (ELD) methods. The water contact angle (WCA) of pristine cotton was 0°. After deposition of copper, the surface became rougher and the WCA rose to 126.4°, and the hierarchical deposition of Ag over Cu further increased the surface roughness, and consequently the WCA was increased to 160.9°. The results of electrical conductivity of as-coated samples showed that pristine cotton was an electro-insulator while the copper coated cotton fabric showed a sheet resistance of 1.15 Ω/sq and bimetal deposited cotton fabric had a sheet resistance of 0.15 Ω/sq. The chemical durability and electrical stability are important for commercial applications. The as-fabricated samples were found to maintain their sheet resistances and contact angle in all pH ranges (Figure 8.9). The coatings were durable and had good electrical stability and superhydrophobicity in all pH solutions. The washing test results showed that the sheet resistance was almost the same after repeated washings. These results indicated that coating had strong adhesion and showed chemical and washing durability of as-coated samples (Figure 8.10).

It should be noted that the development of adhesion between metals and substrates has been a major issue in the industry for decades. Direct coating of cotton fibers/yarns with metals typically resulted in flaking off at bending and rubbing, and cannot withstand washing tests. A novel, facile, and versatile approach for preparing highly durable, electrically conductive cotton yarns by ELD of metal particles onto cotton yarns modified with polyelectrolyte brushes as adhesion layer was reported [15]. Polyelectrolyte brushes, polymers that covalently tether one end on a surface, were first grown from cotton surfaces

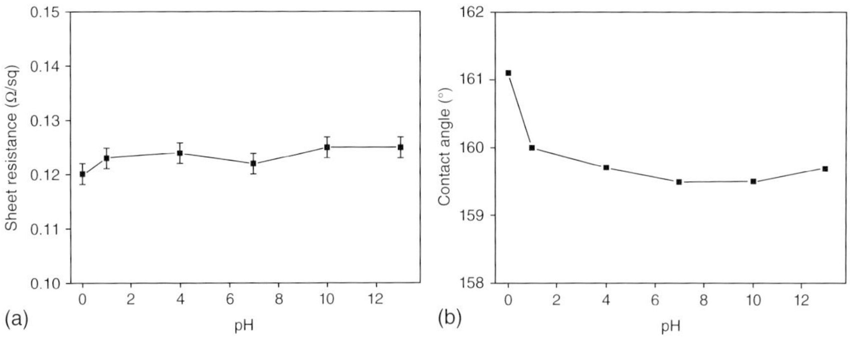


**Figure 8.9** Chemical durability of bimetal coated cotton fabric: (a) sheet resistance and (b) wettability against various pH solutions.

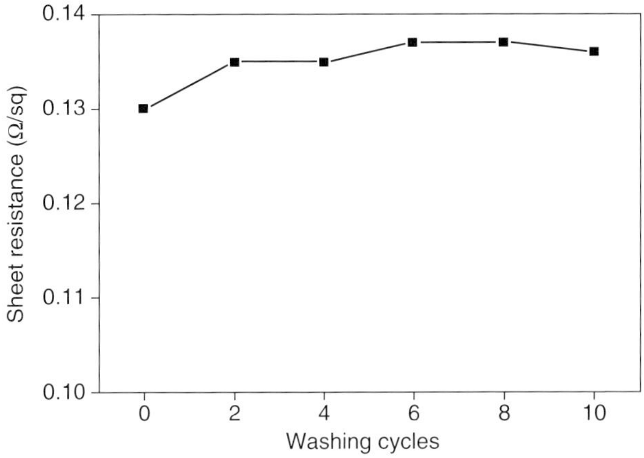


**Figure 8.10** Washing durability of as-coated samples after repeated washings.

by surface-initiated atomic transfer radical polymerization. Subsequent ELD of metal particles onto the brush-modified cotton yarns yielded electrically conductive yarns. Importantly, the formation of polymer brushes bridged metal/cotton hierarchical structures provided robust mechanical and electrical durability under many stretching, bending, rubbing, and washing cycles. Bare cotton is well known to be insulated. In contrast, the Cu coated cotton yarns were electrically conductive. The conductivity of the yarns increased with increasing ELD time (and thus increasing amount of plated Cu particles), reaching as high as ~1 S/cm at 60 minutes, which demonstrated that it could be used as electrical wires for powering light-emitting diodes. After five washing and drying cycles, no obvious degradation of the conductivity was observed in Figure 8.11.

Electrically conductive PET fabric with high endurance and durability by ELD of copper nanoparticle was reported as well [16]. As shown in Figure 8.12, the active interface was created via plasma treatment to generate active chemical groups on the PET substrate, which reacted with atomic transfer radical polymerization initiator for the following polymerization. The Pd ions entrapped

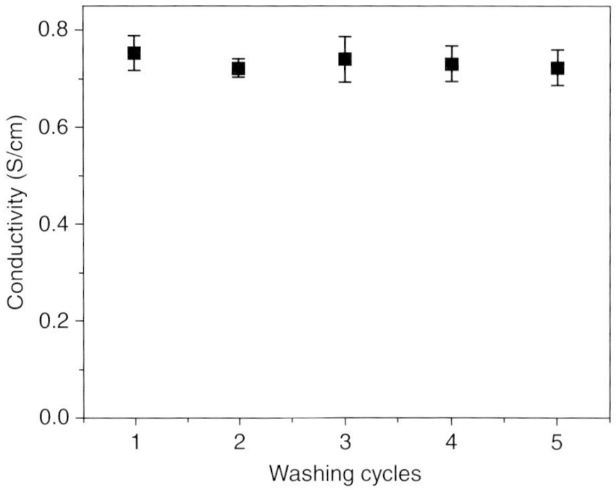


**Figure 8.11** Washability test of the conductive cotton yarns (60 minutes plating).

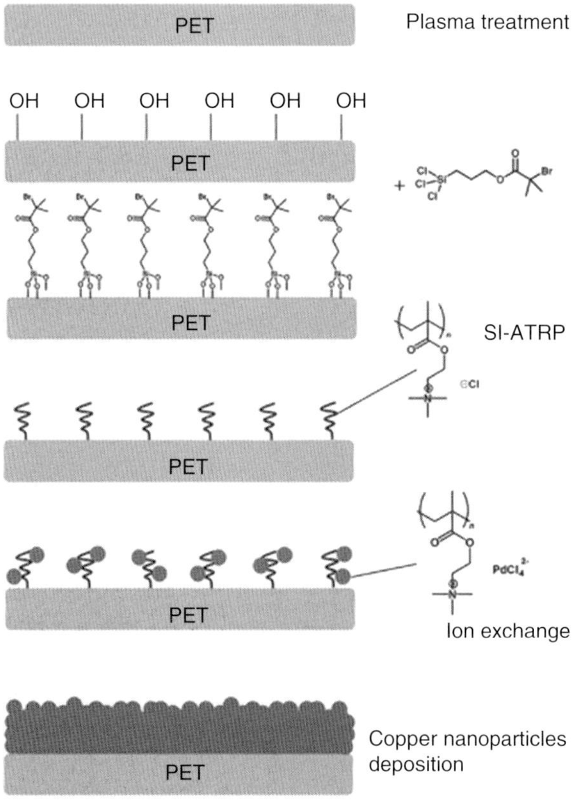


**Figure 8.12** Schematic illustration of the process of preparing copper film on PET substrates.

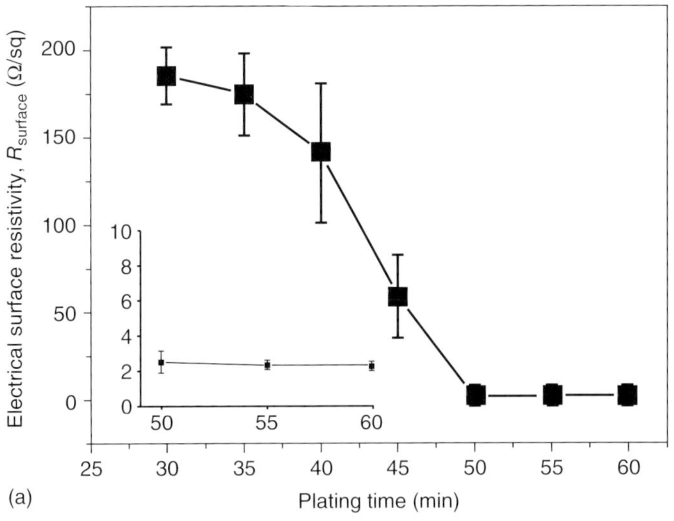


**Figure 8.13** Conductivity tests of the prepared conductive PET fabrics with (a) different plating time and (b) wash fitness test of the conductive PET fabrics (60 minutes plating) and samples of commercial cotton fabrics by thermal evaporation.

by polymer brushes were then reduced in situ and the $Pd^0$ species acted as a catalytic seed layer for subsequent electroless copper nanoparticles growth on the active interface. It could be observed from Figure 8.13a that the electronic surface resistivity of PET fabrics decreased with increasing ELD time, reaching as low as 2 Ω/sq at 60 minutes. It could be attributed to the quantity of copper nanoparticles increasing on the PET fabrics, which made the copper films more continuous. The most critical challenge of conductive fabrics was the washability and long-term durability under multiple cycles. After five washing and drying cycles, there was no obvious increase of surface resistivity on the sample fabricated by ELD (Figure 8.13b). This was basically attributed to the covalent bond between the surface and the grafted polymer brushes as well as to the nanometer-scale mechanical interlocking of the growing electroless plated layer within the brushes, which was provided by the polymer viscoelastic and high-swelling intrinsic properties.

#### 8.2.2.2 Metal Conductive Ink Printing

The powders of copper can be used in conductive inks, but the use of copper ink causes some difficulties, because it is thermodynamically unstable in atmospheric conditions due to oxidation. Since gold ink is very expensive, silver based ink is the best option. Electrical conductive textiles can be obtained by screen printing and inkjet printing deposition.

Monodisperse silver nanoparticles conductive ink was successfully synthesized by in situ synthesis method in an aqueous solution and deposited on textiles by screening printing [17]. The size distribution of the silver nanoparticles was tested and the average diameter was within 10 nm. The corresponding size distribution also proved the continuous coalescence resulting in the large diameter of silver particles to connect into a conductive path. Silver metal nanoparticles were combined together due to the bridging role of HCl. Electrical resistivity of patterns on cotton fabrics and A4 papers with different HCl concentrations when sintering at 60 °C were demonstrated in Figure 8.14. They all had a sudden decrease in resistivity of patterns when adding HCl and it was getting smaller with the increasing HCl concentration. When hydrogen chloride concentration was 50 mM, the minimum resistivity of the printed pattern on cotton fabrics and A4 paper was $2 \times 10^{-5}$ Ω m and $1.9 \times 10^{-6}$ Ω m, respectively. This might be assigned to the fact that the A4 paper's surface was smoother than that of the cotton fabrics, which provided a better printed surface that facilitates conductivity.

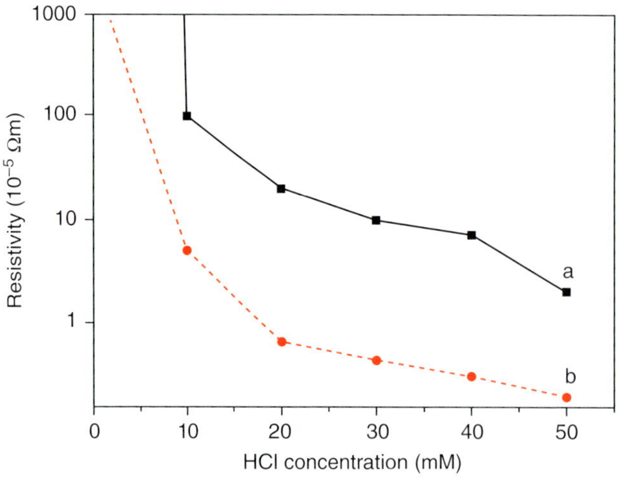


**Figure 8.14** Electrical resistivity of patterns on (a) cotton fabrics and (b) A4 paper with different HCl concentrations and sintering at 60 °C.

The main problem in preparing stable and printable inks for the inkjet printing of textile substrates is to overcome their roughness and high sintering temperature. Reactive silver ink was synthesized by a modified Tollens' process [18]. The choice of this type of ink ensured inkjet printability of textile surfaces with simultaneous sintering at low temperatures, not exceeding 90 °C. Electroconductive layers on flat textile surfaces were printed using the inkjet printing technique. The surface resistance of the inkjet-printed silver layers on all the textile fabrics steadily decreased with the increasing number of silver layers deposited. In the case of eight prints on synthetic textiles made of polyacrylonitrile, the surface resistance of PP, PET, and basalt fibers in the range of 0.155–0.235 Ω/sq was achieved. For some textile substrates made of natural fibers and their blends with synthetic fibers, an interlayer of acrylic copolymer containing resin was applied to achieve proper conductivity. The surface resistance for the same number of prints was in the range of 0.389–0.622 Ω/sq after the washing in distilled water containing surfactant, followed by the dry-cleaning process. All textile samples still retained good electroconductivity, which was important from the point of view of their application in smart wearable textiles.

### 8.2.3 Conductive Fiber Substrates Based on Carbon Material

The application of early carbon materials in textiles mainly includes carbon fiber and carbon black. Current research focuses on low-dimensional carbon materials such as carbon black, carbon nanotubes (CNTs), and graphene. Especially for graphene, a two-dimensional (2D) carbon material with single-atom thickness has attracted increasing attention all over the world in recent years, due to its remarkable electrical and thermal conductivities, great mechanical strength, and a large theoretical specific surface area.

#### 8.2.3.1 Vacuum Filtration

A flexible conductive cotton fabric was fabricated by formulation and deposition of graphene oxide (GO) onto a cotton fabric by vacuum filtration followed by the reduction of GO [19]. As shown in Figure 8.15, the deposited GO amount was controlled by the concentration and volume of the GO dispersion. The GO was reduced by a hot press method at 180 °C for 60 minutes, and no chemical reducing agent was needed in both the deposition and reduction processes. The carbon–oxygen ratio increased from 1.77 to 3.72 after the hot press reduction. The as-prepared flexible conductive cotton fabric showed a sheet resistance as low as 0.9 kΩ/sq. The sheet resistance rises slightly with the number of washing cycles, going from 0.9 to 1.2 kΩ/sq. After 10 washing cycles, it was still lower than 2 kΩ/sq, which indicated that the washing cycles did not significantly affect the conductivity of the reduced graphene oxide (rGO) deposited cotton fabric. Moreover, the as-prepared rGO based conductive cotton fabric showed a reproducible characteristic as a strain sensor.

#### 8.2.3.2 Dip Coating

Multifunctional silk fabrics that were electrically conductive, anti-ultraviolet, and water repellent were successfully prepared by surface modification with GO [20].

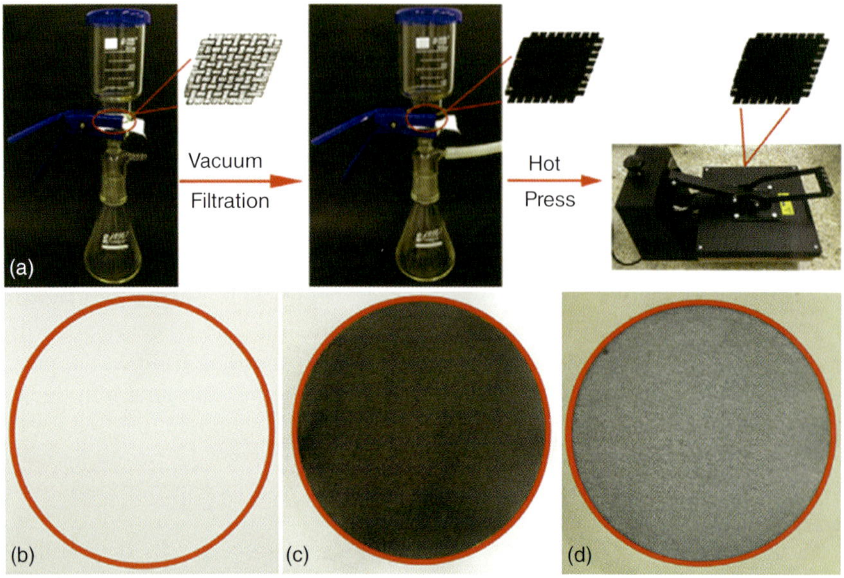


**Figure 8.15** Photographs of (a) the fabrication process of the rGO deposited cotton fabric with vacuum filtration and hot press treatment, and (b) control cotton fabric, (c) GO deposited cotton fabric, (d) rGO deposited cotton fabric hot pressed at 180 °C for 60 minutes.

As shown in Figure 8.16, the yellow-brown GO that is deposited on the surface of the silk fabric was converted into graphitic black rGO by sodium hydrosulfite. The surface resistance decreased with increasing rGO surface modification times, and the low surface resistance value reached 3.24 kΩ/cm after nine times of modification, indicating excellent conductivity of the silk obtained. The electrical surface resistance of rGO coated silk fabrics increased with increasing washing times, indicating that the electrical conductivity slightly decreased with washing times. This can be due to the presence of most of the rGO sheets on the surface of silk fabrics after many times of washing. Because the GO sheets were absorbed on the surface amino group of silk via electrostatic interaction and hydrogen bond, the $\pi-\pi$ interactions and hydrophobic force occurred on the rGO sheets. Only a small amount of unlinked rGO sheets fell off the surface of silk fabrics.

Another method to produce graphene based wearable e-textiles through the chemical reduction of GO is using a simple pad-dry technique (Figure 8.17) [21]. After one padding pass with rGO, the sheet resistance of coated fabric was 361.82 kΩ/sq. There was a slight change in the sheet resistance of rGO coated cotton fabric up to two padding cycles. However, the sheet resistance decreased sharply by ~50–60% for each padding pass up to five padding passes. The sheet resistance of the rGO coated fabric decreased by ~90% from 348.34 kΩ/sq (after two padding passes) to 36.94 kΩ/sq. This could possibly be explained by an absorption and adsorption phenomena, where absorption of rGO dispersion into the fibers is predominant in the first few padding passes. Once it reached the saturation point, it was mainly adsorbed on the fiber surface and formed a continuous conductive film by creating better connections between flakes.

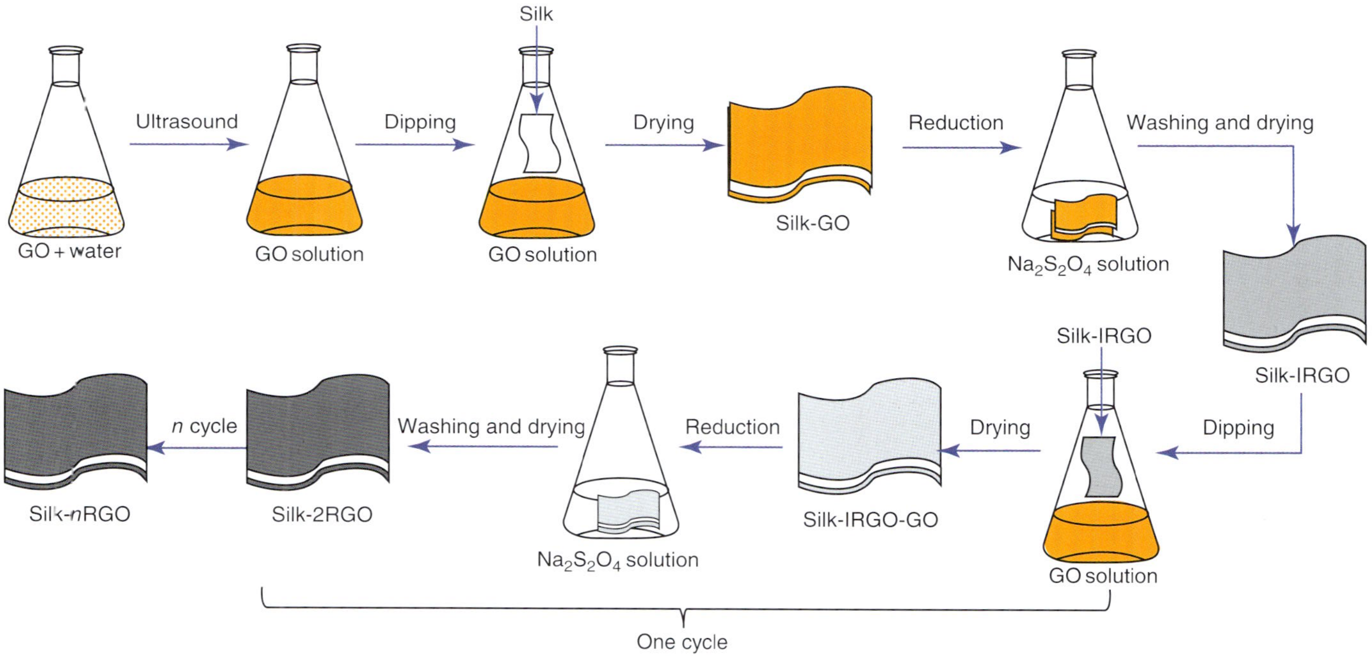


**Figure 8.16** Illustration of the surface modification of silk fabrics by GO.

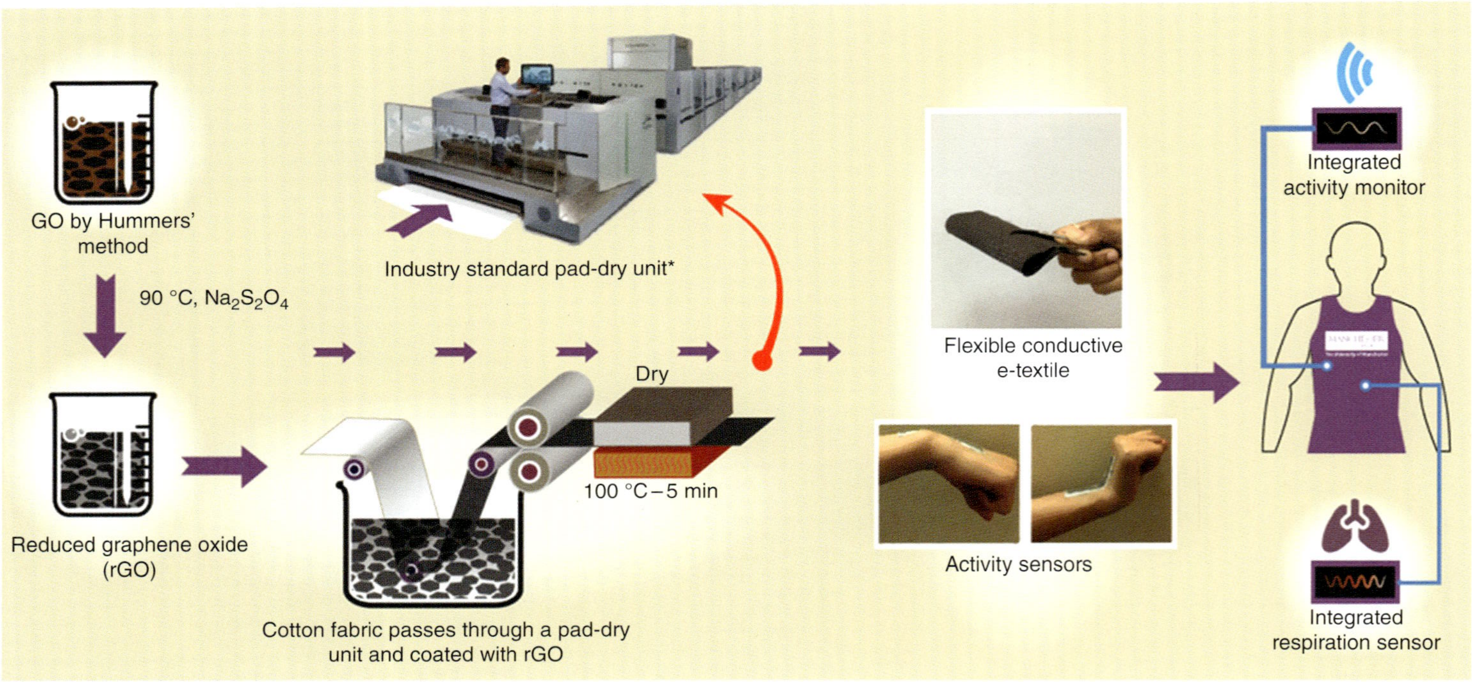


**Figure 8.17** Schematic diagram of the scalable production of graphene-based wearable e-textiles.

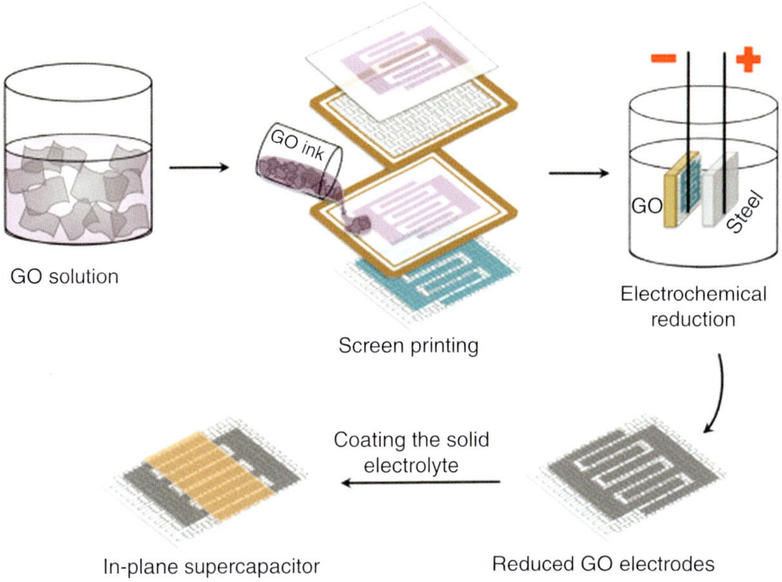


**Figure 8.18** Schematic representation of the printed supercapacitor fabrication process.

The sheet resistance of rGO coated fabric increased significantly from 36.94 to 70.32 kΩ/sq after the first washing cycle, probably due to the removal of unfixed rGO flakes from the fabric surface. After that, the sheet resistance increased moderately to 139.09 kΩ/sq after 10 washing cycles.

The electrical resistance of wool fabric can be reduced significantly by using CNTs in the simple exhaustion method [22]. Using citric acid and sodium hypophosphite helped improve CNTs adsorption on wool fabric surface. Single-wall carbon nanotubes (SWCNTs) were more efficient than multiwall carbon nanotubes (MWCNTs) to produce lower electrical resistance wool fabrics. Also, preparation of wool fabric through oxidation with potassium permanganate helped obtain improved results and better conductivity. The lowest electrical resistance was $1.0 \times 10^4$ Ω/sq, and after washing it was $5.3 \times 10^4$ Ω/sq.

#### 8.2.3.3 Printing

A solid-state flexible supercapacitor device printed on textiles using GO ink and screen-printing technique was reported [23]. As shown in Figure 8.18, after printing, GO was reduced in situ via a rapid electrochemical method avoiding the use of any reducing reagents that may damage the textile substrates. The printed electrodes exhibited excellent mechanical stability due to the strong interaction between the ink and textile substrate. The obtained aerial capacitance was as high as 2.5 mF/cm$^2$ and maintained 95.6% of these values when tested under bending conditions. The device obtained showed outstanding cyclic stability over 10 000 cycles and maintained excellent mechanical flexibility, which was necessary for wearable applications.

Stretchable conductive fabrics were fabricated by screen-printing carbon black ink [24]. Carbon black ink was prepared by stirring a mixture containing

20% (w/w) carbon black and 10% (w/w) surfactant in water at room temperature for one hour and then sonicating for two hours. The fabric was screen-printed three times to ensure full penetration of carbon black through the yarns, and then dried in an oven at 130 °C for two hours to remove any residual water. The polyurethane solution was printed on oven-dried carbon black printed fabric using the same printing method and repeated two times. The fabricated material showed a resistance value of 71 Ω/cm. The polyurethane/carbon black/PET fabric system showed excellent stability, flexibility, and stretchability due to the properties of polyurethane matrix.

Nanoscale carbon particles were stabilized into macro-structured carbon black/polymer composite with the help of an aqueous solution of natural rubber latex, polyvinyl alcohol, and other auxiliary additives under high-speed stirring [25]. The "knife-over-roll" technique was used to apply structured carbon black composite on the plain woven cotton fabric. The lowest resistivity level achieved by this process was less than $60 \pm 5.4\,\Omega/cm^2$.

#### 8.2.3.4 Dyeing

GO as a dyestuff was immobilized on the surfaces of acrylic yarns through a conventional dyeing approach. The GO-dyed yarn or the fabric was immersed in an aqueous sodium hydrosulfite solution at around 90 °C for 30 minutes, which converted the GO into rGO. The rGO created a graphitic-colored and electrically conductive thin layer over each yarn in the fabric. The values of the electrical resistivity ranged from 102 to 109 Ω/cm; it can be verified by the overall graphene content in the conductive thin layer of the graphene-colored fabrics. This was achieved by controlling the total amount of the GO that was immobilized on the surface of the initial untreated fabric [26].

#### 8.2.3.5 Ultrasonic Depositing

A hydrophobic-conductive cotton fabric was produced using methyltrimethoxysilane (MTMS) and MWCNT via sol–gel method [27]. The optimized conditions for preparing the conductive-hydrophobic cotton fabric were as follows: pH = 4, concentration of MTMS 2.00% v/v, sol–gel time of 60 minutes, and concentration of 0.03 g/ml of MWCNT under ultrasound waves. The results obtained suggested that the washing cycle conductivity of the samples was reduced. However, the samples somehow regained their conductive properties after 20 times of washing. The abrasion fastness of the coated samples was good after wear, but their conductivity reduced slightly after wear. The WCA and the resistivity of the padded sample in a 0.03 g/ml solution of MWCNT treated with a 2.00% v/v MTMS were found equal to 146° and $40\,k\Omega/cm^2$, respectively.

Washable e-textiles were fabricated by ultrasonic soldering of CNTs on low-cost and eco-friendly nonwoven fabric (NWF) [28]. When a piece of NWF was subjected to ultrasonication in a CNT dispersion, the NWF could be locally softened so that CNTs were soldered on the fiber surface and led to conductive textiles. The ultrasonic nanosoldering of CNTs onto the polymer surface was attributed to the huge pressure and high temperature produced by the collapsing of bubbles during ultrasonication. The strong mechanical and physical attachment of the soldered CNTs endowed the CNTs–NWF with good washability and durability. The CNTs remained on the fiber surface even after vigorous mechanical washing in water for 40 hours, and the conductance of the textile decreased slightly, and the change arose from the damage of NWF fibers rather than the rinsing away of the CNTs.

#### 8.2.3.6 Brushing Coating

Conductive fabric can be fabricated by a brushing coating method. GO was treated on the surfaces of wool fabrics through the brushing technique [29]. The pH of the solution was adjusted to 4.50 before treating wool by GO. Then, GO was reduced to rGO through a UV reduction method. The surface resistivity of rGO coated wool fabric showed a decrease from 139.5 to 45.0 kΩ/sq.

### 8.2.4 Conductive Fiber Substrates Based on Graphene Composite Materials

As the characteristics of conductive materials used for conductive textiles are different, two or more conductive materials used jointly for textile would improve the textile properties. Composite material conductive textiles can be fabricated via dip coating and in situ polymerization methods using graphene and CNTs, graphene and conducting polymers, graphene and metal composites, and other conductive materials.

#### 8.2.4.1 Dip Coating

In graphene-based supercapacitors, re-stacking of graphene sheets can lead to a decrease of active surface area. To solve this problem, CNTs can be used as spacers to create nanopores among graphene layers, and at the same time provide good conductivity. For example, the hybrid carbon films prepared by the mixture of poly(ethyleneimine) (PEI)-modified rGO sheets and acid-oxidized MWCNTs gave an average specific capacitance of 120 F/g at a high scan rate of 1 V/s [30].

The knittable, flexible, and lightweight fibriform electrodes of rGO-cotton thread (CT) and carbon nanoparticles (CNPs)/rGO-CT were fabricated through dip coating and low-temperature vapor reduction method [31]. The GO nanosheets coated onto cotton fibers strongly due to the attraction of opposite electrostatic charges between GO and bovine serum albumin molecules, which was beneficial to establish stable rGO-CT electrode materials. The rGO nanosheets wrapped cotton fiber resulted in the compact CT, which had impressive conductivity. Meanwhile, the CNPs coated onto rGO-CT uniformly and steadily via implanting into the wrinkles and ripples of rGO nanosheets. The rGO nanosheets with implanted CNPs not only formed the 3D hierarchical nanostructures that were beneficial for ion transmission, but also contacted with each other due to the abundant cotton fiber, which could further enhance the conductivity of electrode materials. Moreover, the two kinds of fibriform supercapacitors (FSCs) based on rGO-CT and CNPs/rGO-CT electrodes had excellent and stable electrochemical performance, which could tolerate severe and long-term bending. The CNPs/rGO-CT based FSC exhibited the merit that could be knitted into textile in perfect union without sacrificing its electrochemical performance, which was attributed to the good mechanical strength and softness of CT. More importantly, the CV of rGO-CT based FSC could be improved about threefold by coating with CNPs.

#### 8.2.4.2 In Situ Polymerization

GO was reduced in the presence of poly(sodium 4-styrenesulfonate) (PSS) to modify the surface of graphene in order to avoid the aggregation of the graphene nanosheets (GNS) [32]. The thickness of the GNS was around 2 nm, possibly due to double or multiple layers of single graphene sheets. After mixing surface modified GNS and Py monomers in ethylene solution, ferric chloride ($FeCl_3$) was added to the mixture to start the polymerization, and GNS/PPy composites were formed. A mixture of GNS/PPy composites could be represented by checking the presence of both peaks of pure GNS and PPy in the Raman spectrum. The GNS/PPy composites had almost twice the specific capacitance and much better cycling performance as the supercapacitor electrode compared to pure PPy film. These results provided evidence that the incorporation of graphene into PPy could facilitate electrochemical utilization of PPy and provided mechanical support to PPy to enhance the structural stability of the composites during the charge and discharge cycles.

Composites of PANI and graphene could be prepared by in situ polymerization of aniline with graphene suspension in acidic solution. Zhang et al. used GO to mix with PANI to form the composite and drop cast onto glassy carbon working electrode for electrochemical characterization [33]. The GO was prepared from graphite by a modified Hummers method and had a layer-like structure with the size of tens of micrometers. Oxidative polymerization method was used to synthesize PANI nanofibers (PANI-F) in 1 M aqueous HCl acidic media while ammonium peroxydisulfate ($(NH_4)_2S_2O_8$) was used as the oxidizing agent. The composite morphology of the GO/PANI composite was affected by the GO concentration, which affected the electrochemical behavior of the supercapacitor electrode. Pure PANI-F had one pair of redox peaks in the CV curves, but the GO/PANI composites exhibited both characteristics of pure GO and pure PANI, which showed two pairs of the redox peaks. Pure PANI-F electrode had a very high specific capacitance of 420 F/g; however, the cycling stability was rather poor, and the specific capacitance reduced by almost 40% after only five cycles. The specific capacitance of the composite decreased as the GO concentration increased. On the other hand, with higher GO concentration in the composites, the capacitance retention over long cycles was greatly improved, since GO nanoparticles could compensate the volumetric change of the PANI during the charge and discharge cycles.

## 8.3 Waterproof Modification for Conductive Fiber Substrates

The waterproof finishing of conductive textiles is the same as the common hydrophobic finishing method of textiles, and it contributes to the washability and durability of conductive textiles during use. Textiles have the advantages of breathability and softness, but the waterproof finishing of textiles would take away these advantages. Hydrophobic textiles not only possess the characteristics of waterproofing, but also retain the original breathability and softness of textiles. Therefore, the waterproof characteristics of conductive textiles are achieved through hydrophobic finishing. The methods for hydrophobic finishing surface are to reduce the surface energy of the textiles and increase the roughness. Low surface energy materials such as silicones, fluorosilanes, and long-chain alkanes are commonly used, which are deposited onto textiles by coating, deposition, or grafting to reduce the surface energy. The ways to increase the roughness of textile surface are etching, nanoparticle deposition, and so on. The methods of textile hydrophobic finishing include chemical bath deposition, spray coating, surface etching, sol–gel, dip coating, etc. Chemical bath deposition means that spherical or rod-shaped inorganic nanoparticles grow on the surface of the fiber under hydrothermal conditions, and are then chemically modified by low surface energy substances to form a micro/nanoscale roughness structure with low surface energy, thereby achieving superhydrophobicity [34]. The spraying method uses a spray gun to spray nanoparticles and low surface energy materials onto the textile surface, and then dried or solidified to obtain a superhydrophobic surface [35, 36]. The surface etching uses plasma or alkali solution to treat the surface of the fiber, causing the surface of the fiber to be corroded to form a micro-nano-like convex structure, and then preparing a superhydrophobic fabric by grafting or coating a low surface energy substance [37, 38].

### 8.3.1 Dip-Coating Method

The dip-coating method is to immerse the fabric into the hydrophobic finishing solution. The hydrophobic finishing agents evenly enter the gap between fibers. The low surface energy substances and nanoparticles exhibit an excellent uniform distribution and cover the fiber surfaces after the solvents are removed, thereby imparting low surface energy and roughness to the fabric [39, 40]. For example, Suryaprabha et al. [14] proposed a conductive and hydrophobic metal textile by two kinds of metal dip-coating reduction methods. Copper acetate was first dipped onto cotton fabric to increase the micro-nano structure of the fabric to obtain hydrophobicity. The silver salt was further deposited on the fabric, resulting in greatly improved conductivity and slightly increased hydrophobicity. The final resistance of the fabric was less than 1 Ω, and the contact angle was above 160°. However, the utilization of hazardous agents in order to reduce the highly toxic hydrazine hydrate was a limitation of this type of metal fabric research.

The superhydrophobic conductive fabric was obtained by impregnating NWF with MWCNT moisture dispersions [41]. The hydrophobicity level obtained was completely determined by the polymethylsilsesquioxane (PMSQ) structure formed on the surface of the sample, while the PMSQ structure was dependent on the initial water content in the fiber. The minimum resistance of the cotton fabric sample with stacked MWCNTs was 0.71 k$\Omega$/cm$^2$.

### 8.3.2 Sol–Gel Method

The sol–gel method is to utilize the hydrolysis of tetraethyl orthosilicate in water or ethanol to obtain spherical or rod-shaped nanoparticles. The fabric is immersed in silica nanoparticle solution to obtain a rough surface. The construction of roughness provides a good foundation of hydrophobicity [42, 43]. Nasirizadeh et al. [27] prepared a hydrophobic conductive cotton fabric using MTMS and MWCNT by sol–gel method. The hydrophobicity and conductivity were evaluated by measuring WCA and surface resistance. The optimum conditions for preparing conductive hydrophobic cotton fabric were pH = 4, MTMS concentration 2.00% v/v, sol time 60 minutes, and MWCNT concentration under ultrasonic 0.03 g/ml. Under the optimum conditions, the WCA and surface resistance of the cotton fabric obtained were 146° and 40 k$\Omega$/cm$^2$, respectively.

### 8.3.3 Chemical Vapor Deposition

Zhu et al. [44] deposited 1H,1H,2H,2H-perfluorooctyltriethoxsysilane (POTS) fluorosilane on PEDOT film by chemical vapor deposition (CVD) to achieve a resistance value of $3.2 \times 10^{-4}$ $\Omega$ m and self-healing superhydrophobic property, which is a more prominent work done in this type of research. M. Skrifvars and coworkers [45] produced relatively high conductive PEDOT coated VS yarns by oxidative chemical vapor deposition (OCVD) techniques. Longer length PEDOT coated conductive yarns could be prepared using OCVD techniques. Surface resistance measurements indicated that the silicon coated conductive yarn had sufficient conductivity because of the thin layer of silicone coating on the surface. The hydrophobicity test results showed that silicone modification could improve the water repellency of the coated VS yarn. The mechanical properties test showed that the strength of the coated PEDOT VS fiber remains basically the same, but the surface modification of the silicone resin could improve its flexibility. Silicone-modified conductive fibers could be used as pressure sensors and tensile sensors in the health-care and sports fields.

## 8.4 Washing Evaluations of Conductive Textiles

Textiles are made of interwoven fibers or coils, and are the indispensable materials for daily life with a wide application in clothing, decoration, and industrial products. Washing is an inevitable procedure of textiles in the process of production and application. For smart wearable textiles, the most important indicator is the conductivity. The conductive textiles in smart wearable are utilized in daily life for special monitors. Like normal textiles, conductive textiles also require washing to remove surface contaminants. In this case, the washing evaluation of textile conductivity is vital and necessary. There are two methods to evaluate the washing properties of conductive textiles.

For the conductive textiles made by coating, depositing, or embedding electrically conductive components on the surface of textiles, the washing method is to wash textiles with some detergents [14, 46] or just wash by hand [16]. C. Wang et al. [47] fabricated a functional textile through silver/reduced GO composites coated on PET fabric after bovine serum albumin functionalization.

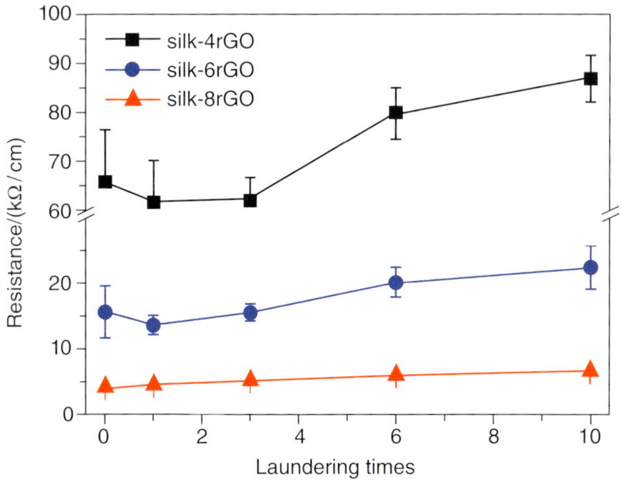


**Figure 8.19** The effects of washing times on electrical surface resistance of silk-*n*rGO fabrics.

Ag/rGO-PET fabric was washed with commercial synthetic detergent (0.37 wt%) at room temperature using a magnetic stirring bar at 900 rpm for 10 minutes to investigate the stability against laundering, and there was no obvious decrease in electrical conductivity after 10 cycles of washing, whereas the conductivity decreased after the fabric was washed in hot water. H. M. Qin et al. [5] fabricated conductive chitosan fabrics by electroless plating of silver nanoparticles onto the surfaces of the fibers. The fabrics were immersed in 4 g/l commercial nonionic detergent at the oscillation rate of 350 rpm in 40 °C for 30 minutes. The conductivity is reduced after washing six times.

Another evaluation method used is according to the standards or a simplified version of the standard text methods. Relative conductivity ($\sigma/\sigma_0$) after standard washing (AATCC test method 61-2007 (1A) without steel balls) and boiling water treatment are measured for the durability of electroconductive PET fabric fabricated by via reduction and cross-linking of GO with $SnCl_2$ [46]. M. Mičušík et al. [9] modified textile surface with a pyrrole-functionalized silane followed by the in situ synthesis of a conducting PPy overlayer. Washing cycles were performed at 40 °C and pH 8.5 according to STN-EN ISO 105 C-06. Electrical conductivity measurements were performed before and after five washing cycles. J. L. Cao et al. [20] treated silk fabric with repeated coating of GO and chemical reduction method to make it electrically conductive and water repellent. The rGO coated silk fabrics were washed in an automatic washer according to the AATCC test method 135-2000. The electrical surface resistance of rGO coated silk fabrics increased with the increasing washing times, as shown in Figure 8.19.

Conductive textiles are important substrates in smart wearables, and the washing durability of the conductive layer is of great significance since the textiles are used in different environment. T. Suryaprabha et al. [14] developed a simple and inexpensive method for the fabrication of superhydrophobic textiles with electrical conductivity and antibacterial activity through bimetallic deposition of copper and silver over cotton fabric. The washing durability is evaluated by repeated washing of as-coated samples with 0.3% of liquid detergent at constant stirring speed of 500 rpm for 30 minutes at room temperature. The electrical conductivity is measured after each washing cycle. The sheet resistance was almost the same after repeated washings.

Standards are also utilized for the durability of conductive textiles. ISO 6330:2000 is used for the washing behavior of conductive textiles [13, 48]. I. Kazani et al. [49] fabricated prototype conductive parts that were screen-printed on several textile substrates using two different silver-based conductive inks with a thermoplastic polyurethane coating. A study of the laundering behavior of the printed textile antennas was performed in a reference washing machine according to the international standard ISO 6330:2000 – Textiles – Domestic washing and drying procedures for textile testing. The washing procedure 8A was chosen, as described in the datasheet of the manufacturer of the thermoplastic polyurethane layer. It washed at a temperature of $30 \pm 3\,°C$, for a period of three minutes, followed by three rinsing cycles, two of three minutes and one of two minutes, and concludes with a spinning time of two minutes. Reference detergents are added as requested by the standard. B. T. Wu et al. [50] prepared a conductive textile via covalently grafting PANI onto cotton by a multistep treatment process. The washing test of conductive textile was conducted with the dry wash test AATCC 132-2004 and AATCC 86-2005. During each washing cycle, the fabrics were washed in capped bottles containing 200 ml of tetrachloroethylene (TTE) detergent solution at $30 \pm 2\,°C$ for 30 minutes with intense stirring. All the samples were conditioned at a relative humidity of $65 \pm 2\%$ and a temperature of $20 \pm 1\,°C$ for one hour before being subjected to the conductivity tests. The resultant conductive fabrics were able to withstand 40 simulated dry-wash cycles, with almost no decay in the electrical conductivity. H. X. Wang et al. [13] prepared a robust, electrically conductive, superamphiphobic fabric by vapor-phase polymerization of EDOT on fabric in the presence of FD-POSS and a FAS. The washing durability of the treated fabric was examined by washing the coated fabrics in a washing machine according to the method specified in the Australian Standard (AS 2001.1.4). Such a standard wash procedure was equivalent to five cycles of home machine launderings. The conductivity only had a small reduction after 100 washing cycles according to the standard.

## 8.5 Conclusions

Washability and durability are important for conductive textiles. Various approaches have been adopted by researchers to prepare washable and durable conductive textiles, and it can be divided into four types – polymer conductive textiles, metal conductive textiles, carbon material conductive textiles, and composite material conductive textiles – due to the difference in conductive components. Conducting polymers, such as PANI, PPy, and PEDOT, are applied to textile materials by solution coating, graft modification, in situ chemical polymerization, in situ electrochemical polymerization, and in situ vapor phase polymerization. Metals such as copper, silver, and iron can be used to fabricate conductive textiles via electroless plating, metal conductive ink printing, metal wire spinning, and weaving. Carbon material conductive textiles can be prepared by brushing coating, vacuum filtration, dip coating, inkjet printing, and dyeing. For composite material conductive textiles, due to various composite types, such as graphene and CNTs, graphene and conducting polymers, graphene and metal, there are many fabricated methods. Washing evaluation was important, especially for washing durability. Conductive textiles with a high washing fastness can integrate advanced functions such as smart clothes, energy storage and conversion, sensors, and color change, or can deliver responses to external stimuli, among others.

## References


1 Sultan, N. (2015). Reflective thoughts on the potential and challenges of wearable technology for healthcare provision and medical education. *Int. J. Inf. Manage.* 35 (5): 521–526.

2 Murali, S., Rincon, F., and Atienza, D. (2016). A wearable device for physical and emotional health monitoring. *Comput. Cardiol. Conf. IEEE*: 121–124.

3 Kim, R.H. (2015). Cure performance and effectiveness of portable smart healthcare wear system using electro-conductive textiles. *Procedia Manuf.* 3 (1): 542–549.

4 Honarvar, M.G. and Latifi, M. (2016). Overview of wearable electronics and smart textiles. *J. Text. Inst. Proc. Abstr.* 108 (4): 631–652.

5 Qin, H., Li, J., He, B. et al. (2018). Novel wearable electrodes based on conductive chitosan fabrics and their application in smart garments. *Materials* 11 (3): 370.

6 Salavagione, H.J., Gómez-Fatou, M.A., Shuttleworth, P.S., and Ellis, G. (2018). New perspectives on graphene/polymer fibers and fabrics for smart textiles: the relevance of the polymer/graphene interphase. *Front. Mater.* 5: 18.

7 Choi, C.M., Kwon, S.N., and Na, S.I. (2017). Conductive PEDOT:PSS-coated poly-paraphenylene terephthalamide thread for highly durable electronic textiles. *J. Ind. Eng. Chem.* 50: 155–161.

8 Ankhili, A., Tao, X.Y., Cochrane, C. et al. (2018). Washable and reliable textile electrodes embedded into underwear fabric for electrocardiography (ECG) monitoring. *Materials* 11: 256.

9 Mičušík, M., Nedelčev, T., Omastová, M. et al. (2007). Conductive polymer-coated textiles: the role of fabric treatment by pyrrole-functionalized triethoxysilane. *Synth. Met.* 157: 914–923.

10 Lin, T., Wang, L., Wang, X., and Kaynak, A. (2005). Polymerizing pyrrole on polyester textiles and controlling the conductivity through coating thickness. *Thin Solid Films* 479: 77–82.

11 Maiti, S., Das, D., and Sen, K. (2012). Studies on electro-conductive yarns prepared by in situ chemical and electrochemical polymerization of pyrrole. *J. Appl. Polym. Sci.* 123: 455–462.

12 Kaynak, A., Najar, S.S., and Foitzik, R.C. (2008). Conducting nylon, cotton and wool yarns by continuous vapor polymerization of pyrrole. *Synth. Met.* 158: 1–5.

13 Wang, H., Zhou, H., Gestos, A. et al. (2013). Robust, electro-conductive, self-healing superamphiphobic fabric prepared by one-step vapour-phase polymerisation of poly(3,4-ethylenedioxythiophene) in the presence of fluorinated decyl polyhedral oligomeric silsesquioxane and fluorinated alkyl silane. *Soft Matter* 9: 277–282.
14 Suryaprabha, T. and Sethuraman, M.G. (2017). Design of electrically conductive superhydrophobic antibacterial cotton fabric through hierarchical architecture using bimetallic deposition. *J. Alloys Compd.* 724: 240–248.
15 Liu, X., Chang, H., Li, Y. et al. (2010). Polyelectrolyte-bridged metal/cotton hierarchical structures for highly durable conductive yarns. *ACS Appl. Mater. Interfaces* 2 (2): 529–535.
16 Liu, X., Guo, R., Shi, Y. et al. (2016). Durable, washable, and flexible conductive PET fabrics designed by fiber interfacial molecular engineering. *Macromol. Mater. Eng.* 301 (11): 1383–1389.
17 Wang, Z., Wang, W., Jiang, Z., and Yu, D. (2016). Low temperature sintering nano-silver conductive ink printed on cotton fabric as printed electronics. *Prog. Org. Coat.* 101: 604–611.
18 Stempien, Z., Rybicki, E., Rybicki, T., and Lesnikowski, J. (2016). Inkjet-printing deposition of silver electro-conductive layers on textile substrates at low sintering temperature by using an aqueous silver ions-containing ink for textronic applications. *Sens. Actuators, B* 224: 714–725.
19 Ren, J., Wang, C., Zhang, X. et al. (2017). Environmentally-friendly conductive cotton fabric as flexible strain sensor based on hot press reduced graphene oxide. *Carbon* 111: 622–630.
20 Cao, J. and Wang, C. (2017). Multifunctional surface modification of silk fabric via graphene oxide repeatedly coating and chemical reduction method. *Appl. Surf. Sci.* 405: 380–388.
21 Karim, N., Afroj, S., Tan, S. et al. (2017). Scalable production of graphene-based wearable e-textiles. *ACS Nano* 11: 12266–12275.
22 Montazer, M., Asghari, M.S.G., and Pakdel, E. (2011). Electrical conductivity of single walled and multiwalled carbon nanotube containing wool fibers. *J. Appl. Polym. Sci.* 121 (6): 3353–3358.
23 Abdelkader, A.M., Karim, N., Vallés, C. et al. (2017). Ultraflexible and robust graphene supercapacitors printed on textiles for wearable electronics applications. *2D Materials* 4: 035016.
24 Pahalagedara, L.R., Siriwardane, I.W., Tissera, N.D. et al. (2017). Carbon black functionalized stretchable conductive fabrics for wearable heating applications. *RSC Adv.* 7 (31): 19174–19180.
25 Sarkar, K., Das, D., Chaki, T.K., and Chattopadhyay, S. (2017). Macro-structured carbon clusters for developing waterproof, breathable conductive cotton fabric. *Carbon* 116: 1–4.
26 Fugetsu, B., Sano, E., Yu, H. et al. (2010). Graphene oxide as dyestuffs for the creation of electrically conductive fabrics. *Carbon* 48 (12): 3340–3345.
27 Nasirizadeh, N., Dehghani, M., and Yazdanshenas, M.E. (2015). Preparation of hydrophobic and conductive cotton fabrics using multi-wall carbon nanotubes by the sol-gel method. *J. Sol-Gel Sci. Technol.* 73 (1): 14–21.

28 Du, D., Tang, Z., and Ouyang, J. (2017). Highly washable e-textile prepared by ultrasonic nanosoldering of carbon nanotubes onto polymer fibers. *J. Mater. Chem. C* 6 (4): 883–889.

29 Javed, K., Galib, C.M.A., Yang, F. et al. (2014). A new approach to fabricate graphene electro-conductive networks on natural fibers by ultraviolet curing method. *Synth. Met.* 193: 41–47.

30 Yu, D. and Dai, L. (2009). Self-assembled graphene/carbon nanotube hybrid films for supercapacitors. *J. Phys. Chem. Lett.* 1: 467–470.

31 Ye, X., Zhou, Q., Jia, C. et al. (2016). A knittable fibriform supercapacitor based on natural cotton thread coated with graphene and carbon nanoparticles. *Electrochim. Acta* 206: 155–164.

32 Bose, S., Kim, N.H., Kuila, T. et al. (2011). Electrochemical performance of a graphene-polypyrrole nanocomposite as a supercapacitor electrode. *Nanotechnology* 22 (29): 295202.

33 Zhang, K., Zhang, L.L., Zhao, X.S., and Wu, J. (2010). Graphene/polyaniline nanofiber composites as supercapacitor electrodes. *Chem. Mater.* 22 (4): 1392–1401.

34 Wang, H., Zhou, H., Yang, W. et al. (2015). Selective, spontaneous one-way oil-transport fabrics and their novel use for gauging liquid surface tension. *ACS Appl. Mater. Interfaces* 7 (41): 22874–22880.

35 Wang, S., Sha, J., Wang, W. et al. (2018). Superhydrophobic surfaces generated by one-pot spray-coating of chitosan-based nanoparticles. *Carbohydr. Polym.* 195: 39–44.

36 Liu, M., Hou, Y., Li, J. et al. (2018). Robust and self-repairing superamphiphobic coating from all-water-based spray. *Colloids Surf., A* 553: 645–651.

37 Yang, J., Pu, Y., Miao, D., and Ning, X. (2018). Fabrication of durably Superhydrophobic cotton fabrics by atmospheric pressure plasma treatment with a siloxane precursor. *Polymers* 10 (4): 460.

38 Xue, C., Li, Y., Zhang, P. et al. (2014). Washable and wear-resistant superhydrophobic surfaces with self-cleaning property by chemical etching of fibers and hydrophobization. *ACS Appl. Mater. Interfaces* 6 (13): 10153–10161.

39 Wang, H., Zhou, H., Gestos, A. et al. (2013). Robust, superamphiphobic fabric with multiple self-healing ability against both physical and chemical damages. *ACS Appl. Mater. Interfaces* 5 (20): 10221–10226.

40 Xu, B., Ding, Y., Qu, S., and Cai, Z. (2015). Superamphiphobic cotton fabrics with enhanced stability. *Appl. Surf. Sci.* 356: 951–957.

41 Makowski, T., Kowalczyk, D., Fortuniak, W. et al. (2014). Superhydrophobic properties of cotton woven fabrics with conducting 3D networks of multiwall carbon nanotubes, MWCNTs. *Cellulose* 21 (6): 4659–4670.

42 Gurav, A.B., Xu, Q., Latthe, S.S. et al. (2015). Superhydrophobic coatings prepared from methyl-modified silica particles using simple dip-coating method. *Ceram. Int.* 41 (2): 3017–3023.

43 Yu, M., Li, P., Feng, Y. et al. (2018). Positive effect of polymeric silane-based water repellent agents on the durability of superhydrophobic fabrics. *Appl. Surf. Sci.* 450: 492–501.

44 Zhu, D., Lu, X., and Lu, Q. (2014). Electrically conductive PEDOT coating with self-healing superhydrophobicity. *Langmuir* 30 (16): 4671–4677.

45 Bashir, T., Skrifvars, M., and Persson, N.K. (2011). Surface modification of conductive PEDOT coated textile yarns with silicone resin. *Mater. Process. Rep.* 26 (3): 135–139.
46 Babaahmadi, V., Montazer, M., and Gao, W. (2018). Surface modification of PET fabric through in-situ, reduction and cross-linking of graphene oxide: towards developing durable conductive fabric coatings. *Colloids Surf., A* 545: 16–25.
47 Wang, C., Guo, R.H., Lin, S.J. et al. (2018). A highly electro-conductive and flexible fabric functionalized with bovine serum albumin for a wearable electronic device. *J. Mater. Sci. - Mater. Electron.*: 1–8.
48 Kazani, I., Hertleer, C., Mey, G.D., and Langenhove, L.V. (2012). Electrical conductive textiles obtained by screen printing. *Fibres Text. East. Eur* 20 (1): 57–63.
49 Kazani, I., Declercq, F., Scarpello, M.L. et al. (2014). Performance study of screen-printed textile antennas after repeated washing. *Autex Res. J.* 14 (2): 47–54.
50 Wu, B.T., Zhang, B.W., Wu, J.X. et al. (2015). Electrical switchability and dry-wash durability of conductive textiles. *Sci. Rep.* 5: 11255.

# 9

# Flexible Microfluidics for Wearable Electronics

*Dachao Li*[1], *Haixia Yu*[2], *Zhihua Pu*[1], *Xiaochen Lai*[1], *Chengtao Sun*[1], *Hao Wu*[1], *and Xingguo Zhang*[1]

[1] *Tianjin University, School of Precision Instruments and Optoelectronics Engineering, Department of Precision Instrument Engineering, State Key Laboratory of Precision Measuring Technology and Instruments, No. 92 Weijin Road, Tianjin 300072, China*
[2] *Tianjin University, School of Precision Instruments and Optoelectronics Engineering, Department of Biomedical Engineering, Tianjin Key Laboratory of Biomedical Detecting Techniques and Instruments, 92 Weijin Road, Tianjin 300072, China*

## 9.1 Introduction

With the development of micro-electro-mechanical-systems (MEMS) technology, microfluidics is developed and widely used in various fields such as environment detection, cell biology, protein analysis, and gene engineering. Conventional microfluidic systems are generally fabricated using rigid materials such as silicon and glass. Rigid microfluidics requires complicated manufacturing procedures and expensive processing equipment; unfortunately, high integration is still very difficult. In recent years, flexible microfluidics is developed rapidly based on the fast development of materials and fabrication technologies. Flexible materials such as polymer, silicone elastomer, and fabric are used to fabricate the microfluidic system instead of conventional rigid materials. New fabrication technologies such as layer transfer and lamination, soft lithography, inkjet printing, and 3D printing are developed for flexible microfluidics. These flexible materials and fabrication technologies simplify the manufacturing procedures, cut the cost, and make flexible microfluidics multifunctional and highly integrated. In addition to flexibility, flexible microfluidics has the advantages of lightweight, conformability, portability, low cost, disposability, and easy integration. Therefore, flexible microfluidics is widely used, especially in wearable applications, including biosensing and motion sensing. In the next sections, more information about flexible materials (Section 9.2), fabrication technologies (Section 9.3), and applications of flexible microfluidics (Section 9.4) will be discussed, respectively.

## 9.2 Materials

Many materials have already been developed and used in flexible microfluidics. These materials can be divided into three classes: polymer, silicone elastomer,

*Flexible and Wearable Electronics for Smart Clothing*, First Edition.
Edited by Gang Wang, Chengyi Hou, and Hongzhi Wang.

and fabric. Each of these materials possesses inherent qualities compatible for different applications. For example, polymeric substrates are flexible, robust, and generally possess strong resistance to chemicals. Silicone elastomers are stretchable and conformable. The fabric is soft, absorbent, and breathable [1].

As the most popular polymer in microfluidics, polydimethylsiloxane (PDMS) has been widely used in flexible microfluidics due to its special characters, such as elasticity, transparency, biocompatibility, gas permeability, and ability to form conformal contact with surfaces. The Young's modulus of PDMS is approximately 2 MPa, although it can be modified by altering the ratio of PDMS base and curing agent or by changing the curing temperature. Consequently, a thin layer of PDMS can be soft, flexible, and deformable, similar to that of the skin. Furthermore, the structures fabricated with PDMS are hydrophobic, which could be altered by surface modifications. As such, PDMS substrates can be bonded easily and quickly to different kinds of materials without the need for adhesives.

Another thermoset polymer, which is usually used as the substrate of a flexible microfluidic chip, is polyimide (PI)[2]. Thermosetting PI is known for its thermal stability, good chemical resistance, excellent mechanical properties, and characteristic orange/yellow color. Polyimides compounded with graphite or glass fiber reinforcements have flexural strengths of up to 50 000 psi (340 MPa) and flexural moduli of 3 000 000 psi (21 000 MPa). Thermoset polymer matrix polyimides exhibit very low creep and high tensile strength. These properties are maintained during continuous use to temperatures of up to 232 °C (450 °F) and for short excursions, as high as 400 °C (752 °F) [3].

Other polymeric substrates include polyethylene terephthalate (PET), polyethylene (PE), polystyrene (PS), polyurethane (PU), polypropylene (PP), polycarbonate (PC), and cyclic olefin (co)polymers (COC/COP). These materials are thermoplastics, and they have been widely used in flexible printed electronics [4]. With fabrication techniques to achieve metallization of these polymeric substrates, these structures can serve as soft wires or even serve as sensors in various microfluidic chips. Another special kind of polymers is photosensitive resin, which can be cured under UV light. These photosensitive polymers are the fundamental materials of 3D printing, which is suitable for the fabrication of complex structures and multilayers or compound foils with more functionality.

Silicone rubber represents another class of materials for flexible microfluidics. It has already been well utilized in wearable devices due to its durability and excellent chemical resistance. Furthermore, its viscoelasticity matches that of the skin and can get the perfect fit with people's skin. For example, Ecoflex silicone rubber possesses an elastic modulus of approximately 125 kPa, which is similar to that of the skin. It can be stretched over 100% without plastic or permanent deformation. Similarly, the soft lithography technique can be used for silicone rubber substrates. To ensure mold release, precast silicone rubber should be peeled off the mask before it is fully cured. Electrical components can also be embedded in the silicone substrate.

Also, paper-based microfluidic devices fall within our definition of flexible microfluidics. In fact, not only is paper a potential candidate for flexible microfluidic chips but also it has already been successfully used as a substrate material for lateral flow assays, which are sold in billions of units in point-of-care diagnostics [5]. Patterned paper can be used to manipulate micro-flow using capillary forces [6]. Paper is used when mainly capillary matrices are required. Suitable grades are often made of nitrocellulose such as chromatographic paper.

As an emerging extension, the fabric or textile-based microfluidics utilizes a wicking force similar to that in the paper that is produced by hydrophilic yarns (e.g. cotton yarns) to direct biological reagents along with the fibrous structure. This structure affords the aforementioned operational capacities of interfacial microfluidics while providing a low-cost and scalable solution based on well-established traditional textile manufacturing techniques such as automatic weaving, knotting, and stitching. In addition, textile-based microfluidics is inherently more durable and could be easily woven onto clothes and accessories. However, in order to control bio-fluidic transport, specialized channels have to be fabricated. Chemical treatment and photolithographic means are usually used for surface modification to alter textile hydrophobicity. Taking advantage of the difference in wetting properties, liquid droplets can be self-propelled in an open-channel configuration under Laplace pressure gradients. Other than fluidic transport, electrical coating of textile or e-textile is necessary for providing electrochemical reactions and data transmission. Here, secondary processes such as dip coating and stencil printing have been employed to achieve electrically conductive fabric.

## 9.3 Fabrication Technologies

The fabrication technologies used for microfluidic systems have evolved to improve functionality. Early microfluidic systems were manufactured from glass and silicon. However, glass/silicon fabrication requires complex etching processes and is fundamentally limited to very planar, passive devices. To form an integrated microfluidic system, a micro/nanostructured layer is bonded to a microchannel layer. Various micro/nanofabrication techniques can be used for manufacturing integrated microfluidic systems on flexible substrates, such as a polymer, silicone elastomer, and fabric. In the first part of this section, fabrication methods of traditional microfluidic devices made by polymer and silicone elastomer are introduced, including layer transfer and lamination, soft lithography, inkjet printing, and 3D printing; in the second part, the processing methods of open-surface microfluidics made by fabric are introduced.

### 9.3.1 Layer Transfer and Lamination

The layer transfer and lamination technique can be used to fabricate flexible microfluidic channels in various shapes and with a wide range of dimensions. High bond strengths can be achieved by cure cycle adaptation, and the microchannels can be combined with metallization layers to fabricate electrodes inside and outside the channels.

Metz et al. present a method to fabricate flexible microfluidic devices using layer transfer and lamination technique [2]. The device is micromachined on a carrier substrate. The resulting structures can be removed from the carrier substrate by using a release technique based on the anodic metal dissolution of aluminum. On the aluminum surface, a 5–20 μm thick layer of photosensitive polyimide is applied, photostructured and cured (Figure 9.1a). Subsequently, a layer of titanium-platinum is sputtered on the substrate and structured by dry etching using a positive photoresist as etch mask (Figure 9.1b). The second layer of photosensitive polyimide is spin-coated to define the channel walls and insulate the metallization layer where embedded metallic conductors are desired (Figure 9.1c). On the surface of Mylar™ foil on the second substrate, a 5–20 μm thick layer of photosensitive polyimide is spin-coated (Figure 9.1d). Following a soft bake step on a hotplate, the second substrate is flipped over and bonded to the open channel structures on the first wafer by lamination (Figure 9.1e). After peeling off the Mylar™ foil (Figure 9.1f), the top layer of polyimide is photostructured and the devices are again cured under nitrogen at 350 °C for one hour (Figure 9.1g). During the anodic metal dissolution process of aluminum, the gold remains on the substrate due to the difference in the electrochemical potential compared with aluminum and ensures electrical contact until the structures are released (Figure 9.1h). The fabricated device

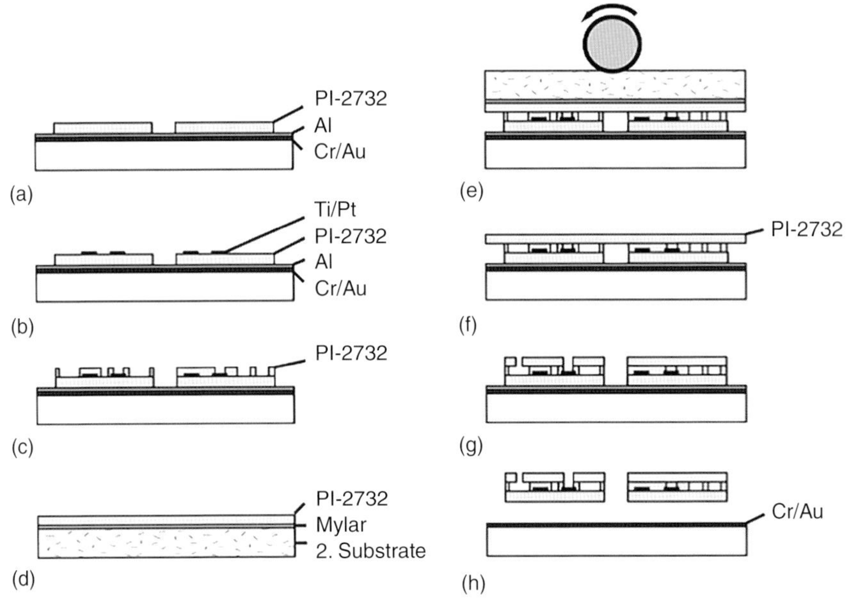


**Figure 9.1** Fabrication steps for polyimide-based microfluidic devices. (a) Photolithography. (b) Sputtering and dry etch. (c) Photolithography. (d) Cover layer preparation. (e) Cover layer bonding. (f) Substrate and Myler removal. (g) Photolithography and Cure. (h) Final release. Source: (a–h) Metz et al. 2001 [2]. Copyright 2001, Reproduced with permission of Wiley-VCH Verlag GmbH & Co. KGaA; Liaw et al. 2012 [3]. Copyright 2012, Reproduced with permission of Wiley-VCH Verlag GmbH & Co. KGaA; Wang et al. 2016 [4]. Copyright 2016, Reproduced with permission of Wiley-VCH Verlag GmbH & Co. KGaA; Schubert-Ullrich et al. 2009 [5]. Copyright 2009, Reproduced with permission of Wiley-VCH Verlag GmbH & Co. KGaA; Martinez et al. 2007 [6]. Copyright 2007, Reproduced with permission of Wiley-VCH Verlag GmbH & Co. KGaA.

can be used for flexible fluidic and electrical connectors, and for many other applications in medical, chemical, and biological research.

### 9.3.2 Soft Lithography

Soft lithography is a popular method for rapid prototyping of microfluidic devices. The majority of reported laboratory-scale prototypes were fabricated with this method. Soft lithography has provided a low-expertise route toward micro/nanofabrication and plays an important role in microfluidics, ranging from simple channel fabrication to the creation of micropatterns onto a surface or within a microfluidic channel. A low-cost and simple method is to fabricate a solid matrix by using lithography to construct the SU-8 layer, as shown in Figure 9.2a. Using the SU-8 master mold, a PDMS microfluidic device could be fabricated following the procedures in Figure 9.2b. The PDMS mixture is poured into the master batch and then solidified for several hours at relatively low temperatures (60–80 °C). After stripping and surface treatment with oxygen plasma, the structured PDMS film is contacted with clean glass, silica, or another surface activated PDMS [7]. Instead of drilling the reservoirs and entry holes in the PDMS layer, glass posts could be placed on the SU-8 mainframe to determine the access to the holes and reservoirs.

Soft lithography is also a direct pattern transfer technique. The term "soft" refers to an elastomeric stamp with patterned relief structures on its surface. PDMS has been used successfully as the elastomeric material. Two soft lithographic methods are introduced to fabricate micro/nanopatterns onto a surface

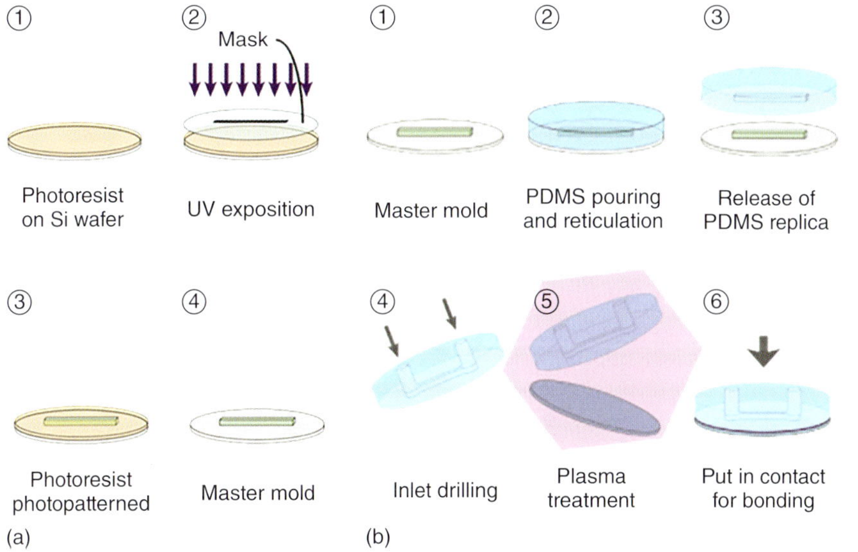


**Figure 9.2** Fabrication method of the microfluidic device by soft lithography. (a) The fabrication process of the master mold and the (b) fabrication process of the microfluidic device using the master mold. Source: (a, b) Nguyen 2008 [7]. Copyright 2008, Reproduced with permission of Elsevier.

**Table 9.1** Soft lithographic methods for the fabrication of micro/nanopatterns.

| | Contact printing | Capillary molding |
|---|---|---|
| Mold | Soft mold<br>PDMS<br>Hard PDMS(h-PDMS) | PDMS<br>PUA |
| Resolution | ~500 nm<br>~50 nm with h-PDMS | ~50 nm with PUA mold |
| Application | 1. Channel fabrication<br>2. Direct printing of biological molecules | 1. Channel fabrication<br>2. Fabrication of micro/nanostructures |
| Process | 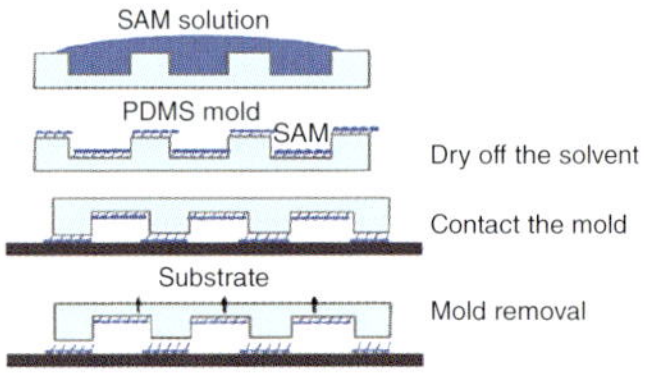 | 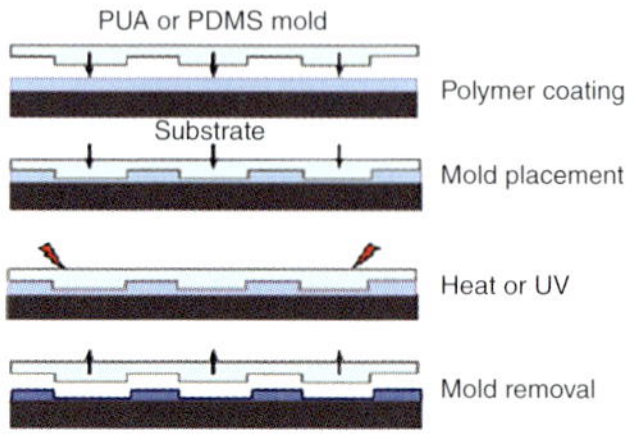 |

Source: Kim et al. 2008 [8]. Copyright 2008, Reproduced with permission of WILEY-VCH Verlag GmbH & Co. KGaA.

or within a microfluidic channel: contact printing and capillary molding. Contact printing generates a nonstructured, chemically modified surface, while capillary molding fabricates a topographically modified physical micro/nanostructure [8] (Table 9.1).

### 9.3.3 Inkjet Printing

Photolithography relies on photomasks and photosensitive materials, which limits the flexibility of lithography. Printing technology, as a kind of high-efficiency technology, has shown tremendous application prospects in pattern-based fabrication and functional materials fabrication of electronic devices, plate making, and microchannels.

A simple method for fabricating microchannels is presented based on inkjet printing liquid templates [9]. By using a temperature-sensitive ink, the shape of the liquid template can be well controlled. Figure 9.3 shows the fabrication process of a typical microfluidic reactor by inkjet printing. The first step is to pour the prepolymer mixture into the container. The Y pattern, which acts as a template, is printed onto the surface of the prepolymer mixture and spontaneously wrapped in the prepolymer mixture. The prepolymer mixture is then thermally cured. At the same time, the liquid template evaporates, leaving the Y microchannel in the matrix. A typical microfluidic reactor is obtained by stripping the matrix from the container.

### 9.3.4 3D Printing

The advent of soft lithography allowed for an unprecedented expansion in the field of microfluidics. However, the vast majority of microfluidic devices are still

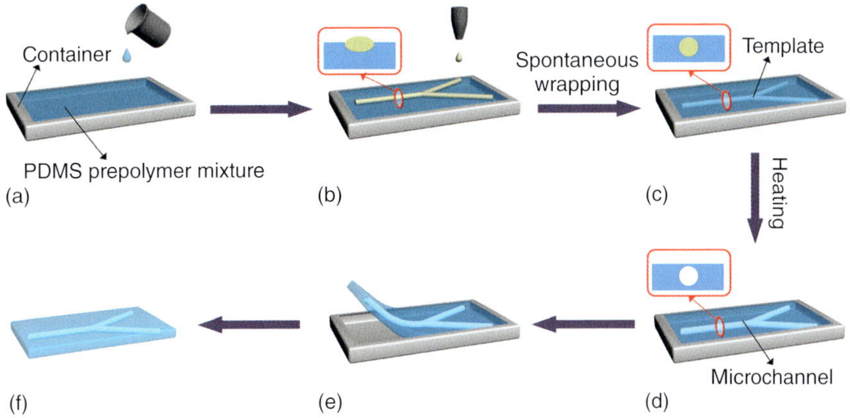


**Figure 9.3** Fabrication process of a typical microfluidic reactor by inkjet printing. (a) Pouring the PDMS prepolymer mixture into a Container. (b) Printing a Y-shape pattern on the surface of the prepolymer mixture. Inset: relative position sketch of the pattern and prepolymer mixture surface. (c) Standing for a few seconds to allow the pattern to be wrapped spontaneously. Inset: relative position sketch of the pattern and prepolymer mixture surface. (d) After heating the prepolymer mixture with the liquid template, the prepolymer mixture thermally cures and the liquid template evaporates, leaving the microchannel in the PDMS matrix. Inset: section sketch of the microchannel in the matrix. (e) Peeling off the fabricated PDMS matrix with microchannel. (f) The typical microfluidic reactor in the PDMS matrix. Source: Guo et al. 2015 [9]. Copyright 2015, Reproduced with permission Royal Society of Chemistry.

made with extensive manual labor, are tethered to bulky control systems, and have cumbersome user interfaces, which all render commercialization difficult. On the other hand, 3D printing has begun to embrace the range of sizes and materials that appeal to the developers of microfluidic devices. For 3D printing, prior to fabrication, a design is digitally built as a detailed 3D computer-aided design (CAD) file. The design can be assembled in modules by remotely collaborating teams, and its mechanical and fluidic behavior can be simulated using finite-element modeling. As structures are created by adding materials without the need for etching or dissolution, the processing is environmentally friendly and economically efficient.

#### 9.3.4.1 3D Printing Sacrificial Structures

3D printing technology is suitable for many kinds of materials. For example, soluble materials could be printed and act as the sacrificial template for microfluidic device fabrication. Here, an easy two-step acrylonitrile butadiene styrene (ABS) scaffold-removal method is presented for achieving 3D, multilayer, intricate, micrometric channels in a single block of PDMS. Using the scaffold-removal fabrication method, external components, such as heating elements, electronics, or radio frequency (RF) circuitry, can be embedded directly in microfluidic devices.

Figure 9.4 shows the schematic representation of the ABS scaffold-removal fabrication method for manufacturing microfluidic devices. An ABS plastic scaffold is modeled or 3D printed in the desired shape. Consecutively, it is suspended in PDMS with or without the addition of external components and then the polymer is cured. Finally, the scaffold is removed by immersion in acetone, creating the microfluidic channels [10].

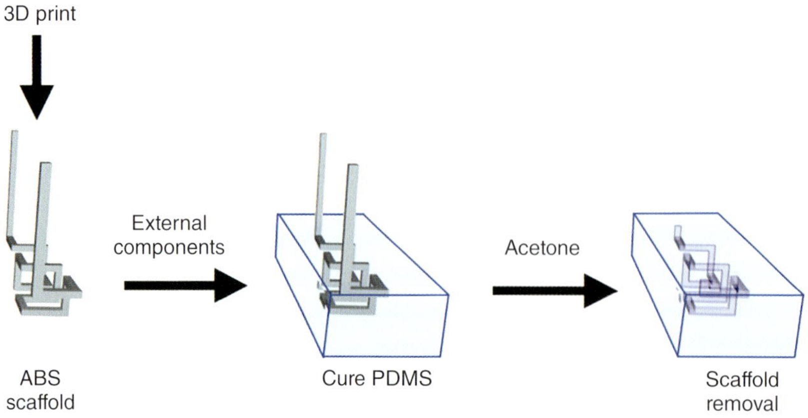


**Figure 9.4** Schematic representation of the ABS scaffold-removal fabrication method for microfluidic devices. Source: Saggiomo and Velders 2015 [10]. Copyright 2015, Reproduced with permission John Wiley & Sons.

#### 9.3.4.2 3D Printing Templates

Direct printing of flexible microfluidic device, implemented by micro-stereolithography, has traditionally required expensive infrastructure, whereas the emergence of affordable 3D printers has the potential to grant broader access to custom-made microstructures, by minimizing resources and fabrication skills.

As shown in Figure 9.5, Comina et al. present a fabrication method of conventional lab-on-a-chip devices made by 3D printer [11]. The template is designed by CAD and printed with a commercial micro-stereolithography 3D printer. The printed templates replace clean room and photolithographic fabrication resources and deliver resolutions of 50 μm. Whereas the templates are smooth enough, it can direct transfer and proper sealing to substrates. After protecting the surface of templates, the polymer is directly cast on the 3D printed templates with no additional processing cost or effort. This simplified method promotes access to build flexible device configurations with minimal fabrication requirements and provides a versatile flexible device development platform with a 3D printer.

### 9.3.5 Fabrication of Open-Surface Microfluidics

Open-surface and interfacial microfluidics, where one or more gas/liquid interfaces exist as boundary conditions, is the emerging trend in microfluidics from which several new and flexible operations have been established, including self-propelled motion, three-dimensional connectivity, open sample accessibility, direct reactivity, and readability, in addition to conventional microflow manipulations.

#### 9.3.5.1 Fabrication of Paper-Based Microfluidic Device

The paper-based microfluidics diagnostic device is inexpensive, robust, lightweight, and independent of supporting infrastructure. Martinez et al.

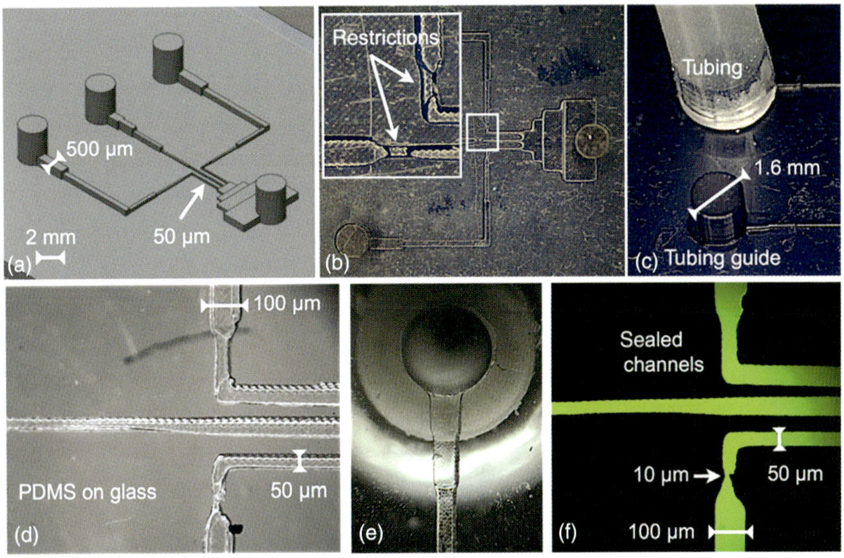


**Figure 9.5** Lab-on-a-chip device fabricated using templates printed with 3D printer. (a) CAD with 50 μm and 2 mm features combined in the same design. (b) 3D printout of the structure in (a). The inset highlights hindrances developed at the 50–100 μm channel step. (c) Image of the printed pillar guides used to integrate silicone tubing in the PDMS structure. (d) PDMS device assembled on a glass substrate showing no texture at the PDMS–glass interface. (e) Microscopic image of tubing connected to a 500 μm channel. (f) Epi-fluorescence microscopic image of PDMS structure filled with fluorescein solution. Source: (a–f) Comina et al. 2014 [11]. Copyright 2014, Reproduced with permission of Royal Society of Chemistry.

described a new type of paper-based device, which makes the paper layer into the hydrophilic channel, hydrophobic wall, and ribbon structure, and connects the channels on different paper layers with tape patterned holes [12]. These devices extend the paper-based analysis from simple one-dimensional lateral flow systems to 3D devices with complex microfluidic paths, and significantly expand the capabilities of low-cost analysis systems.

As shown in Figure 9.6, the 3D paper-based flow system is produced by stacking alternating paper and waterproof double-sided adhesive tape. Patterned hydrophobic polymers on paper delineate channels for fluid transverse movement, impermeable double-sided tape layers separate channels in adjacent paper layers, and holes cut into the tape allow fluid to flow vertically. The paper is patterned with SU-8 photoresist by being impregnated with photoresist, dried, exposed to ultraviolet light through a transparent mask, and developed with acetone and isopropanol. The adhesive tape is patterned with a laser cutting machine, which can also be manually patterned with a puncher. Then the layers of paper and adhesive tape are stacked together. A layer of adhesive tape is attached to the patterned paper layer. The holes on the adhesive tape are filled with a paste made from cellulose powder and water. The protective film is removed from the adhesive tape and second layers of paper placed on the second surface. This stacking-paper-tape-paper-tape process is repeated as needed to complete the device, and a reproducible method for manufacturing the device is provided.

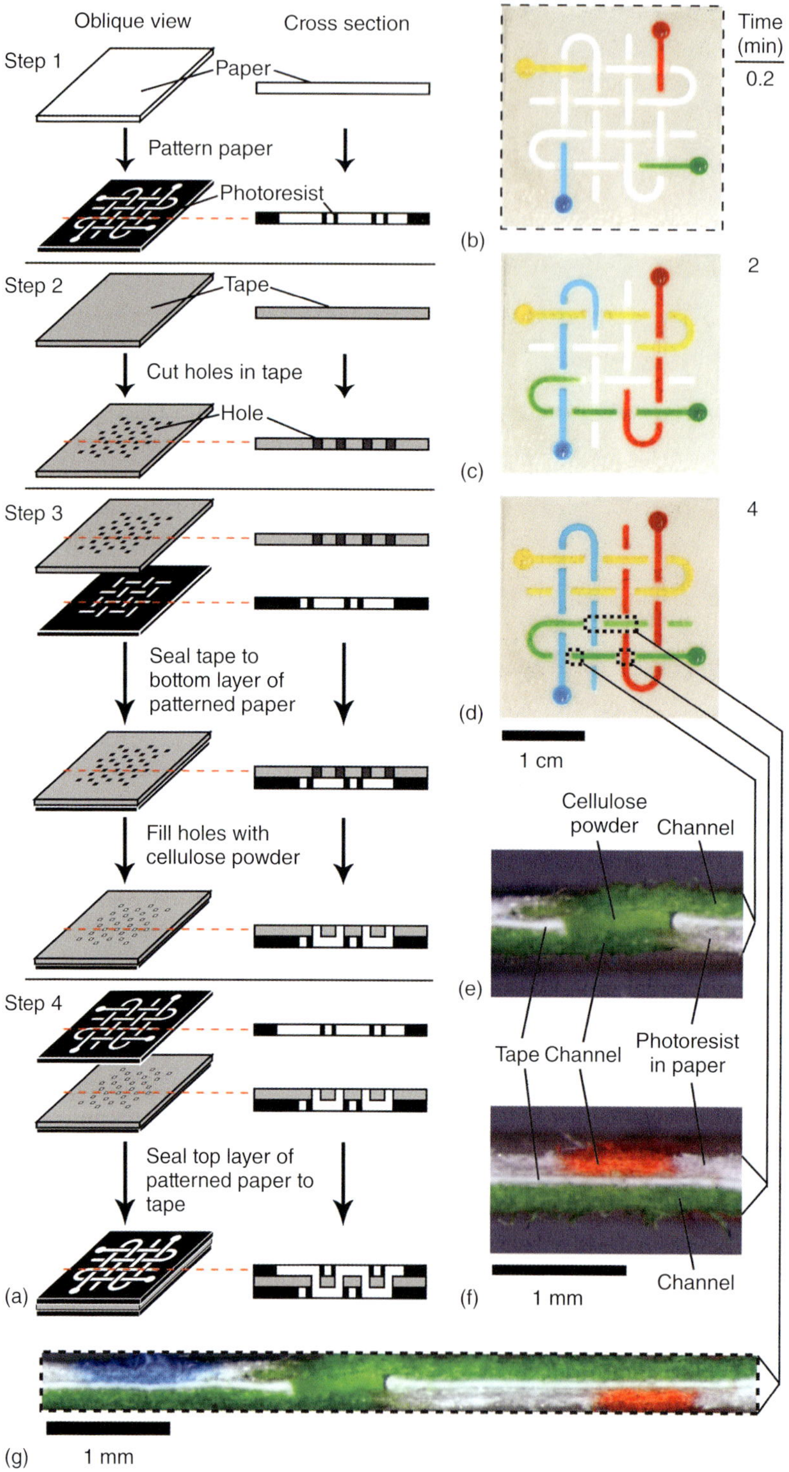


**Figure 9.6** Three-dimensional microfluidic devices fabricated in layered paper and tape. Source: Martinez et al. 2008 [12]. Copyright 2008, Reproduced with permission of PNAS.

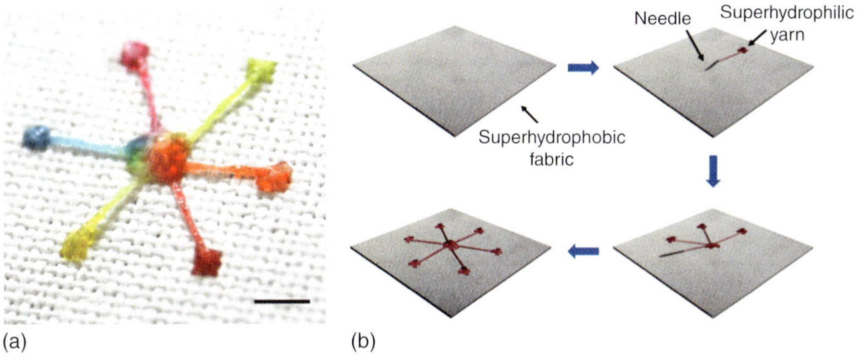


**Figure 9.7** The fabrication process of MST with multi-inlet-single-outlet design. (a) A multi-inlet-single-outlet design on the MST platform using the autonomous interfacial transport concept. (b) The fabrication process of MST with multi-inlet-single-outlet design. Source: Xing et al. 2013 [13]. Copyright 2013, Reproduced with permission of Royal Society of Chemistry.

#### 9.3.5.2 Fabrication of Textile-Based Microfluidic Device

The interfacial microfluidic concept has been extended to textile-based structures and surfaces. The textile-based microfluidics utilizes a similar wicking force to that in paper that is produced by hydrophilic yarns (e.g. cotton yarns) to direct biological reagents along the fibrous structure, which affords the aforementioned operational capacities of interfacial microfluidics while providing a low-cost and scalable solution based on well-established traditional textile manufacturing techniques such as automatic weaving, knotting, and stitching.

In a recent study, a novel interfacial microfluidic transport principle is used to drive 3D liquid flow on a micropatterned superhydrophobic textile (MST) platform in a more autonomous and controllable manner [13]. Specifically, the MST system utilizes the surface tension-induced Laplace pressure to facilitate liquid motion along the hydrophilic yarn, in addition to the capillarity present in the fibrous structure. The fabrication of MST is simply accomplished by stitching hydrophilic cotton yarn into a superhydrophobic fabric substrate, from which well-controlled wetting patterns are established for interfacial microfluidic operations. The geometric configurations of the stitched micropatterns, e.g. the lengths and diameters of the yarn and bundled arrangement, can all influence the transport process (Figure 9.7).

## 9.4 Applications

With the fast development of flexible materials and fabrication methods, the combination of fluid manipulation and flexible electronics promotes the application of functional microfluidics, e.g. wearable microfluidics for biosensing and health care. Typically, wearable microfluidics detects biofluids by the following procedures: (i) collect the biofluids, (ii) transfer biofluids to detection site, and (iii) perform detection. Blood, serum, sweat, interstitial fluid (ISF), saliva, tears, mucus, semen, and stools are usually the source of the biofluids [14]. Especially, sweat is the analyte most commonly processed. In addition, wearable microfluidics could also be used for motion sensing of human bodies.

### 9.4.1 Wearable Microfluidics for Sweat-Based Biosensing

Sweat contains plenty of physiological information. Sweat sensors integrated into wearable microfluidics offer methods for gleaning molecular-level insight into the dynamics of our bodies. Many wearable sweat sensors have been developed in recent years and combined with different form factors, substrates, and detection mechanisms.

Kim et al. used superabsorbent polymer (SAP) valves and colorimetric sensing reagents as the enabling components of soft, skin-mounted microfluidic devices designed to capture, store, and chemically analyze sweat released from eccrine glands [15]. The valving technology guided the flow of sweat from an inlet location into a collection of isolated reservoirs, in a well-defined sequence. Figure 9.8 shows the schematic illustration of the structure of the chrono-sampling epidermal microfluidic device for sweat chloride monitoring. The channel system was designed into a double layer geometry that separately supported the microfluidic channels and valves. A collection of active valves based on SAP materials and

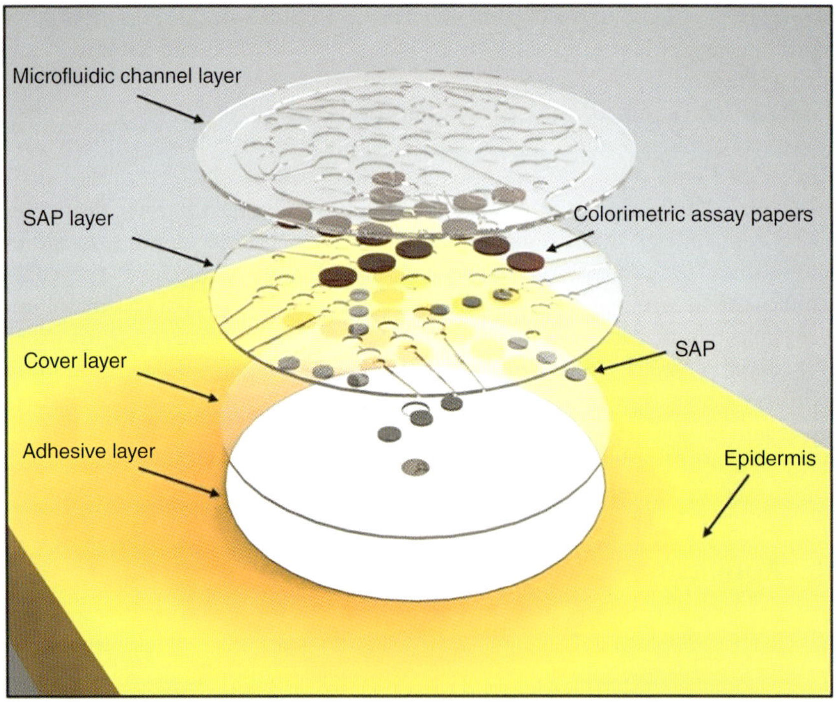


**Figure 9.8** The schematic illustration of the structure of the chrono-sampling epidermal microfluidic device for sweat chloride monitoring. Source: Kim et al. 2018 [15]. Copyright 2018, Reproduced with permission of John Wiley & Sons.

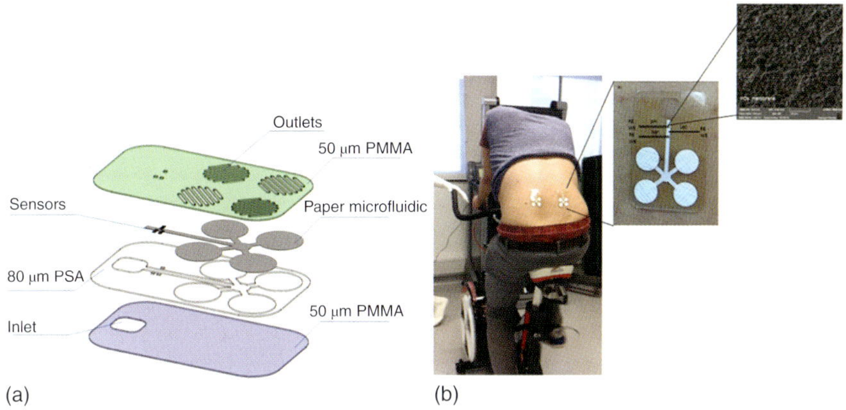


**Figure 9.9** (a) Schematic representation of the fabrication steps of the microfluidic chip; (b) photograph of the chip attached to the body and scanning electron microscopic (SEM) image (magnification 2.00 k×) photograph of IrOx pH sensor membrane on top of a 50 μm Pt wire . Source: (a, b) Anastasova et al. 2017 [17]. Copyright 2017, Reproduced with permission of Elsevier.

passive valves exploited tailored hydrophobic and hydrophilic surface chemical functionalization on the surfaces of the channels. The active valves are used to close the inlets and outlets of each reservoir. The passive valves are used to guide the flow directionality.

Koh et al. presented a series of materials and device designs for soft, flexible, and stretchable microfluidic systems to establish sweat monitoring technologies [16]. The sweat can spontaneously get through a microfluidic network and a set of reservoirs by access points. Colorimetric fashions were used to analyze markers such as chloride and hydronium ions, glucose, and lactate. In addition, the wireless communication electronics integrated into the wearable microfluidic systems can intimately and robustly bond to the surface of the skin without chemical and mechanical irritation.

Anastasova et al. presented a fully integrated flexible microfluidic system for continuous, simultaneous, and selective measurement of sweat metabolites, electrolytes, and temperature [17]. The flexible microfluidic system can transmit information wirelessly for ease of collection and storage, with the potential for real-time data analytics. Figure 9.9 shows the schematic representation of the fabrication steps of the microfluidic chip and a photograph of the chip attached to the body.

Benito-Lopez et al. presented a wearable, robust, flexible, and disposable microfluidic chemical barcode platform [18]. The platform incorporated novel ionic liquid polymer gels (ionogels). The device can monitor the sweat pH in real time during an exercise period. The microfluidic structure makes the fresh sweat continuously pass through the sensing area and that ensures continuous real-time analysis and immediate feedback regarding sweat composition.

Yang et al. reported a wearable microfluidic platform to measure the flow rate on a patterned fabric surface referred to as digital droplet flowmetry (DDF). The platform was made by using conventional fabric materials and laser

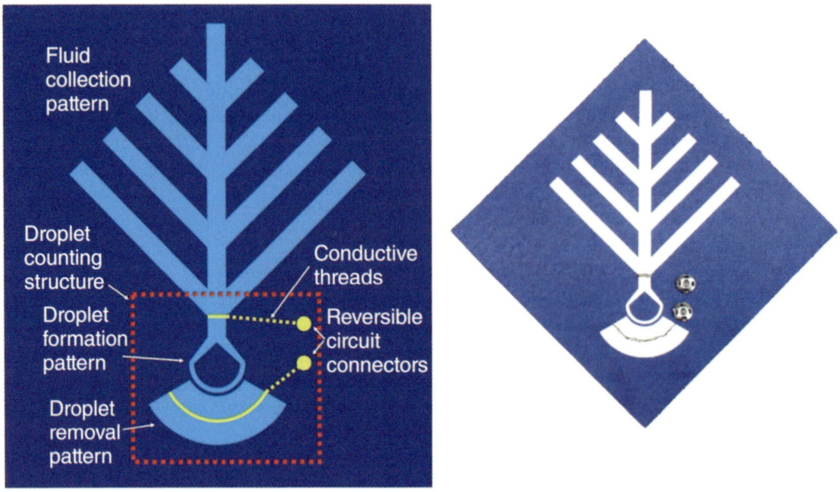


**Figure 9.10** (a) Illustration and (b) a prototype of the wearable digital droplet flowmetry. Source: Yang et al. 2017 [19]. Copyright 2017, Reproduced with permission of Royal Society of Chemistry.

micromachining [19]. The proposed wearable DDF can collect and measure continuous perspiration with high precision (96% on average) in a real-time fashion. Moreover, the proposed DDF platform can be conveniently integrated with regular apparel or a wearable device and has potential to be applied to dynamic removal, collection, and monitoring of biofluids for various physiological and clinical processes. Figure 9.10 shows the illustration and prototype of wearable DDF.

### 9.4.2 Wearable Microfluidics for ISF-Based Biosensing

ISF is another source for molecular-level detection of physiological information of the human body. Li's group has done a great deal of work on transdermal extraction of ISF by flexible microfluidics providing a new way for continuous blood glucose monitoring [20–24]. Yu et al. fabricated a five layers PDMS microfluidic chip for ISF transdermal extraction, collection, and measurement toward the application of continuous and real-time glucose monitoring [20]. The photograph of the chip integrated a fiber-optic surface plasmon resonance (SPR) sensor for glucose concentration measurement is shown in Figure 9.11 [21]. Wu et al. proposed a new method of manufacturing stable microelectrodes on PDMS by inkjet printing and fabricated a three-electrode electrochemical sensor for glucose detection in a microfluidic system based on PDMS [22]. Figure 9.12 shows the optical image of the three-electrode electrochemical sensor fabricated on PDMS. In order to improve the performance of the electrochemical sensors, Pu et al. modified the working electrode of the sensor with graphene and gold nanoparticles by inkjet printing to form a composite nanostructure [23]. Pu et al. also reported an all-inkjet-printed flexible epidermal microfluidic system, which

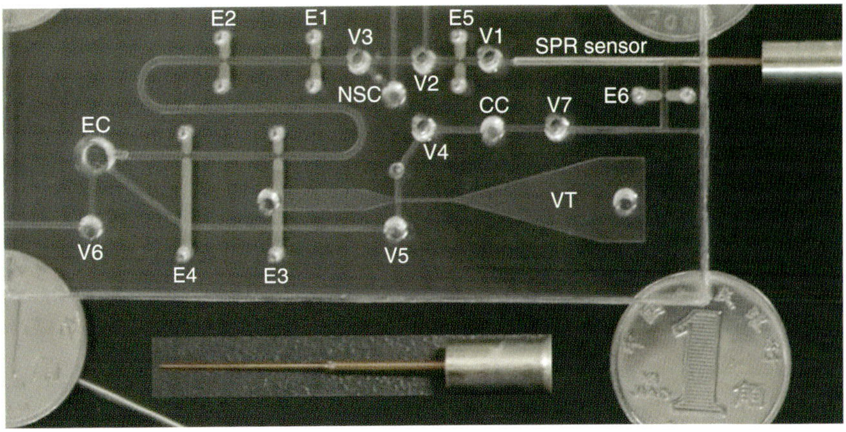


**Figure 9.11** Optical image of the flexible microfluidic chip integrated with a fiber-optic SPR sensor for continuous ISF transdermal extraction and glucose concentration monitoring. Source: Li et al. 2016 [21]. Copyright 2016, Reproduced with permission of American Institute of Physics.

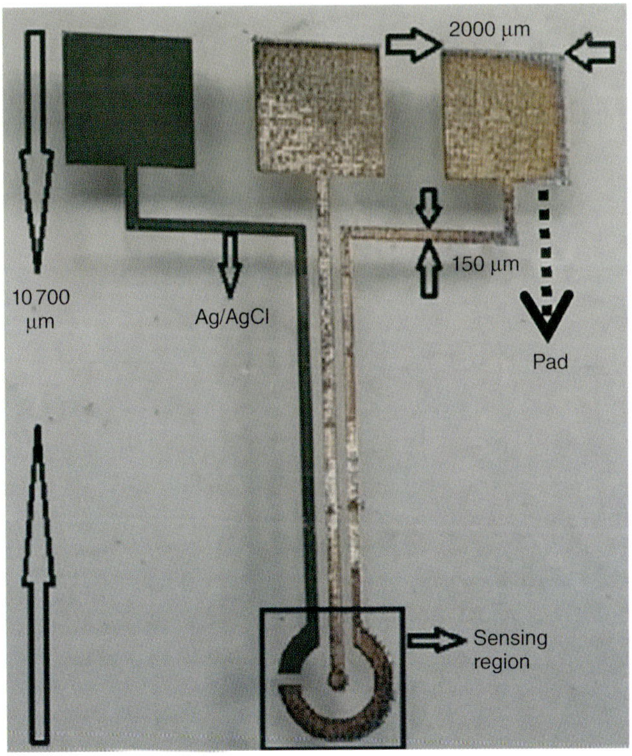


**Figure 9.12** Optical image of a three-electrode electrochemical sensor on PDMS after silver chloride (black) formation. Source: Wu et al. 2015 [22]. Copyright 2015, Reproduced with permission of Royal Society of Chemistry.

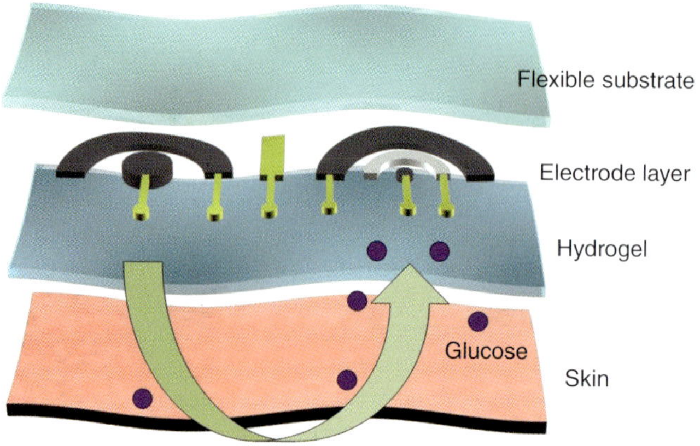


**Figure 9.13** Photographs of the flexible epidermal microfluidic system attached to the skin [24].

consists of seven disposable four-electrode measuring array units [24]. Two of the electrodes are used to form epidermal microfluidic channels through the skin to transdermally extract ISF with glucose molecules by ions transportation. Figure 9.13 shows photographs of the flexible epidermal microfluidic system attached on the skin.

### 9.4.3 Wearable Microfluidics for Motion Sensing

Wearable microfluidics is also used for motion sensing of human bodies. Liquid metal in the microchannel is used as pressure sensor. Yeo et al. developed a liquid-based thin-film microfluidic tactile sensor, which has high flexibility, robustness, and sensitivity [25]. A highly conductive metallic liquid, eutectic gallium indium (eGaIn), functionalized the microfluidic sensor. The principle of the sensor is that the micro-deformation of the pressure sensor results in fluid displacement and that changes the electrical resistance. Figure 9.14 shows the photographs of the sensors. The sensors were multiplexed to detect forces at multiple regions of the hand.

Jiao et al. developed all-flexible strain sensors made of graphene, microfluidic liquid metal, and stretchable elastomer. Liquid metal was introduced into

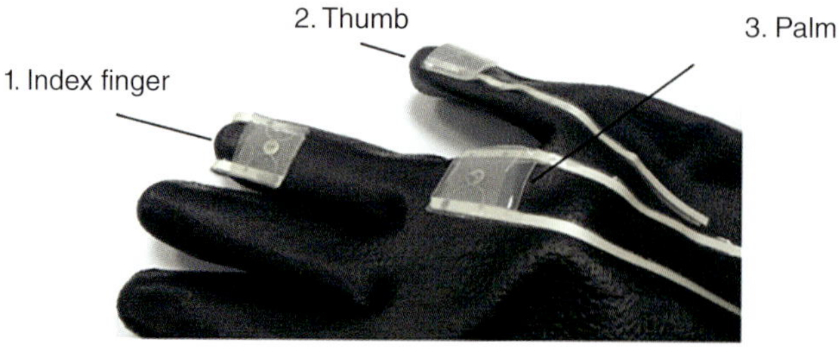


**Figure 9.14** Positions of the sensors: index finger, thumb, and palm region. Source: Yeo et al. 2016 [25]. Copyright 2016, Reproduced with permission of Royal Society of Chemistry.

microfluidic channels for wiring inside the devices [26]. Graphene was patterned inside of the liquid metal containing elastomer. The fluid eGaln made the microfluidic electrical wires to flow and reshape without cracks and fatigue in response to the sensor deformation during mechanical sensing. The graphene sensor was able to track the angular motion of a human wrist.

Chossat et al. designed and manufactured a kind of soft artificial skin with an array of embedded soft strain sensors. The sensor can detect various hand gestures by measuring joint motions of the five fingers [27]. Two different liquid conductors, an ionic liquid and a liquid metal, were filled into microchannels, which were made of a hyperelastic elastomer material.

### 9.4.4 Other Flexible Microfluidics

#### 9.4.4.1 Soft Robotics

Apart from the abovementioned sensing applications, flexible microfluidics has been employed in soft actuating devices, namely the soft robotics. Pioneered by the Whitesides group, soft robotics has expanded to a wide range of use from giant robots for industrial applications to miniaturized robots for surgery and drug delivery. The soft robots have several advantages over rigid robots, such as the shock-free contact to objects, safe human–machine interaction, and simple gripping mechanism.

The dominant actuation strategy of soft robotics is pneumatic actuation. Figure 9.15 illustrates the operating mechanisms of soft robots that are pneumatically actuated by a fluid. Adding materials to introduce asymmetries enables the design of structures that deform into specific shapes [28].

The soft robotics has a large range of applications. For instance, soft robots could be used in the food industry for handling brittle objects. In clinical applications, soft robotic assistance is beneficial for delivering therapy and for restoring motor functionality [29]. In various works, the soft robots, integrated with sensing components, are used as rehabilitation assistance [30] and prostheses [31]. The combined abilities of sensing and actuating built on an entirely soft and flexible platform are advantageous for stroke and rehabilitation patients. Apart from macroscopic applications, soft robots are also reported to be capable of handling delicate tasks, such as soft robots for surgery [32].

#### 9.4.4.2 Drug Delivery

Drug delivery is another fascinating application of flexible microfluidics. Although drug patches have been successfully developed previously, the efficacy of drug release over time remains a challenge due to its concentration effect. Microfluidics-enabled drug patch may overcome this problem by ensuring consistent drug dispensing over prolonged periods. Furthermore, the possibility of fabricating microneedles allows painless drug administration and ease of use. Continual monitoring of the patient's physiological state with capabilities to provide drug intervention will greatly benefit the biomedical community. For example, Xiang and coworkers designed a flexible microfluidic drug patch for inflammation treatment [33]. The device comprises a micropump, microneedles, and microchannels preloaded with diclofenac. Even more recently, Lee et al.

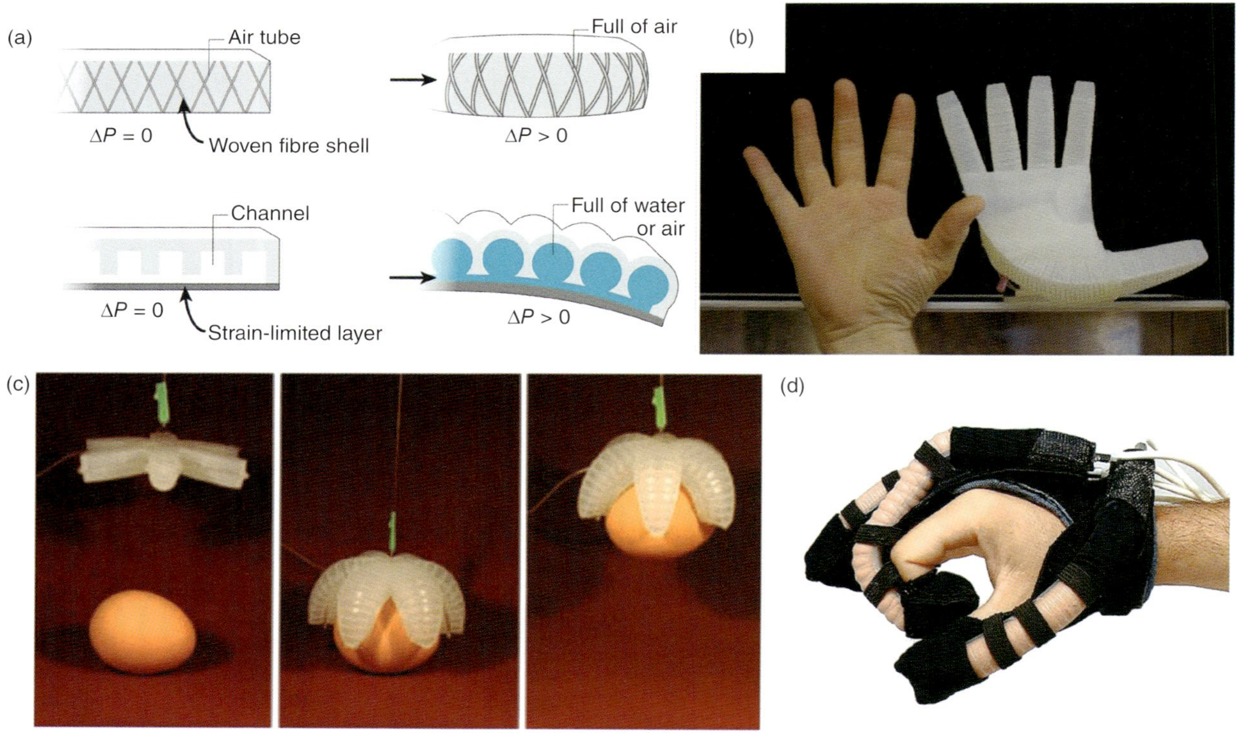


**Figure 9.15** (a) Schematic illustration of the actuation mechanisms of microfluidics-based soft robotics [28]. (b) A compliant and underactuated robotic hand for dexterous grasping. (c) Simple gripper fabricated by soft lithography for handling fragile objects. (d) A soft robotic glove for combined assistance and at-home rehabilitation. (a–d) Rus and Tolley 2015 [28]. Copyright 2015, Reproduced with permission of Springer Nature.

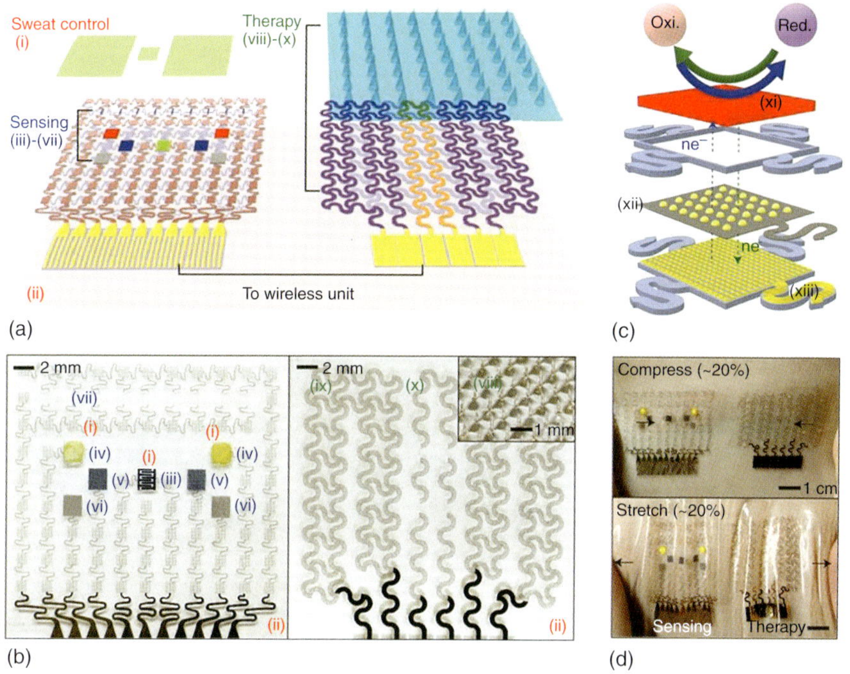


**Figure 9.16** (a) Schematic drawings of the microfluidic drug patch. (b) Optical camera image of the electrochemical sensor array (left), therapeutic array (right), and magnified view of the drug-loaded microneedles (inset). (c) Schematic of the GP-hybrid electrochemical unit. (d) Optical camera images of the diabetes patch laminated on human skin under mechanical deformations. Source: (a–d) Lee et al. 2016 [34]. Copyright 2016, Reproduced with permission of Springer Nature.

developed a wearable patch that combines monitoring and intervention of diabetes through sweat analysis [34]. Figure 9.16a–c shows the schematic of the device layout. The patch comprises different functional elements, including temperature, humidity, glucose, and pH sensors, which allow complete glucose monitoring solutions. The device mounted on the skin could sustain conformability even under compression and tension (Figure 9.16d). Accordingly, glucose can be tracked continuously together with other variables and when abnormally high levels were detected, metformin may be administered to the skin through thermal activation via the microneedles on the sensor patch. Such proactive monitoring drug patches may potentially reduce catastrophic side effects and even save lives.

#### 9.4.4.3 Implantable Devices

Flexible microfluidics is also suitable for implantable applications such as drug injection and status monitoring. Soft microfluidics could allow pharmacological infusions while minimizing tissue damage and inflammation, making long-term implantation possible. For instance, Jeong et al. developed a wireless optofluidic system that enabled programmable in vivo pharmacology and optogenetics, as shown in Figure 9.17 [35]. A neural probe composed of sufficiently thin and

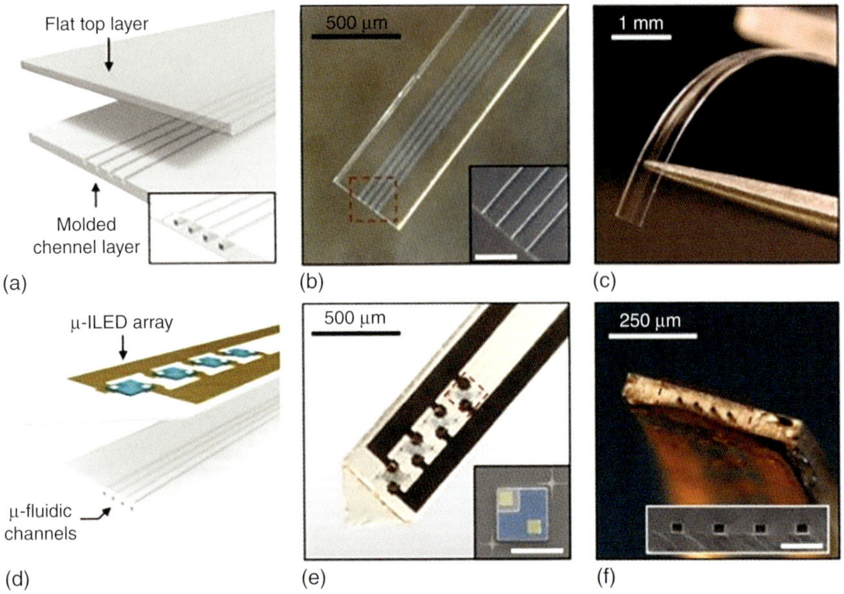


**Figure 9.17** Schematic illustration and photographs of the optofluidic neuro probe. (a) Schematic diagram of the assembly of a soft microfluidic probe. (b) Optical micrograph of a microfluidic probe formed in this way. (Inset) Magnified view of the channels. Scale bar, 100 μm. (c) The picture that illustrates the soft, compliant mechanics of the device. (d) Schematic diagram of the integration of a soft microfluidic probe with a flexible array of m-LEDs (light emitting diode) and metal interconnect traces on a film of PET. (e) Optical micrograph of an integrated probe, which was referred to as an optofluidic system. (Inset) Colorized SEM of a representative m-ILED (inorganic light emitting diode). (f) Tilted view of an optofluidic probe that shows the tip end. (Inset) SEM of the outlets of the microfluidic channels. Source: (a–f) Jeong et al. 2015 [35]. Copyright 2015, Reproduced with permission of Elsevier.

flexible microfluidics could be implanted into the deep brain tissue region of rodents, which enabled wireless programmed spatiotemporal control of fluid delivery and photostimulation while having a size that is orders of magnitude smaller than conventional counterparts. The minimally invasive operation of these probes forecasts utility in other organ systems and species, with potential for broad application in biomedical science, engineering, and medicine.

#### 9.4.4.4 Flexible Display

Flexible microfluidics has also been used for flexible display recently. Kobayashi and Onoe reported a multicolored flexible display based on sequentially introducing colored water droplets and air pockets into a microfluidic device fabricated with PDMS, as shown in Figure 9.18 [36].The microfluidic device based on a flexible polymer enables a display with zero power consumption. Single water droplets define individual pixels, which are separated by air gaps; pumping water and air through the device in sequence allows colored patterns to be displayed, which can be controlled by varying the sequence of water and air, as well as water color. Additionally, retaining the pattern does not require the input of energy,

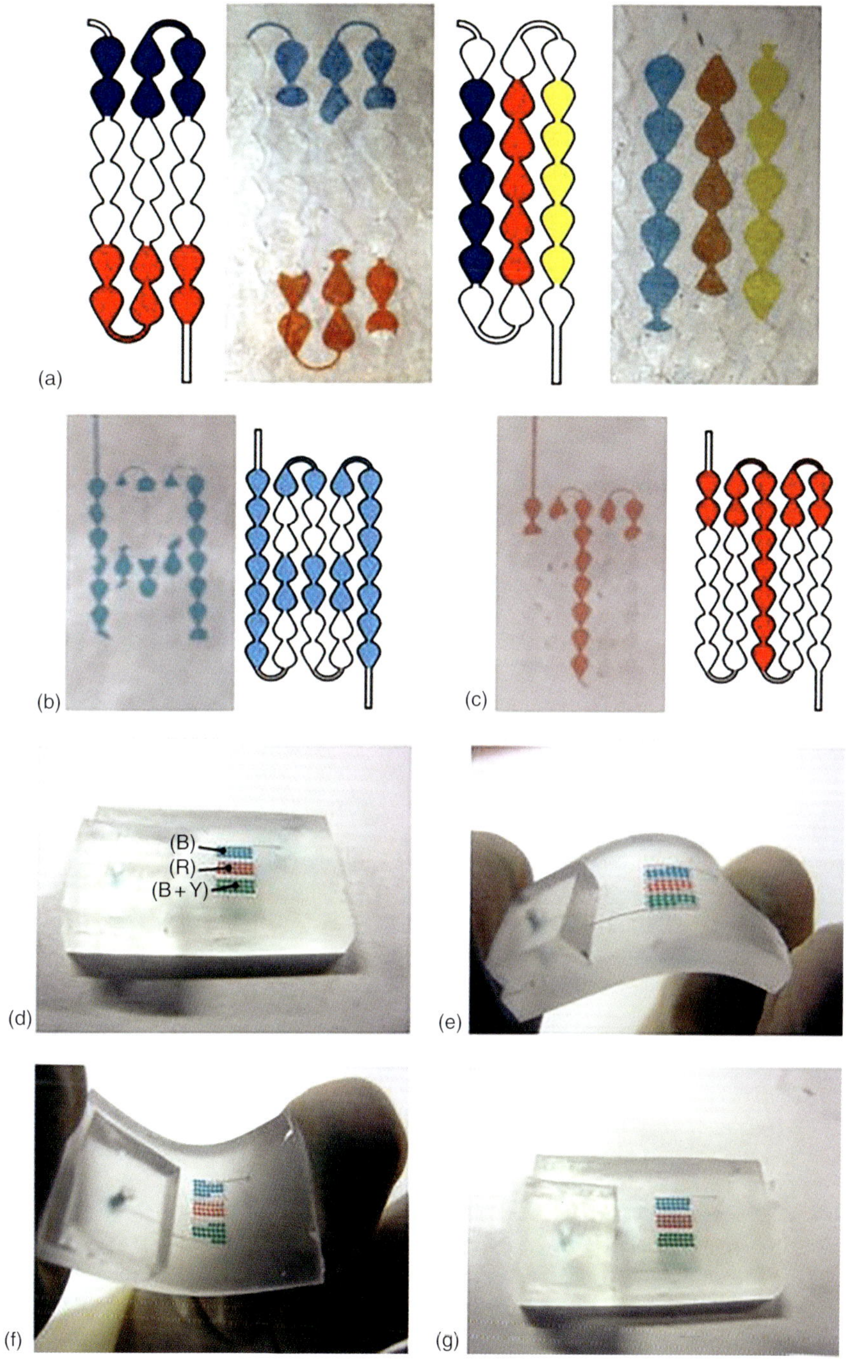


**Figure 9.18** Demonstrations of the microfluidics-based flexible display. (a) Multicolored stripe patterns (vertical and horizontal) displayed on meandering microchannel. (b–c) Bitmap characters "A" and "T" on microfluidics-based reflective display. (d–g) Bending test of the display. The display image was maintained after the bending. Source: (a–g) Kobayashi and Onoe 2018 [36]. Copyright 2018, Reproduced with permission of Wiley-VCH Verlag GmbH & Co. KGaA. https://www.nature.com/articles/s41378-018-0018-1#rightslink. http://creativecommons.org/licenses/by/4.0/. Licensed under CCBY 4.0.

while PDMS makes the device flexible. Such a device could find application in displays on clothes and the skin of robots.

## 9.5 Challenges

In the future, more and more flexible materials with great physical properties, chemical stability, and biocompatibility will be developed. The emergence of novel materials and the development of fabrication technologies will promote the applications of flexible microfluidics in life science, health care, chemical analysis, and so on.

Further research and development are required to find better ways of packaging rigid, fragile sensors into the flexible microfluidic system. Flexible microfluidics can also benefit from the performance improvements of the sensors on flexible substrates. In addition, more robust and stable fluid manipulating and sensing components must be designed to enable long-term use.

## References

1 Yeo, J.C., Kenry, K., and Lim, C.T. (2016). *Lab Chip* 16: 4082.
2 Metz, S., Holzer, R., and Renaud, P. (2001). *Lab Chip* 1: 29.
3 Liaw, D.-J., Wang, K.-L., Huang, Y.-C. et al. (2012). *Prog. Polym. Sci.* 37: 907.
4 Wang, S.Q., Chinnasamy, T., Lifson, M.A. et al. (2016). *Trends Biotechnol.* 34: 909.
5 Schubert-Ullrich, P., Rudolf, J., Ansari, P. et al. (2009). *Anal. Bioanal. Chem.* 395: 69.
6 Martinez, A.W., Phillips, S.T., Butte, M.J., and Whitesides, G.M. (2007). *Angew. Chem. Int. Ed.* 46: 1318.
7 Nguyen, N.T. (2008). *Micromixers*. Elsevier.
8 Kim, P., Kwon, K.W., Park, M.C. et al. (2008). *Biochip J.* 2: 1.
9 Guo, Y.Z., Li, L.H., Li, F.Y. et al. (2015). *Lab Chip* 15: 1759.
10 Saggiomo, V. and Velders, A.H. (2015). *Adv. Sci.*: 2.
11 Comina, G., Suska, A., and Filippini, D. (2014). *Lab Chip* 14: 424.
12 Martinez, A.W., Phillips, S.T., and Whitesides, G.M. (2008). *Proc. Natl. Acad. Sci. U.S.A.* 105: 19606.
13 Xing, S.Y., Jiang, J., and Pan, T.R. (2013). *Lab Chip* 13: 1937.
14 Heikenfeld, J. (2016). *Electroanalysis* 28: 1242.
15 Kim, S.B., Zhang, Y., Won, S.M. et al. (2018). *Small* 14: e1703334.
16 Koh, A., Kang, D., Xue, Y. et al. (2016). *Sci. Transl. Med.* 8: 366ra165.
17 Anastasova, S., Crewther, B., Bembnowicz, P. et al. (2017). *Biosens. Bioelectron.* 93: 139.
18 Benito-Lopez, F., Coyle, S., Byrne, R., and Diamond, D. (2001). Sensing sweat in real-time using wearable microfluidics. *Proceedings of the 7th International Workshop on Wearable and Implantable Body Sensor Networks*, Singapore (26–28 June 2010).
19 Yang, Y., Xing, S., Fang, Z. et al. (2017). *Lab Chip* 17: 926.

20 Yu, H., Li, D., Roberts, R.C. et al. (2012). *J. Microelectromech. Syst.* 21: 917.
21 Li, D., Lu, B., Zhu, R. et al. (2016). *Biomicrofluidics* 10: 011913.
22 Wu, J., Wang, R., Yu, H. et al. (2015). *Lab Chip* 15: 690.
23 Pu, Z., Wang, R., Wu, J. et al. (2016). *Sens. Actuators, B* 230: 801.
24 Pu, Z., Yu, H., Lai, X. et al. (2017). Flexible epidermal microfluidic system for continuous glucose monitoring. *The 21st International Conference on Miniaturized Systems for Chemistry and Life Sciences (MicroTAS 2017)*, Savannah, Georgia, USA (22–26 October 2017).
25 Yeo, J.C., Yu, J., Koh, Z.M. et al. (2016). *Lab Chip* 16: 3244.
26 Jiao, Y., Young, C.W., Yang, S. et al. (2016). *IEEE Sens. J.* 16: 7870.
27 Chossat, J., Yiwei, T., Duchaine, V., and Park, Y. (2015). Wearable soft artificial skin for hand motion detection with embedded microfluidic strain sensing. *2015 IEEE International Conference on Robotics and Automation (ICRA)*, Seattle, Washington, USA (26–30 May 2015).
28 Rus, D. and Tolley, M.T. (2015). *Nature* 521: 467.
29 Cianchetti, M., Laschi, C., Menciassi, A., and Dario, P. (2018). *Nat. Rev. Mater.* 3: 143.
30 Polygerinos, P., Wang, Z., Galloway, K.C. et al. (2015). *Rob. Autom. Syst.* 73: 135.
31 Deimel, R. and Brock, O. (2015). *Int. J. Rob. Res.* 35: 161.
32 Noh, Y., Sareh, S., Back, J. et al. (2014). *IEEE Int. Conf. Rob.*: 6388.
33 Wang, H., Xiang, Z., Giorgia, P. et al. (2016). *Nano Energy* 23: 80.
34 Lee, H., Choi, T.K., Lee, Y.B. et al. (2016). *Nat. Nanotechnol.* 11: 566.
35 Jeong, J.W., McCall, J.G., Shin, G. et al. (2015). *Cell* 162: 662.
36 Kobayashi, K. and Onoe, H. (2018). *Microsyst. Nanoeng.* 4: 1–11.

# Part IV

# Integrating and Connecting

# 10

# Piezoelectric Materials and Devices Based Flexible Bio-integrated Electronics

*Xinge Yu*

*City University of Hong Kong, Department of Biomedical Engineering, 83 Tat Chee Avenue, Kowloon Tong, Hong Kong, China*

## 10.1 Introduction

Recent development in materials science, chemistry, physics, mechanics, and advanced fabrication technologies allow electronic devices in formats that are lightweight, ultrathin, flexible, and even stretchable [1–3]. These electronics are well known as wearable devices and in some cases, epidermal electronics, which have attracted great attention from both academia and industry because of their unique characteristics and practical applications in long-term health monitoring, disease diagnosis, and treatment [2, 4–7]. The key to the success of such kind of electronics is the choice of suitable materials as desired functional components in flexible electronics [8–10]. Piezoelectric materials have attracted growing interest in flexible and stretchable electronics due to their ability to interconvert mechanical and electrical forms of energy [11–13]. For instance, electrical power could be generated from movements and vibrations associated with nature, such as human body activities, environmental power sources, and the friction that exists in operating machinery [14]. In turn, applying electrical energy to piezoelectric materials can also generate controllable material deformation and thus lead to mechanical forces that are typically utilized in robotics actuation, medical imaging instruments, and nano/micro-scale indentation in metrology tools [14–16].

Organic piezoelectric materials, such as polyvinylidene fluoride (PVDF) and poly(vinylidenefluoride-*co*-trifluoroethylene) (PVDF-TrFE), have the advantage of inherent flexibility and are thus easy to integrate into soft electronic devices [17–20]. However, improving the power conversional efficiency of organic based piezoelectric materials is still a challenge. Inorganic piezoelectric materials, such as lead zirconate titanate (PZT) [14, 21, 22] and zinc oxide (ZnO) [23, 24], exhibit much better performance than organic piezoelectric materials, and have been applied in various commercialized devices. In recent years, piezoelectric materials based electronics have made a great progress due to materials choices and device designs, and provide the foundations for a large number of embodiments, particularly focusing on epidermal and implantable biomedical devices.

*Flexible and Wearable Electronics for Smart Clothing*, First Edition.
Edited by Gang Wang, Chengyi Hou, and Hongzhi Wang.

This chapter first discusses the methodologies and techniques to create unconventional electronics in a flexible or stretchable format, which is critical for application in biomedical related and clinically relevant devices. Various routes of realizing flexible piezoelectric materials and devices are summarized in the first part, and mechanics and engineering strategies in device construction are highlighted. Then, a summary of the latest piezoelectric based electronics that combines the use of piezoelectric materials and class of engineering for applications in biomedical engineering are discussed in the second part (Sections 10.2 and 10.3).

## 10.2 Piezoelectric Materials

Organic electronics, such as organic light-emitting diodes (OLEDs) [25], organic photovoltaics (OPVs) [26], and organic thin-film transistors (OTFTs) [27, 28], have proved successful in next-generation electronics due to low fabrication costs, lightweight, mechanical flexibility, and compatibility with a wide range of applications. Indeed, the flexibility and stretchability associated with organic electronics are due to the intrinsic flexibility of organic materials [29]. Besides organic semiconducting materials, recently developed organic and composite piezoelectric materials and devices exhibit the potential for various applications in flexible bio-electronics. PVDF and its copolymer, PVDF-TrFE, are two most popular examples, which typically exhibit better piezoelectric properties when deposited as a well-aligned fiber film using a particular method, such as electrospinning. The well-aligned fibers facilitate crystallization in the β-phase; thus, the dipoles can align in parallel and lead to surface charge separation for greater piezo response than that of thin films [19, 30]. According to literature reports, the piezoelectric constants of fiber based composites are about four times greater than those of controlled thin-film based ones [31]. Another advantage of the fiber format of organic piezoelectric materials is the potential of integration into textile industry to form a wearable platform, such as cloth, gloves, or other wearable items [32] (Figure 10.1).

Compared to organic piezoelectric materials, inorganic materials such as PZT [33] and ZnO [23], and $Ba(Zr_xTi_x)O_x$ (BZT) [34], also known as piezoelectric

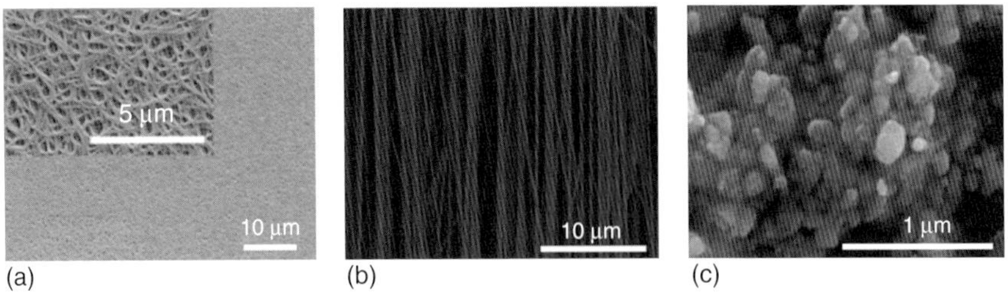


**Figure 10.1** Scanning electron microscope (SEM) of various piezoelectric materials and composites. (a) PVDF-TrFE thin film, (b) PVDF-TrFE fibers, (c) PZT composites. Source: (a) Sharma et al. 2012 [32]. Reproduced with permission of Elsevier; (b) Persano et al. 2013 [19]. Reproduced with permission of Springer Nature; (c) Chaipanich et al. 2007 [33]. Reproduced with permission of Elsevier.

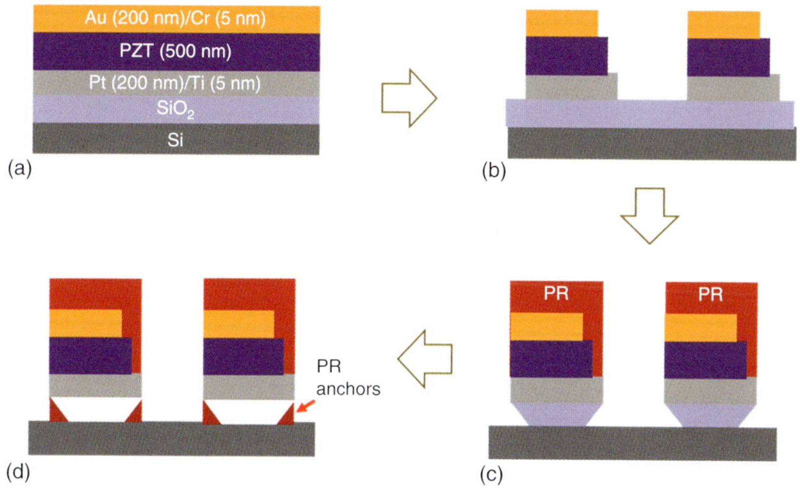


**Figure 10.2** Schematic demonstration of the processes for fabricating PZT nanomembranes used for transfer printing. (a) PZT wafer with structure Au/Cr/PZT/Pt/Ti/$SiO_2$/Si. (b) The PZT membranes were patterned and defined layer by layer through photolithography technique, (c) partially under cutting of the sacrificial layer. (d) Final PZT membranes formed by fully undercut etching of the sacrificial layer.Source: Yu et al. 2018 [7]. Reproduced with permission of Springer Nature.

ceramics, have attracted more attention, and proved to be the most successful perovskite metal oxides, because both the piezoelectric coefficients and constants of the inorganic piezoelectric materials are much better than those of organic materials. Despite the great performance that inorganic materials exhibit, the brittle and rigid nature of bulk inorganic ceramics limits their direct application on soft curvy surfaces. However, advanced processing technologies enable these rigid materials to form into sub-micrometer thin films, membranes, ribbons, nanowires, or hemisphere structures blending with polymers, which could significantly increase the flexibility or even stretchability [35, 36]. Overall, decreasing thicknesses in both active materials and substrates is the key to developing flexible electronics. As the champion piezoelectric material, PZT is the most used active layer in piezoelectric based devices (Figure 10.2).

The widely used processing technique for PZT is based on sol–gel chemistry, where lead, zirconium, and titanium precursors are mixed together and dissolved in solvent, spin-coated (or similar solution deposition methods) after hydrolysis to form a thin film, and then pre-baked to remove organic residuals [9, 37]. After repeating this step until achieving the desired film thickness, high-temperature annealing follows to facilitate the formation of crystal structures. In contrast with the solution processing deposition methods, physical vapor deposition, such as sputtering, is the low-temperature deposition route of piezoelectric ceramics [38]. However, the deposition durations and limited chamber spaces of those vacuum methods restrict the film thickness, which is typically less than 1 μm, and device areas [39]. On the other hand, the dielectric constants and coefficients of the vapor deposited films are typically lower than the solution processed ones [38]. Similarly to organic piezoelectric materials, inorganic composites can also be formed in formats of nanowire based fibers, processed by electrospinning. The resulting fiber based piezoelectric materials exhibit enhanced performance due to the highly oriented dipole separation [40]. Meanwhile, the fiber based piezoelectric ceramics also exhibit favorable flexibility under controlled mechanical strains [41].

Optimized mechanical design of the thin inorganic piezoelectric films as well as device architectures is extremely important for building the platform on a flexible substrate, such as plastics [35]. One very successful example is processing PZT thin films into small pieces of nanoribbon or nanomembrane and then transfer printing on a flexible substrate [42]. This process typically starts on a wafer with preformed high-performance piezoelectric thin film, for example, PZT, with a thickness of 100 nm [7]. The PZTs and electrodes with a capacitor architecture (Pt/PZT/Au) are coated on a Si wafer, with a thin oxide sacrificial layer, such as $SiO_2$ or MgO, in between the PZT and Si [43]. After a series of patterning and etching by photolithography techniques, undercut etching of the sacrificial layers enables the formation of PZT nanomembranes, which is also called "inks" for transfer printing. These "inks" can be picked up by silicone based stamps and transfer printed on to a receiving semi-cured polyimide (PI) as a flexible substrate. Then, the PI typically used as the top encapsulation layer for PZT is spin-coated and cured, followed by opening of the electrode contacts. Interconnects and wiring of the PZT membranes can be realized by lift-off, and then a final encapsulation layer will be deposited on top. The devices fabricated by this method exhibit the advantage of superior flexibility because of the reduced thickness and device area of the rigid PZT thin film [42].

## 10.3 Piezoelectric Devices for Biomedical Applications

Based on the nature of the energy exchange between electrics and mechanics in piezoelectric materials, piezoelectric devices show the possibility for diverse applications in various fields, ranging from microelectromechanical systems (MEMS), to mechanical energy harvesters, to robotics, to biomedical engineering [35]. The application of piezoelectric systems in biomedical devices has attracted great attention in recent years [44]. Latest studies show that PZT based electronics enable the measurement of cells mechanics, generating energy from the motions of the inner organs, monitoring blood pulse waves, and testing tissue elastomer modulus, etc. (Figure 10.3).

The use of piezoelectric devices as interfaces in energy harvesting associated with human body is the most straightforward application of biomedical engineering. Piezoelectric materials with high piezoelectric coefficients are usually preferred for use in energy-harvesting devices. Proper formats of the piezoelectric materials, such as ribbons and membranes, help enhance the flexibility and even stretchability. The electromechanical behavior of these flexible piezoelectric formats is basically designed and optimized by a shape buckling model [45]. The mechanical deformations of these ribbons or membranes by

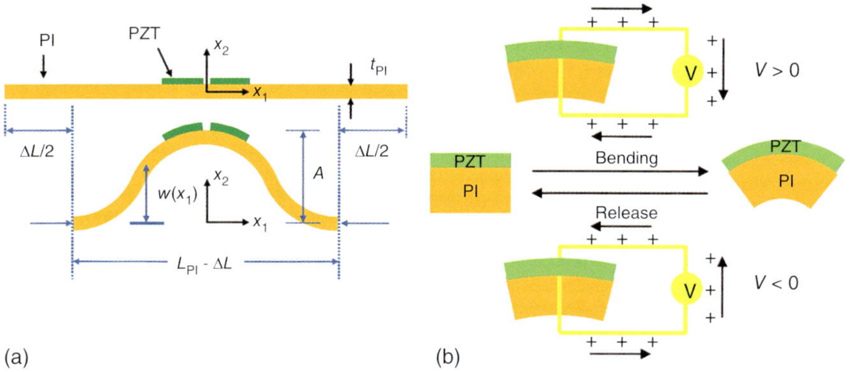


**Figure 10.3** Illustrations of coupling of the PZT nanomembrane under mechanical strains [35, 45]. (a) Demonstration of the shape deformation under bending. (b) Electric signal generation along with bending and release. Source: Adapted from Dagdeviren et al. 2016 [35] and Su et al. 2015 [45].

external mechanical forces yield an out-of-plane buckling, thus generating electrical power [46]. The output power of these piezoelectric films is usually proportional to the mechanical strain. In most cases, the area of the supporting substrate is larger than that of the piezoelectric materials, so the strain of the membranes or the ribbons can be calculated by the following equation:

$$\varepsilon_{\mathrm{m}} = 4\pi \frac{\overline{EI}_{\mathrm{PI}}}{\overline{EI}_{\mathrm{comp}}} \frac{h}{L} \sqrt{\frac{\Delta L}{L}}$$

where $\overline{EI}_{\mathrm{PI}}$ and $\overline{EI}_{\mathrm{comp}}$ are the plane-strain bending stiffness of the PI substrate without and with PZT, respectively, and $h$ is the distance from the midpoint through the thickness of the PZT to the neutral mechanical plane [47]. According to further calculations of the relations among applied stress, strain, and induced electrical field and displacement, the output voltage of this system can be defined as

$$\frac{\mathrm{d}V}{\mathrm{d}t} + \frac{N t_{\mathrm{pzt}}}{A_{\mathrm{pzt}} R \bar{k}} = -\frac{N \bar{e} t_{\mathrm{pzt}}}{\bar{k}} \frac{\mathrm{d}\varepsilon_{\mathrm{m}}}{\mathrm{d}t}$$

where $V$ is the measured output voltage, $R$ is the loaded resistance of the voltmeter, $N$ is the number of the ribbon groups electrically connected in series, $\bar{e}$ is the effective piezoelectric constant, $A_{\mathrm{pzt}}$ is the area of the PZT, $t_{\mathrm{pzt}}$ is the thickness of the PZT, and $\frac{\mathrm{d}\varepsilon_{\mathrm{m}}}{\mathrm{d}t}$ is the strain rate. It can be observed from this equation that the output voltage of the system is highly dependent on the strain of the PZT component.

Mechanical stress exists in every part of the human body during physical activities, due to both external body part motions and inner organ contractions. So, various kinds of piezoelectric devices with the capability of converting mechanical energy into electrical power are associated with biomedical applications. ZnO nanowires have been widely studied as piezoelectric components for energy harvesting [47]. Compared to the bulk or thin-film ZnO, ZnO nanowires exhibit a greater piezoelectric constant and thus can result in higher energy output values [48]. For instance, flexible energy harvesters using single ZnO nanowire have been reported with an output voltage up to 0.1 V when worn by small animals [49]. Flexible devices with ZnO nanowire arrays can further improve the energy-harvesting efficiency with voltage up to 2 V and current of 0.1 μA [50]. This micro-level power is sufficient to operate a single light-emitting diode (LED). Compared to ZnO, perovskite type piezoelectric materials such as PZT exhibit even greater piezoelectric properties. One representative example is an implantable flexible PZT energy-harvesting device, using a capacitor based PZT membrane as the piezoelectric component, which can be conformally laminated on the inner organ tissues, including the heart and lung of a cow, and generate power during the motions of these organs [42]. Thus, these devices are able to harvest energy without placing any excessive loads, which provides a new thought on power supply for implanted biomedical devices. To further improve the output power, multiple PZT flexible devices can be stacked in series, which can result in output voltage of ~4 V and a power density of 1.2 $\mu W/cm^2$.

Since piezoelectric materials have the ability of bidirectional signal conversion between mechanics and electrics, pressure or strain sensing is another promising area for the application of such materials. As is well known, pressure monitoring is a very important topic in biomedicine, because all the pressures associated with the human body, such as blood pressure and intracranial pressure, are highly clinically relevant. Continuous monitoring of blood pressure transients or pulse waves of arterial blood flow can provide extremely important information for physicians, who can use such information to improve the accuracy of diagnosis. Traditional cuff type sphygmomanometer is not practicable for continuous monitoring of blood pressures. Therefore, development of flexible wearable electronics that can be mounted on patients daily and monitoring of blood pressures in real time is an interesting but challenging topic. In 2014, PZT based devices in an epidermal electronics format were developed. In such devices, the pressure wave associated with blood flow appears as a mechanically induced strain on PZT membranes, and a field-effect transistor is integrated to amplify the signal, thus realizing real-time pulse wave monitoring [51]. However, more efforts, not only on the hardware part, but also on the models and algorithms are still needed for precise continuous monitoring of blood pressure. PZT can also be processed into nanocomposites, such as nanorods and nanoribbons, to apply for sub-micro or even nanoscale sensing. Attributed to the high sensitivity of PZT associated with mechanical stress, even very small deformations and deflections of a single cell at several nanometers range can still be detected by nanocomposite based PZTs. So, piezoelectric materials also exhibit much potential for the application of cell mechanics. One representative embodiment is the processing of PZT into nanoribbons format and integrating onto a flexible silicone substrate. This platform enables detecting the mechanical deformations of single PC12 cells via a standard whole-cell patch clamp method [52]. The output voltage response is up to 10 mV of a PZT nanoribbon sensor, indicating the capability of nanometer scale detection of cell deformations (Figure 10.4).

In addition to energy harvesting and sensing, mechanical actuation of piezoelectric devices is another very important function for the application

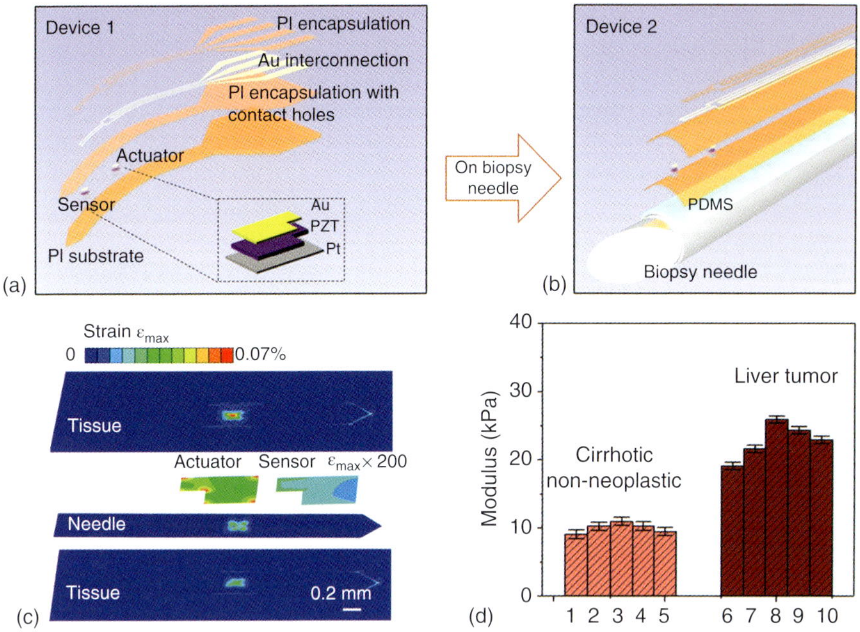


**Figure 10.4** Needle-shaped tissue modulus sensor based on membrane PZT actuators and sensors. (a) Expanded-view schematic illustrations of a flexible device. (b) An updated device based on device 1 (integrated on a biopsy needle). (c) Finite element analysis demonstration of the working process of modulus sensor when injected into a tissue. (d) Clinical measurement of liver tumor using this modulus sensor. Source: Yu et al. 2018 [7]. Reproduced with permission of Springer Nature.

of biomedical devices and instruments. One of the most related medical instruments that strongly depend on actuation is the mechanical assessment of biological tissue. The use of the mechanical properties of tissues, generally speaking, stiffness, for detecting tissue pathology can be traced back to 400 BCE. The development of many currently widely used bioinstruments, such as imaging tools, were strongly motivated by the elastographic difference between normal organ tissues and those with different pathologies [53]. The existing clinical tools are irritating, invasive, and have limited spatial resolution, and thus it is difficult to directly use these tools on patients for wearable continuous monitoring of skin health [54]. Combining the unique mechanical–electrical properties and advanced processing technologies of PZT mentioned in the previous Section 10.2, a lightweight piezoelectric device that enables the measurements of skin tissue viscoelasticity in the near-surface regions of the epidermis was reported in 2015 by Rogers and coworkers [43]. The thickness of these devices is in the 100 μm scale, and thereby can be conformally mounted on nearly every part of our skin, as well as the surface of other organ tissues. The reported device consists of seven actuators (lateral dimensions of $200 \times 1000\ \mu m^2$) and six sensors (lateral dimensions of $100 \times 500\ \mu m^2$), which are all based on 500 nm thick PZT nanomembranes, a low modulus silicone as a substrate, and serpentine Au wires as interconnects. Those epidermal electronic devices can be conformally attached on nearly every part of the human body and distinguish elastomer modulus difference between normal tissues and lesion areas.

The abovementioned epidermal platforms are designed for detection of skin lesions via modulus measurement of the shallow skin layers. However, the mechanical properties of the tissue of the inner organs typically attract the physician's attention more since elastic modulus, in particular, can serve as the basis for noninvasive medical diagnostics of soft tissues, especially the inner organs of the human body, such as the liver, kidney, lung, and brain. As normal tissue becomes afflicted with disease, such as chronic inflammation or malignancy, resultant changes in the modulus can be used to locate and identify pathology such as tumors, which exhibit abnormally high stiffness and density. In this way, modulus may be considered as a means for tumor detection. Existing characterization methods rely on bulk measurements of displacement (i.e. strain) as a function of applied force (i.e. stress) delivered with instruments that apply vacuum suction, tools that impart compressive forces, or fine tips that induce small-scale indentations, all typically ex vivo. The associated instruments often involve large characteristic dimensions, thereby preventing use in direct evaluation of organ tissues in actual patients. On the other hand, percutaneous biopsy of tumor tissue resulting in insufficient malignant cells can lead to misdiagnosis and delay of care. Ultrasound and computed tomography (CT) are helpful in guiding needles to the tumor tissue; however, accuracy and utility of these imaging tools can be limited especially when lesions are small. Despite best imaging practices and operator experience, high rates of inadequate tissue sampling of lesions <1 mm in the lungs and of lesions <3 cm in the abdomen are common. Very recently, we developed a clinical standard tool that adapts and extends the PZT actuators/sensors combination for characterizing the mechanical properties of the skin to a percutaneous biopsy needle based system for pathology tissue targeting [7]. This platform is able to clearly distinguish the in depth modulus difference of biology tissues, thus providing the foundations for modulus based tools capable of rapid tissue characterization with broad potential for clinical use. These studies establish the foundations for minimally invasive sensors for improving tissue targeting, where Young's modulus serves as the basis for guiding accurate specimen collection for histological and genetic testing.

Based on the processing, transfer printing, and integration techniques of the PZT nanomembranes into flexible and bio-electronics, we recently also developed three-dimensional mesostructures for MEMS applications, and demonstrated the capabilities for integrating functional high-performance ceramic PZT into complex 3D structures, thus creating opportunities in unusual classes of systems with active, programmable mechanical function [55]. This technology is compatible with well-established in-plane (2D) fabrication process, so diverse materials besides PZT, including PI encapsulation layer, metal interconnects, and electrodes, and SU8 epoxy substrate can be chosen as functional layers for the purposes of realizing and extending design options of 3D MEMS electronics. The overall results bring the full versatility of heterogeneous combinations of 2D planar technologies into the field of 3D MEMS. The implications extend beyond applications demonstrated here, to include biosensing, energy harvesting, and other possibilities.

## 10.4 Conclusion

In this chapter, we have introduced the piezoelectric materials and the basic technologies of processing of these materials into high-performance thin-film structure with a flexible format. In particular, it is demonstrated that the brittle but high-performance piezoelectric ceramic is difficult to adapt in flexible platform, while this issue can be resolved by mechanical design and novel processing techniques, such as transfer printing. Then, we discussed the applications of these piezoelectric materials in soft bio-integrated electronics for energy harvesting, biological related mechanical signals, i.e. blood pulse wave sensing, tumor diagnosis through detection of tissue mechanical properties, and active 3D MEMS devices actuations. Advances in materials science, mechanical engineering, and integration strategies mentioned in this chapter have provided the possibility for developing soft bio-integrated electronics in a board range of applications including clinical trials. We believe that such kind of electronics can create more opportunities for practical applications in biomedical engineering.

## References


1 Wong, W.S. and Salleo, A. (2009). *Flexible Electronics: Materials and Applications*, vol. 11. Springer Science & Business Media.
2 Rogers, J.A., Someya, T., and Huang, Y. (2010). Materials and mechanics for stretchable electronics. *Science* 327: 1603–1607. https://doi.org/10.1126/science.1182383.
3 Wang, S., Wang, S., Xu, J. et al. (2018). Skin electronics from scalable fabrication of an intrinsically stretchable transistor array. *Nature* 555: 83. https://doi.org/10.1038/nature25494, https://www.nature.com/articles/nature25494#supplementary-information.
4 Kim, D.-H., Lu, N., Ma, R. et al. (2011). Epidermal electronics. *Science* 333: 838–843. https://doi.org/10.1126/science.1206157.
5 Rogers, J.A., Bao, Z., Baldwin, K. et al. (2001). Paper-like electronic displays: large-area rubber-stamped plastic sheets of electronics and microencapsulated electrophoretic inks. *Proc. Natl. Acad. Sci.* 98: 4835–4840. https://doi.org/10.1073/pnas.091588098.
6 Gao, W., Emaminejad, S., Nyein, H.Y.Y. et al. (2016). Fully integrated wearable sensor arrays for multiplexed in situ perspiration analysis. *Nature* 529: 509. https://doi.org/10.1038/nature16521, https://www.nature.com/articles/nature16521#supplementary-information.
7 Yu, X., Wang, H., Ning, X. et al. (2018). Needle-shaped ultrathin piezoelectric microsystem for guided tissue targeting via mechanical sensing. *Nat. Biomed. Eng.* 2: 165–172. https://doi.org/10.1038/s41551-018-0201-6.
8 MacDonald, W.A., Looney, M.K., MacKerron, D. et al. (2007). Latest advances in substrates for flexible electronics. *J. Soc. Inf. Disp.* 15: 1075–1083. https://doi.org/10.1889/1.2825093.

9 Yu, X., Marks, T.J., and Facchetti, A. (2016). Metal oxides for optoelectronic applications. *Nat. Mater.* 15: 383. https://doi.org/10.1038/nmat4599.
10 Ray, T., Choi, J., Reeder, J. et al. (2019). Soft, skin-interfaced wearable systems for sports science and analytics. *Curr. Opin. Biomed. Eng.* https://doi.org/10.1016/j.cobme.2019.01.003.
11 Kong, X.Y. and Wang, Z.L. (2003). Spontaneous polarization-induced nanohelixes, nanosprings, and nanorings of piezoelectric nanobelts. *Nano Lett.* 3: 1625–1631. https://doi.org/10.1021/nl034463p.
12 Choi, M.-Y., Choi, D., Jin, M.J. et al. (2009). Mechanically powered transparent flexible charge-generating nanodevices with piezoelectric ZnO nanorods. *Adv. Mater.* 21: 2185–2189. https://doi.org/10.1002/adma.200803605.
13 Park, K.-I., Son, J.H., Hwang, G.T. et al. (2014). Highly-efficient, flexible piezoelectric PZT thin film nanogenerator on plastic substrates. *Adv. Mater.* 26: 2514–2520. https://doi.org/10.1002/adma.201305659.
14 Jaffe, B. (2012). *Piezoelectric Ceramics*, vol. 3. Elsevier.
15 Uchino, K. (1998). Piezoelectric ultrasonic motors: overview. *Smart Mater. Struct.* 7: 273.
16 Seminara, L., Capurro, M., Cirillo, P. et al. (2011). Electromechanical characterization of piezoelectric PVDF polymer films for tactile sensors in robotics applications. *Sens. Actuators, A* 169: 49–58. https://doi.org/10.1016/j.sna.2011.05.004.
17 Hadimani, R.L., Bayramol, D.V., Sion, N. et al. (2013). Continuous production of piezoelectric PVDF fibre for e-textile applications. *Smart Mater. Struct.* 22: 075017. https://doi.org/10.1088/0964-1726/22/7/075017.
18 Harris, G.R., Preston, R.C., and DeReggi, A.S. (2000). The impact of piezoelectric PVDF on medical ultrasound exposure measurements, standards, and regulations. *IEEE Trans. Ultrason. Ferroelectr. Freq. Control* 47: 1321–1335. https://doi.org/10.1109/58.883521.
19 Persano, L., Dagdeviren, C., Su, Y. et al. (2013). High performance piezoelectric devices based on aligned arrays of nanofibers of poly(vinylidenefluoride-co-trifluoroethylene). *Nat. Commun.* 4: 1633. https://doi.org/10.1038/ncomms2639. https://www.nature.com/articles/ncomms2639#supplementary-information.
20 Shirinov, A.V. and Schomburg, W.K. (2008). Pressure sensor from a PVDF film. *Sens. Actuators, A* 142: 48–55. https://doi.org/10.1016/j.sna.2007.04.002.
21 Hammer, M., Monty, C., Endriss, A., and Hoffmann, M.J. (1998). Correlation between surface texture and chemical composition in undoped, hard, and soft piezoelectric PZT ceramics. *J. Am. Ceram. Soc.* 81: 721–724. https://doi.org/10.1111/j.1151-2916.1998.tb02397.x.
22 Pulskamp, J.S., Polcawich, R.G., Rudy, R.Q. et al. (2012). Piezoelectric PZT MEMS technologies for small-scale robotics and RF applications. *MRS Bull.* 37: 1062–1070. https://doi.org/10.1557/mrs.2012.269.
23 Wang, Z.L. and Song, J. (2006). Piezoelectric nanogenerators based on zinc oxide nanowire arrays. *Science* 312: 242–246. https://doi.org/10.1126/science.1124005.

24 Gullapalli, H., Vemuru, V.S.M., Kumar, A. et al. (2010). Flexible piezoelectric ZnO–paper nanocomposite strain sensor. *Small* 6: 1641–1646. https://doi.org/10.1002/smll.201000254.

25 Tang, C.W. and VanSlyke, S.A. (1987). Organic electroluminescent diodes. *Appl. Phys. Lett.* 51: 913–915.

26 Tang, C.W. (1986). Two-layer organic photovoltaic cell. *Appl. Phys. Lett.* 48: 183–185.

27 Liao, C., Zhang, M., Yao, M.Y. et al. (2015). Flexible organic electronics in biology: materials and devices. *Adv. Mater.* 27: 7493–7527. https://doi.org/10.1002/adma.201402625.

28 Dimitrakopoulos, C.D. and Malenfant, P.R. (2002). Organic thin film transistors for large area electronics. *Adv. Mater.* 14: 99–117.

29 Klauk, H. (2006). *Organic Electronics: Materials, Manufacturing, and Applications.* Wiley.

30 Kim, H., Kim, S.M., Son, H. et al. (2012). Enhancement of piezoelectricity via electrostatic effects on a textile platform. *Energy Environ. Sci.* 5: 8932–8936. https://doi.org/10.1039/C2EE22744D.

31 Soin, N., Shah, T.H., Anand, S.C. et al. (2014). Novel "3-D spacer" all fibre piezoelectric textiles for energy harvesting applications. *Energy Environ. Sci.* 7: 1670–1679. https://doi.org/10.1039/C3EE43987A.

32 Sharma, T., Je, S.-S., Gill, B., and Zhang, J.X.J. (2012). Patterning piezoelectric thin film PVDF–TrFE based pressure sensor for catheter application. *Sens. Actuators, A* 177: 87–92. https://doi.org/10.1016/j.sna.2011.08.019.

33 Chaipanich, A., Jaitanong, N., and Tunkasiri, T. (2007). Fabrication and properties of PZT–ordinary Portland cement composites. *Mater. Lett.* 61: 5206–5208. https://doi.org/10.1016/j.matlet.2007.04.031.

34 Li, B., Blendell, J.E., and Bowman, K.J. (2011). Temperature-dependent poling behavior of lead-free BZT–BCT piezoelectrics. *J. Am. Ceram. Soc.* 94: 3192–3194. https://doi.org/10.1111/j.1551-2916.2011.04758.x.

35 Dagdeviren, C., Joe, P., Tuzman, O.L. et al. (2016). Recent progress in flexible and stretchable piezoelectric devices for mechanical energy harvesting, sensing and actuation. *Extreme Mech. Lett.* 9: 269–281. https://doi.org/10.1016/j.eml.2016.05.015.

36 Dagdeviren, C., Li, Z., and Wang, Z.L. (2017). Energy harvesting from the animal/human body for self-powered electronics. *Annu. Rev. Biomed. Eng.* 19: 85–108.

37 Budd, K.D., Dey, S., Payne, D. et al. (1985). *British Ceramic Proceedings.* Inst of Ceramics.

38 Pandey, S.K., James, A.R., Prakash, C. et al. (2004). Electrical properties of PZT thin films grown by sol–gel and PLD using a seed layer. *Mater. Sci. Eng., B* 112: 96–100. https://doi.org/10.1016/j.mseb.2004.06.011.

39 Zheng, X., Li, J., and Zhou, Y. (2004). X-ray diffraction measurement of residual stress in PZT thin films prepared by pulsed laser deposition. *Acta Mater.* 52: 3313–3322. https://doi.org/10.1016/j.actamat.2004.02.047.

40 Chen, X., Xu, S., Yao, N. et al. (2009). Potential measurement from a single lead ziroconate titanate nanofiber using a nanomanipulator. *Appl. Phys. Lett.* 94: 253113. https://doi.org/10.1063/1.3157837.

41 Qi, Y. and McAlpine, M.C. (2010). Nanotechnology-enabled flexible and biocompatible energy harvesting. *Energy Environ. Sci.* 3: 1275–1285.

42 Dagdeviren, C., Yanga, B.D., Su, Y. et al. (2014). Conformal piezoelectric energy harvesting and storage from motions of the heart, lung, and diaphragm. *Proc. Natl. Acad. Sci.* 111: 1927–1932. https://doi.org/10.1073/pnas.1317233111.

43 Dagdeviren, C., YanShi, Joe, P. et al. (2015). Conformal piezoelectric systems for clinical and experimental characterization of soft tissue biomechanics. *Nat. Mater.* 14: 728. https://doi.org/10.1038/nmat4289, https://www.nature.com/articles/nmat4289#supplementary-information.

44 Ciofani, G. and Menciassi, A. (2012). *Piezoelectric Nanomaterials for Biomedical Applications.* Springer.

45 Su, Y., Dagdeviren, C., and Li, R. (2015). Measured output voltages of piezoelectric devices depend on the resistance of voltmeter. *Adv. Funct. Mater.* 25: 5320–5325. https://doi.org/10.1002/adfm.201502280.

46 Dagdeviren, C., Hwang, S.W., Su, Y. et al. (2013). Transient, biocompatible electronics and energy harvesters based on ZnO. *Small* 9: 3398–3404. https://doi.org/10.1002/smll.201300146.

47 Yuan, J., Dagdeviren, C., Shi, Y. et al. (2016). Computational models for the determination of depth-dependent mechanical properties of skin with a soft flexible measurement device. *Proc. R. Soc. A* 472: 20160225.

48 Kang, M., Park, J.H., Lee, K.I. et al. (2015). Fully flexible and transparent piezoelectric touch sensors based on ZnO nanowires and $BaTiO_3$-added $SiO_2$ capping layers. *Phys. Status Solidi A* 212: 2005–2011. https://doi.org/10.1002/pssa.201431829.

49 Li, Z., Zhu, G., Yang, R. et al. (2010). Muscle-driven in vivo nanogenerator. *Adv. Mater.* 22: 2534–2537. https://doi.org/10.1002/adma.200904355.

50 Yang, R., Qin, Y., Li, C. et al. (2009). Converting biomechanical energy into electricity by a muscle-movement-driven nanogenerator. *Nano Lett.* 9: 1201–1205.

51 Dagdeviren, C., Su, Y., Joe, P. et al. (2014). Conformable amplified lead zirconate titanate sensors with enhanced piezoelectric response for cutaneous pressure monitoring. *Nat. Commun.* 5: 4496. https://doi.org/10.1038/ncomms5496. https://www.nature.com/articles/ncomms5496#supplementary-information.

52 Nguyen, T.D., Deshmukh, N., Nagarah, J.M. et al. (2012). Piezoelectric nanoribbons for monitoring cellular deformations. *Nat. Nanotechnol.* 7: 587. https://doi.org/10.1038/nnano.2012.112. https://www.nature.com/articles/nnano.2012.112#supplementary-information.

53 Park, E. and Maniatty, A.M. (2006). Shear modulus reconstruction in dynamic elastography: time harmonic case. *Phys. Med. Biol.* 51: 3697–3721. https://doi.org/10.1088/0031-9155/51/15/007.

54 Jenkyn, T.R., Ehman, R.L., and An, K.-N. (2003). Noninvasive muscle tension measurement using the novel technique of magnetic resonance elastography (MRE). *J. Biomech.* 36: 1917–1921.

55 Ning, X., Yu, X., Wang, H. et al. (2018). Mechanically active materials in three-dimensional mesostructures. *Sci. Adv.* 4: eaat8313. https://doi.org/10.1126/sciadv.aat8313.

# 11

# Flexible and Printed Electronics for Smart Clothes

*Yu Jiang and Nan Zhu*

*Zhang Dayu School of Chemistry, Dalian University of Technology, No.2 Linggong Road, Dalian, Liaoning 116024, China*

## 11.1 Introduction

Impressive progress has been achieved in recent years in flexible and printed electronics with a promise to provide miniaturized, low-cost, high-performance, good comfort wearable devices for acquiring real-time data for a broad range of applications. In this chapter, we aim to summarize the fields of printing technology and flexible substrates, and the application of flexible and printed electronics in electrochemical sensors/biosensors, biofuel cells (BFCs), supercapacitors, and batteries. In addition, challenges and opportunities, along with the developments in this research direction, are discussed.

## 11.2 Printing Technology

Currently, the frequently used technologies for fabricating flexible electronics are lithography (e-beam, ion-beam, and photolithography) and printing (non-template printing and template-based printing). Compared with lithography, which needs a clean-room and complex production process, printing methods have great superiorities such as low cost, good reproducibility, and easy procedures. In this section, we will provide a brief introduction about all kinds of printing methods and their application.

### 11.2.1 Non-template Printing

Non-template printing methods include inkjet and 3D printing. Because of the features of template-free methods, these printing technologies highly depend on printing inks, which can uniformly distribute onto the receiving substrates. Therefore, non-template printing methods involved a series of advanced technologies, such as piezoelectric, pneumatic, and electrohydrodynamic [1]. To

*Flexible and Wearable Electronics for Smart Clothing*, First Edition.
Edited by Gang Wang, Chengyi Hou, and Hongzhi Wang.

date, inkjet and 3D printing as high-precision fabrication methods have been widely used in many fields. For example, Ulrich S. Schubert and coworkers reported an inkjet printing method to fabricate narrow conductive tracks without predefined patterns by selecting transparent polyarylate substrate with moderate surface energy, which can further be used in flexible circuits and electrodes [2]. Anni Määttänen' team prepared three electrode systems (gold working and counter electrodes, Ag/AgCl reference electrode) on commercial paper substrate via well-established inkjet printing technology [3]. A potentiometric pH sensor with long lifetime was fabricated by further modifying the working electrode surface with polyaniline (PANI), whereas a highly sensitive glucose sensor was demonstrated by using working electrodes composed of glucose oxidase, poly-3,4-ethylenedioxythiophene (PEDOT) films, and gold layer. This work provided the example for paper-based electrochemical sensors by simple inkjet printing methods. When integrated with flexible printed circuit board (PCB), Bluetooth technology, this paper-based chip can ideally achieve the application for wearable electronics. To drive the innovation and development of lithium microbatteries, J. Le Bideau and coworkers reported the research on solid electrolyte with mesoporous silica host structure [4]. By inkjet printing the prepared ionogel precursor on top of half cells and coating ionogel solid electrolyte between positive and negative composite electrodes, the assembled full battery (carbon-coated lithiated iron phosphate [$LiFePO_4$] as the positive electrode and lithium titanate [$Li_4Ti_5O_{12}$] used for negative electrode) exhibited a high surface capacity (300 μAh/cm$^2$) after 100 cycles and would have the potential to fabricate all-solid-state printed microbatteries. Recently, a rapid development of inkjet-printed triboelectric or piezoelectric nanogenerators was observed for energy-harvesting devices due to the capability of converting mechanical energy to electrical energy. Taesung Kim and coworkers proposed flexible triboelectric nanogenerators (TENGs), which were a composite of high transmittance, low sheet resistance, high flexibility, and high power outputs [5]. The schematic of the partial fabrication process of TENGs was illustrated in Figure 11.1. Firstly, homemade poly(methyl methacrylate) (PMMA) inks were used as a sacrificial patterned layer to inkjet print on various substrates using a commercial inkjet printer and further modified Ag film to obtain Ag-based conductive electrodes with regular empty hole patterns on the corresponding substrate. Secondly, through the particle lithography method, a pyramid-patterned polydimethylsiloxane (PDMS) can be prepared as the interlayer for TENGs. Then, TENGs can be fabricated by using the as-prepared Ag films (as the top and bottom electrodes), patterned PDMS interlayer, and a polyethylene terephthalate (PET) space layer (to further separate the two triboelectric surfaces). These TENGs can not only efficiently power light-emitting diodes (LEDs) in the presence of external electrical resistance loads, but also power commercial electronic sensors and apply to smart window systems. Jong Hee Kim and coworkers designed a facile flexible piezoelectric generator via fully inkjet printing methods based on $BaTiO_3$ hybrid film and Ag electrode layers, which also can successfully harvest output voltage (7 V) and current (2.8 μA) from the repetitive deformations and be further developed for self-powered wearable electronics [6]. In the solid-state microsupercapacitors field, inkjet printing technology has been widely used.

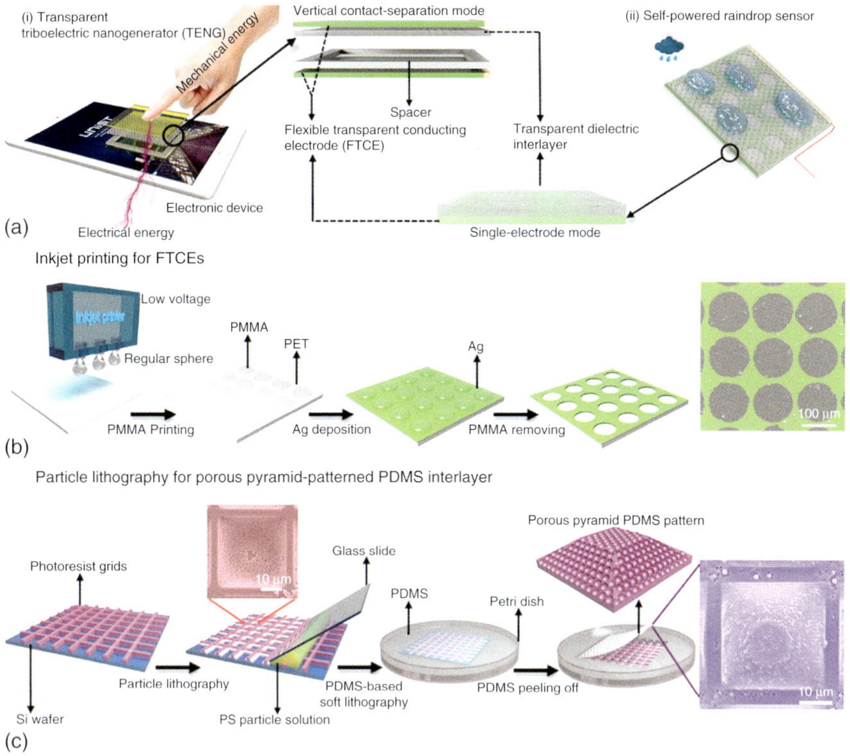


**Figure 11.1** Schematic of fabrication process for transparent triboelectric nanogenerator (a), transparent Ag film with regular empty hole patterns (b), porous pyramid-patterned PDMS interlayer (c). Source: Zhou et al. 2018 [5]. Copyright 2018, Reproduced with permission of Elsevier.

For example, Johan E. ten Elshof and coworkers fabricated a symmetrical microsupercapacitor by preparing optimized 2D $\delta$-$MnO_2$ nanosheet inks to print on polyimide (PI) film substrates, which showed excellent cycling stability (after 3600 charge–discharge tests, capacitance only reduced to 88%), high volumetric capacitance, high energy density, and good mechanical property [7]. The optimization methods of inks would provide new possibilities for other wearable electronics via inkjet printing to obtain patterns without "coffee-ring" effect. Although much progress has been achieved in flexible electronics based on inkjet printing technology, there are still many gaps such as improving the substrate versatility of inks, introducing more new functional materials (stretchable semiconductors, self-healing materials, biodegradable materials), and further improving the flexibility to develop wearable electronic skin.

In 2013, Jennifer A. Lewis and coworkers prepared a Li-ion battery via 3D printing technology [8]. Roughly speaking, there are three steps to fabricate this battery. First, $Li_4Ti_5O_{12}$ nanoparticle inks (anode) and $LiFePO_4$ nanoparticle inks (cathode) were prepared and printed on glass. Second, as-prepared electrode arrays were annealed by a tube furnace to get highly porous structures. Finally, a thin-walled PMMA and PDMS were used for packaging the battery, which was filled with $LiClO_4$ liquid electrolyte. By investigating the electrochemical performance of this 3D Li-ion microbattery with different layers, it was concluded that the height of the electrodes would have an effect on the kinetics of the reaction. Besides, both the $LiFePO_4$ or $Li_4Ti_5O_{12}$-based half-cell and a packaged battery exhibited a good cycle life and a high areal energy density. Chunzhong Li and coworkers devised a highly stretchable PDMS/carbon nanotubes/graphene conductor by taking advantage of 3D printed porous polylactic acid (O-PLA) as the template, which showed good electrical conductivity before and after 5000 bending cycles [9]. Moreover, 3D printing technology has also been applied to the fabrication of TENGs. Recently, Yunming Wang and coworkers explored the effects (such as filling rates, contact angles, excitation frequencies, and the width of the zigzag pattern) on the output voltage of 3D printing TENGs, which would enable the application of 3D printing technology in the energy field.

### 11.2.2 Template-Based Printing

Screen printing, imprinting, flexography, and gravure printing were called template-based printing. Because screen printing was the most widely used method for printed electronics among all of the template-based printing, we would highlight about screen-printing technology and give some representative examples.

Screen-printing technology includes two key components, pattern-based screen mesh and printing inks. For patterned screen mesh, the size of the mesh and the pattern (designed by Auto CAD software) are closely related to the thickness of the pattern, further influencing the mechanical property of printing electronics. It is necessary to use high viscosity ink for screen printing, which would facilitate penetration through the screen mesh to print on various substrates. Owing to the low cost, high efficiency, and simple fabrication process, screen-printing technology was successfully applied for printed batteries, sensors, as well as supercapacitors. For instance, Vivek Subramanian and coworker designed a primary battery by printing the silver current collector, Zn electrode (anode), a photopolymerized poly(acrylic acid) (PAA) separator, monovalent silver oxide ($Ag_2O$) as cathode, and silver flake/poly(ethylene oxide), respectively [10]. After encapsulation by lamination in polyethylene pouches, this printed battery showed high areal capacities (5.4 $mAh/cm^2$), volumetric capacities (7.1 $mAh/cm^3$), and excellent bending performance. Recently, by employing two functional screen-printing inks (Ag@ polypyrrole @$MnO_2$ as cathode and activated carbon as anode), a stretchable asymmetric all-printed supercapacitor was fabricated by Wei Wu' group [11]. Owing to the combination of as-prepared electrode materials with high pseudocapacitance (polypyrrole), high conductivity (activated carbon), and high theoretical capacitance ($MnO_2$), this supercapacitor exhibited capacity retention of 90.8% after 5000 discharge–charge cycles and capacity retention of 86.2% through 40% stretching strain. An electrochemical immunosensor for the detection of *Helicobacter pylori* based on screen-printed Au electrode was reported by Utkarsh Jain and coworkers [12]. To improve the electroconductivity, screen-printed Au electrodes were modified by low-energy ion beam irradiated ZnO tetrapods. The electrochemical measurements and biosensing studies revealed that this immunosensor had a large linear range and low limit of detection for *Helicobacter pylori*. This work would be further developed for other screen-printed electrochemical immunosensors.

In addition, water-assisted transfer printing method [13] and template-stripping nanotransfer printing [14] also have been used for fabricating flexible electronics. Significantly, by employing roll-to-roll gravure printing technology, multifunctional electrochemical sensors (pH sensor, potassium sensor, sodium sensor, glucose sensor, copper sensor, and caffeine sensor) for real-time and in situ sweat analysis during exercise were achieved recently [15]. This roll-to-roll gravure printing method not only can obtain large-area compatible electronics with low cost, but also provide faster printing speeds with high resolution, which was another choice for realizing industrialized production of noninvasive wearable biosensors, as well as wearable electronics.

## 11.3 Flexible Substrates

With the development of Internet-of-Things (IoTs), printed electronics have received more attention in combination with real-time data acquisition for various applications in environment analysis, health management, disease diagnosis, entertainment, and so on. However, conventional electronics are bulky and heavy, which cannot satisfy the requirement of wearable applications. Thus, flexible and printed electronics have huge market competitiveness due to their excellent properties, low cost, flexibility, comfort, and lightweight. Although flexible and printed electronics have various sensing mechanisms (physical, chemical, or biological), active material (metal, metallic oxide, polymer, carbon materials), and signal transformation (light, electricity, magnetism, heat), the flexible substrate can achieve universality and determine the flexibility by not directly acting on the sensing mechanism. Commercially available polymers, papers, tattoo papers, and fiber textile-based flexible platforms have long been employed in wearable devices. To provide an overview of the development actuality, on the basis of these flexible substrates, recent advances in flexible and printed electronics have been highlighted in this section.

### 11.3.1 Commercially Available Polymers

#### 11.3.1.1 Polyethylene Terephthalate (PET)

Owing to the advantages of high transmittance, low cost, moderate coefficient of thermal expansion, good chemical resistance, and moderate moisture absorption [16], PET has become the most popular flexible platform for a series of wearable electrochemical sensors. For example, in 2013 Craig E. Banks and coworkers fabricated screen-printed graphite sensors for the determination of antimony (III) by employing the PET substrate [17]. Joseph Wang's group successively reported noninvasive salivary lactate mouthguard biosensors [18], uric acid mouthguard biosensors [19], eyeglasses-based wireless sweat electrolyte, and metabolite sensors [20] that used PET as platform via screen-printing technology. In 2016,

Ali Javey and coworkers demonstrated a wearable multiplexed heavy metal (Zn, Cd, Pb, Cu, Hg) monitoring PET-based microsensor with high selectivity and repeatability through the photolithography method [21]. Ali Javey and coworkers also presented a fully integrated sensing system for real-time monitoring of $Ca^{2+}$, pH, and skin temperature, which was fabricated on a PET substrate and can ensure the accuracy of measurements by validating with conventional detection methods [22]. To achieve personalized health management, a fully integrated PET-based sensor for simultaneously and selectively detecting glucose, lactate, and sodium and potassium ions, and skin temperature was fabricated by Wei Gao et al., as shown in Figure 11.2 [23]. In 2017, Ronald W. Davis and coworkers devised a wearable PET-based platform through integrated iontophoresis electrodes, $Na^+$ and $Cl^-$ selective sensors, glucose sensors, and sensing circuits for sweat analysis [24]. Previous reports showed that the sweat chloride level is directly related to cystic fibrosis (CF) [25], and sweat glucose concentration can reflect blood glucose level [26]. Accordingly, this perspiration-based wearable system would be able to help noninvasive CF diagnosis and population health monitoring. PET was one of the oldest substrates for flexible electronics, which was not able to adapt well to the nonplanar features characteristic of human anatomy and would be possibly replaced by other alternative platforms with highly stretchable property in the future.

#### 11.3.1.2 Polydimethylsiloxane (PDMS)

PDMS, a silicon-based polymer, is a widely used substrate for flexible and printed electronics due to the advantages of large stretchability, tunable Young's modulus, chemical inertness, ease of processing, no toxicity, and high transparency (transmittance is up to 95%). By mixing Ag flakes and fluorine surfactant/water solution with organic solvents to prepare the elastic conductor inks for printing on PDMS substrates, a highly conductive film was fabricated by Takao Someya' group. This elastic conductor exhibited high initial conductivity (738 S/cm) and had considerable conductivity (182 S/cm) even after 215% stretching test [27]. For disease management of cystic fibrosis, a wearable PDMS-based chloride sweat sensor was devised by Peter C. Searson and coworkers [28]. This chloride sweat detection system was composed of a PDMS-based reference chamber, a salt bridge, a PDMS-based test chamber, reference and test electrodes. The detection mechanism involved making a cell potential directly relate to the ion concentration by using a salt bridge to build equilibration between the reference solution and the test solution. This design can achieve long-time accurate measurements and successfully apply to on-body tests. Yanlin Song and coworkers designed a nanocurves-based strain sensor by employing PDMS platforms through hybridization of pillar-patterned template induced printing, nanoparticle (NP) assembly technology, and vacuum thermal evaporation method [29]. This PDMS-based sensor can detect imperceptible changes in human facial expression by monitoring the resistance changes at four different orientations. Ying-Chih Liao and coworker fabricated flexible conductive tracks with good mechanical stability and high conductivity by screen printing Ag inks on pre-treated PDMS surfaces [30]. When further applied for pressure sensors, these PDMS-based conductive tracks exhibited fast response and excellent

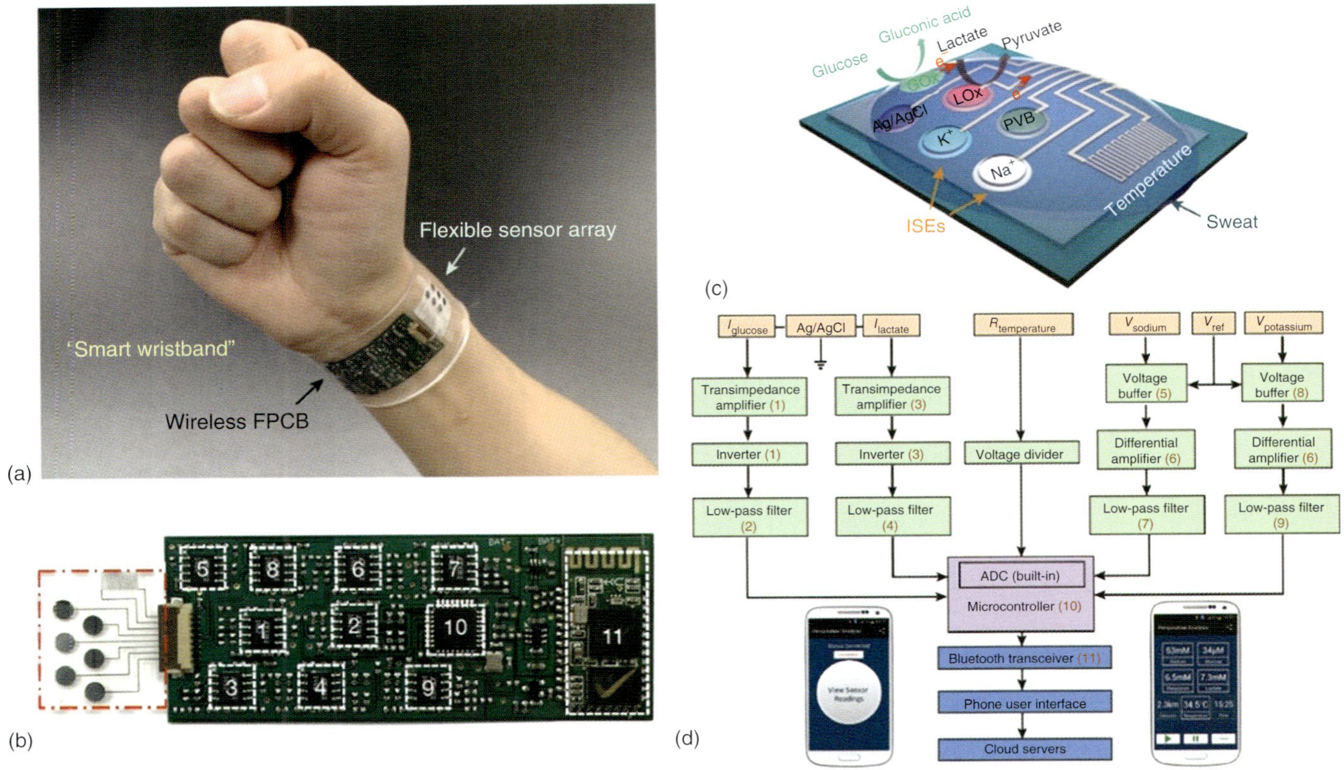


**Figure 11.2** Images and schematic illustrations of the flexible integrated sensing array for multiplexed perspiration analysis. (a,b) Digital images of a wearable integrated device. (c) Schematic of the sensor array for multiplexed perspiration analysis. (d) System-level block diagram of the flexible integrated device including the signal transduction, conditioning, processing, and wireless transmission modules. Source: Gao et al. 2016 [23]. Copyright 2016, Reproduced with permission of Springer Nature.

sensitivity. It boosts the development of conductive tracks for flexible pressure sensors, even all kinds of flexible electronics. Through the integration of a PDMS-based microfluidic device and a PDMS-based electrochemical biosensor, an epidermal detection system for real-time collecting and monitoring of sweat glucose or lactate was designed by Joseph Wang' group [31]. In addition, PDMS also can be mixed with other conductive materials or polymers to fabricate multifunctional flexible electronics with self-healable and highly stretchable performance. You Yu and coworkers prepared a self-healing PDMS film by bringing the reversibly dynamic imine bond (bis(3-aminopropyl)-terminated poly(dimethylsiloxane)) to PDMS network [32]. This as-prepared PDMS film can be fabricated as self-healed sensor for gas sensing with high stretchability (up to 700%) and self-healing performance at room temperature. Recently, Zhenan Bao' group devised an integrated self-healable PDMS-based electronic skin with self-bonding property and high stretchability by embedding the conductive carbon nanotube (CNT) network (Figure 11.3) [33]. This work would introduce new possibilities for self-healable artificial skin, and hence hold considerable promise for the development of wearable multifunctional electronics by taking PDMS as a self-healing platform.

#### 11.3.1.3 Polyimide (PI)

As a thermally stable polymer, polyimide (PI) has a high glass-transition temperature (360–410 °C) and good bendability, and was used for flexible electronics at high work temperatures. In 2008, Joseph Wang's group employed PI films to establish a miniaturized electrochemical norepinephrine and glucose biosensor, which provided examples for the development of minimally invasive biosensors to monitor biomarkers in the tear [34]. In addition, PI substrates were also used by Takao Someya's group to design flexible organic transistors and circuits, which showed excellent bending stability and flexibility due to the good bendability of PI [35]. Similarly, Takao Someya and coworkers designed heat-resistant and flexible organic thin-film transistors, which used an aluminum layer as the gate electrode and dinaphtho[2,3-b:2′,3′-f]-thieno[3,2-b]thiophene (DNTT) as the organic semiconductor; the aluminum oxide (AlOx) layer and an alkylphosphonic acid self-assembled monolayer (SAM) formed the gate dielectric, 50 nm-thick Au to define the source or drain [36]. This PI-based organic thin-film transistor also exhibited mechanical flexibility and high thermal stability. Shalini Prasad and coworkers demonstrated wearable biochemical Au sensors and ZnO sensors on PI substrates for monitoring alcohol consumption by detection of ethyl glucuronide (EtG) [37]. Based on the chemiresistive sensing mechanism, these Au sensors or ZnO sensors can measure the impedance (directly related to the concentration of EtG) when an AC voltage was applied. Both Au sensors and ZnO sensors were proved to have good stability, sensitivity ranges, and specificity, which can be imitated for other wearable biosensing. In addition, PI-based glucose biosensors [38], sweat pH sensors, and skin temperature sensors using PI as substrates [39] have also been reported. However, owing to the limitation of inherent color and not very good stretchability, PI cannot be used as a flexible platform for highly stretchable and transparent electronics.

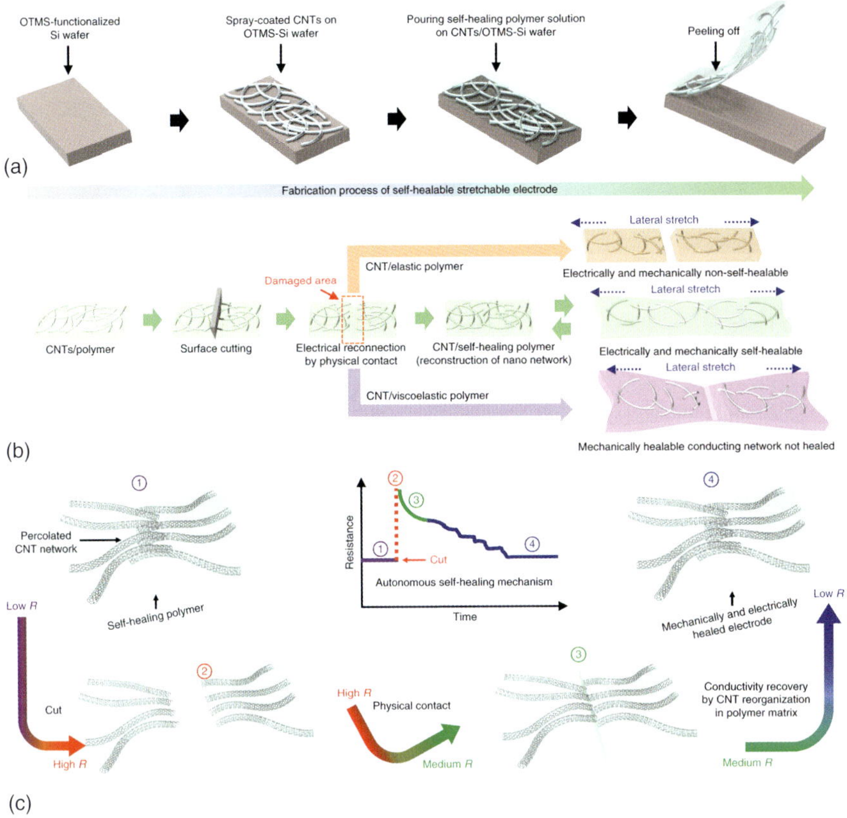


**Figure 11.3** Schematic drawings of dynamic reconstruction of conductive nanonetwork in the tough and stretchable self-healing polymer matrix. (a) The fabrication process for the self-healable stretchable electrode. (b) Sequential schematic illustration of different types of substrate and their response to damage and reconstruction. (c) Proposed recovery mechanism for the self-healing polymer matrix. Source: Son et al. 2018 [33]. Copyright 2018, Reproduced with permission of Springer Nature.

#### 11.3.1.4 Polyurethane (PU)

Polyurethane (PU) macromolecules include a lot of strong polar groups with flexible segment of polyether or polyester, which display high mechanical strength and oxidation stability, excellent flexibility and resilience, and good oil, solvent, and water resistance. Based on the aforementioned performance, PU would be a good choice for the platforms of wearable electronics. In 2013, C.P. Wong's group developed new PU-based electrically conductive adhesives by using poly(ethylene glycol) (PEG) as curing agents of the PU resin and silver flakes as fillers, which provided a simple method to fabricate conductive adhesives, as well as to design highly conductive and stretchable devices by combining with printing technology [40]. Commercial PU sponges can also be used as platforms to establish pressure-sensitive sensors due to their porous property. For example, Shu-Hong Yu and coworkers reported a graphene-wrapped PU sponge [41]. As shown in Figure 11.4, the fabrication process was followed by dip-coating graphene oxide (GO) to modify the pristine PU sponge, followed by hot hydrogen iodide (HI) solution to deal with GO-based sponge, hydrothermal treatment, and compression. The result showed that this sponge can be used as a highly sensitive pressure sensor with good cycling stability and long cycling life. In 2016, Joseph Wang and coworkers designed all-printing electrochemical ion-selective $NH_4^+$ sensors, glucose sensors, and glucose-based BFCs by employing PU as platform [42]. Both electrochemical sensors and BFCs exhibited noteworthy performance, such as stability after repeated 180° torsion, indentations, 300% stretching cycles, good linear relationships ($NH_4^+$ sensors, glucose sensors), and stable power density (BFCs), which would be applied to other sensing or energy-harvesting and storing fields. On the whole, based on the excellent property, not only PU films but also PU sponges have huge potential for the development of next-generation highly stretchable electronics.

#### 11.3.1.5 Others

Besides the abovementioned flexible substrates, other commercially available polymers, such as polyethylene naphthalate (PEN), poly(methylmethacrylate) (PMMA), and Ecoflex, have been applied in flexible electronics. In 2013, Takao Someya and coworkers presented a design for ultrathin tactile sensor arrays by vacuum evaporation and chemical vapor deposition methods, which composed of PEN foil (substrate), a hybrid anodic aluminum oxide-based circuit, a phosphonic acid SAM (gate dielectric), DNTT (organic semiconductor), and parylene (a separating layer between sensing layers and the active-matrix backplane) [43]. As illustrated in Figure 11.5, this ultrathin array can withstand repeated bending and stretching tests, even up to 230% of stretching. After being crumpled, this foil can still keep the completeness like a piece of paper. Besides, highly stretchable Ecoflex was used by Joseph Wang's group as another supporting layer of the substrate to further improve the stretching performance of electrochemical electronics [42, 44]. Jeong Sook Ha and coworkers also employed Ecoflex as the supporting layer for fabricating PET-based PANI nanofiber temperature sensor arrays with excellent stretchable property, which would have great potential to be used for electronic skin devices [45]. In addition, by using multilayer lamination methods, a pH barcode PMMA-based device was fabricated by Fernando Benito-Lopez's group [46]. This pH barcode can be encapsulated into an adhesive plaster and further integrated in the clothes or the wrist to detect pH during exercise.

### 11.3.2 Printing Papers

Papers, an environmentally friendly, recyclable, cheap, highly flexible platform, can be used as the medium for storage and transmission of information, and have the potential capability for developing low-cost wearable systems. In consequence, many paper-based devices, especially printed energy electronics, have been successfully demonstrated over the past decades [47]. Zhong Lin Wang and coworkers reported an all-solid-state flexible supercapacitor on A4 paper substrates, which can store energy coming from a piezoelectric generator [48]. Specifically, the A4 paper was in turn modified with polyvinyl alcohol (PVA), 80 nm thick Au films, and PANI to prepare PANI/Au/paper electrodes.

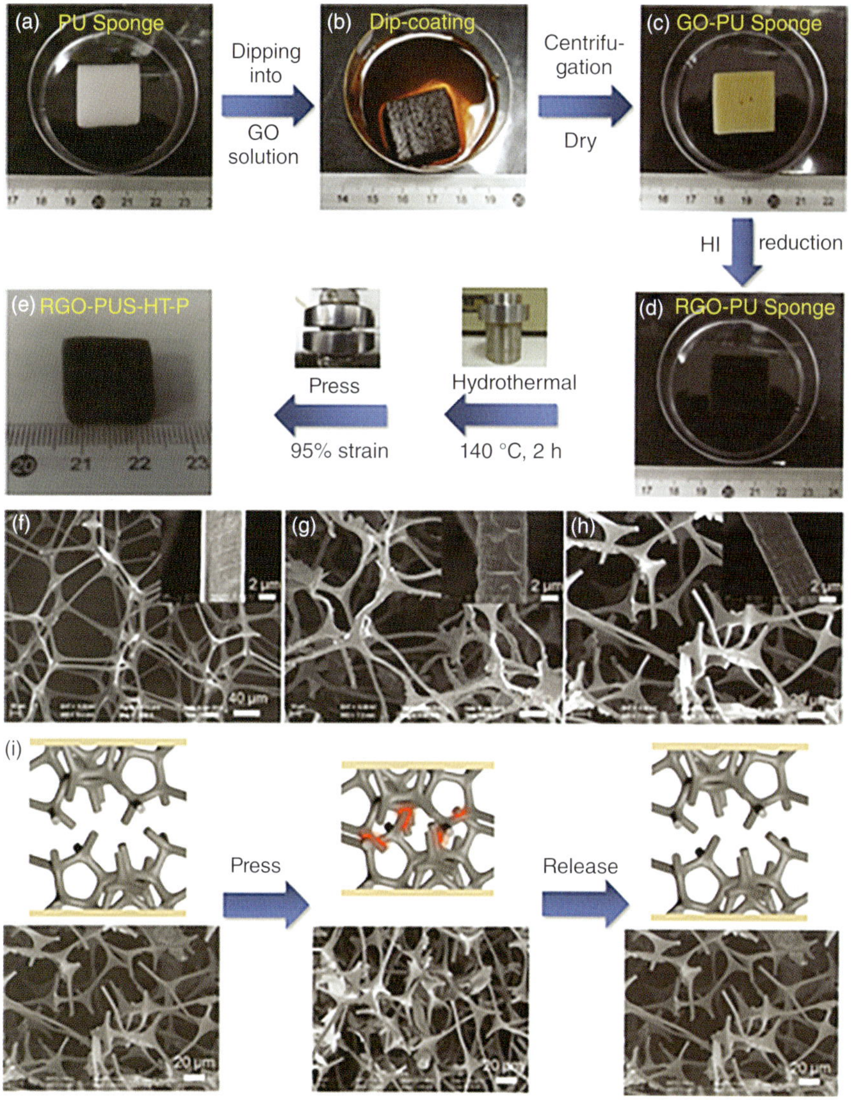


**Figure 11.4** Fabrication process, microstructures, and pressure-sensing models of graphene-wrapped PU sponges. (a–e) Schematic illustration of the fabrication procedure of fractured rGO–PU sponges. (f–h) SEM images of the rGO–PU sponge, hydrothermally treated rGO–PU sponge (rGO–PU–HT), and the compression treated rGO–PU–HT sponge (rGO–PU–HT–P). (i) Sensing mechanism of as-prepared pressure sensors. Source: Yao et al. 2013 [41]. Copyright 2013, Reproduced with permission of John Wiley & Sons.

Then, two as-prepared electrodes were immersed into $H_3PO_4$/PVA solutions to obtain these all-solid-state devices. In 2013, Seiya Tsujimura and coworkers designed a paper-based glucose BFC with high performance via well-established screen-printing technology to print porous carbon inks, which can output a power density up to 0.12 mW/cm$^2$ (at 0.4 V) [49]. To gain high resolution and

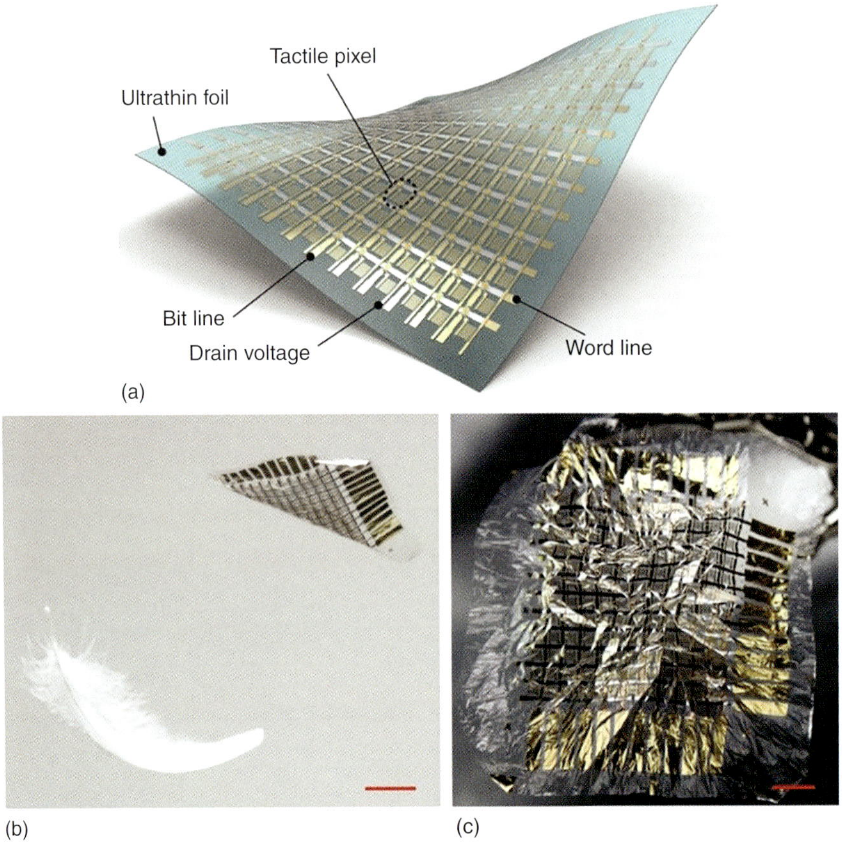


**Figure 11.5** Illustration of a thin large-area and lightweight tactile sensor. (a) Illustration of a thin large-area active-matrix sensor. (b, c) Digital images of ultrathin plastic electronic foils. Scale bar, 1 cm. Source: Kaltenbrunner et al. 2013 [43]. Copyright 2013, Reproduced with permission of Springer Nature.

high conductivity for wearable electronics, Jun Yang and coworkers presented a convenient and versatile fabrication method that combines inkjet technology with electroless deposition process to prepare palladium salt ink and copper-based patterns on flexible photopaper substrates [50]. Sang-Young Lee and coworkers reported all-inkjet-printed supercapacitors on pretreated A4 paper substrates, which displayed excellent cycling performance at constant charge/discharge current density and stability before and after 1000 bending cycles [51]. The highlights of this work were introducing cellulose nanofiber (CNF) as a primer layer on A4 paper to improve the electrical conductivity, printing home-made single-walled carbon nanotube (SWCNT)/active carbon (AC) ink and Ag nanowire (AgNW) ink as supercapacitor electrodes, preparing an ionic liquid (1-butyl-3-methylimidazolium [BMIM][BF4]) for inkjet printing, and UV irradiation to produce solid electrolytes. Overall, the fabrication process of these electric double-layer supercapacitors would provide some references for other flexible power sources, as well as wearable sensors. Similarly, a flexible asymmetric paper-based supercapacitor was reported by Shantanu Bhattacharya's group [52]. In this paper, the authors pre-prepared 40 layers of conducting GO ink on the commercial A4 paper to gain low-resistance substrates. The positive electrode was fabricated by using GO-$MnO_2$ nanocomposite ink for inkjet printing, whereas AC inks were printed as negative electrodes and poly(vinyl alcohol) (PVA)-LiCl was employed as the gel electrolyte. This supercapacitor exhibited noteworthy electrochemical performances (cycle stability, rate capability, energy density) that would promote the development of paper-based printing electronics.

### 11.3.3 Tattoo Papers

Tattoos papers were used as body art in the past years, until the tattoo-based electrochemical sensor was first reported in 2012 [53]. Since then, it brought new vitality for the development of wearable body-compliant electronics. Tattoos papers as platforms were gradually employed for various electrochemical sensors, biosensors, and energy-harvesting and storage devices. Tattoo-based electrochemical sensors were all successfully achieved via screen-printing technology and transfer process to fabricate electrode arrays for monitoring ammonium [54], epidermal pH [55], and sodium [56] in sweats. The sensing mechanism of the abovementioned three kinds of sensors takes advantage of the measurement of the potential (direct rely on the concentration of analytes) generated between the corresponding ion-selective electrode and reference electrode. In 2015, the Zn temporary tattoo sensor, which uses bismuth as working electrode and is based on the stripping-voltammetric determination technology, has been used for on-body noninvasive real-time detection [57]. In addition, tattoo biosensors for dynamic monitoring of sweat lactate [58], noninvasive glucose monitoring [59], and noninvasive monitoring of alcohol [60] were demonstrated, which was promising for other biosensing systems. It is noteworthy that both the glucose tattoo-based biosensor and alcohol tattoo-based biosensor combined iontophoresis with biosensing processes to collect sweat for improving sensing performance. In the microballistic delivery field, tattoo-based devices also have made progress. As shown in Figure 11.6, Joseph Wang and coworkers fabricated a tattoo-based epidermal patch with thousands of microcannons for delivery of the microdose (μD) and perfluorocarbon (PFC) emulsion [61]. Owing to the favorable mechanical properties, tattoos papers have also been used for energy harvesting and storage devices, such as epidermal BFC [62] and printable-tattoo alkaline batteries [63], which can meet the demands (high flexibility, directly covering with highly extensible skins, and tunable discharge capacity) of next-generation wearable power electronics.

### 11.3.4 Fiber Textiles

Fiber textiles have been used as clothes for many years, on account of durability, breathability, deformability, and launderability. Based on the aforementioned advantages, fiber textiles have become a potential alternative to other traditional substrates (such as silicon, glass, and ceramics) for new wearable electronics. For

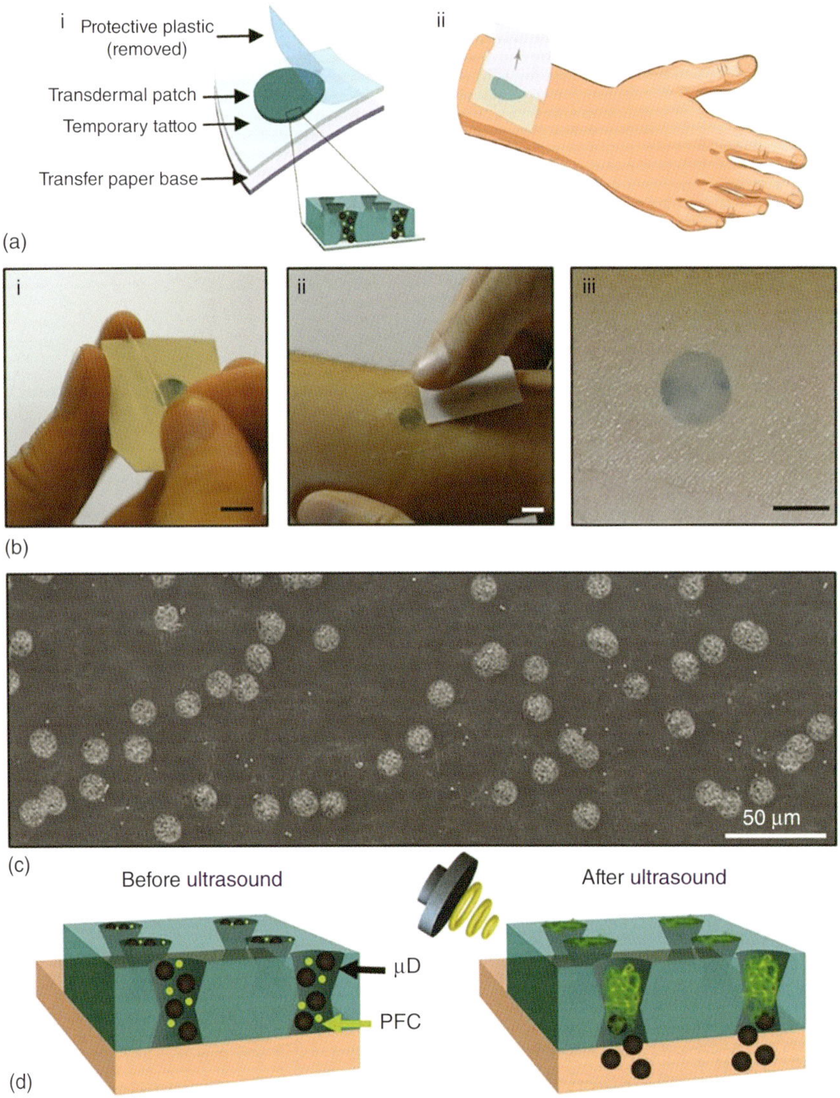


**Figure 11.6** Schematic, photographs, and mechanism of the tattoo-patch acoustic microballistic delivery. (a) Schematic and (b) photographs of the tattoo patch. (C) SEM showing a section of the top view of the loaded micropores. (d) Schematic of the microdose ejection mechanism. Source: Soto et al. 2017 [61]. Copyright 2017, Reproduced with permission of John Wiley & Sons.

example, Michael C. McAlpine and coworkers presented a fully bio-interfaced graphene-based sensor for detection of bacteria on tooth enamel, which would provide a versatile strategy for biochemical sensing by printing onto water-soluble silk [64]. Combined with CNT ink, commercial cotton yarns were also employed for fabricating various ion-selective electrodes to detect pH, $K^+$, and $NH^{4+}$, which confirmed the feasibility of cotton yarns-based potentiometric

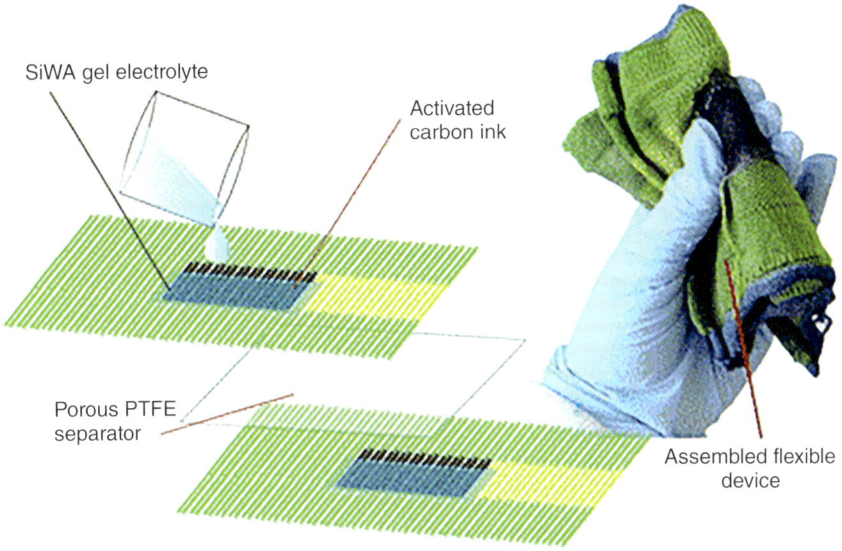


**Figure 11.7** Schematic of the fabrication process for assembled devices. Source: Jost et al. 2013 [68].Copyright 2013, Reproduced with permission of Royal Society of Chemistry.

sensors and offered demonstration for other electrochemical sensors and biosensors, as well as wearable energy electronics [65]. Christina Bini and coworkers designed a sweat analysis system, which is an integrated device with the ability to collect sweat, monitor physiological parameters (sweat rate, electrocardiogram, blood oximetry, and respiration), and analyze sweat composition (pH, $Na^+$, conductivity) by employing fabrics as platform [66]. Genevieve Dion and coworkers presented a universal fabrication method for fabric-based supercapacitor electrodes by combining dip-coating and screen-printing technology to guarantee a sufficient number of active materials impregnated into the fabric structure [67]. In 2013, knitted and screen-printed textile supercapacitors were designed by Genevieve Dion's group [68]. The schematic of the fabrication process is shown in Figure 11.7; activated carbon was screen-printed on knitted carbon fiber clothes and solid polymer electrolytes were modified on electrode arrays; polytetrafluoroethylene (PTFE) was used as a porous membrane separator to prevent short circuit between double electrodes. This textile supercapacitor exhibited high capacitance per area (0.5 $F/cm^2$) and excellent mechanical performance (stretching and bending). Hong Hu and coworkers designed polypyrrole @ carbon nanotubes @ urethane elastic fiber core spun yarn (PPy @ CNTs @ UV) electrodes by using poly(vinyl alcohol)/$H_3PO_4$ (PVA/$H_3PO_4$) as gel electrolyte for fabricating stretchable yarn-based supercapacitors. The electrical resistance of PPy @ CNTs @ UV composites displayed little change (less than 18%) after 100% stretching test, and the symmetric supercapacitors with sandwiched structure showed good capacitance behaviors [69]. In 2016, Xiangkang Meng's group reported all-solid-state wearable textile-based supercapacitors by using ternary metal sulfides' composite ($FeCo_2S_4$-$NiCo_2S_4$) modified on a silver-sputtered textile cloth [70]. Moreover, the fabrics have been employed in thermoelectric

(TE) energy devices with the aid of scalable screen-printing methods. For instance, Renkun Chen and coworkers reported a fabrics-based thermoelectric film with high electrical conductivity and low thermal conductivity through the optimization of the printing inks (including TE particles, binders, and organic solvents) [71]. Overall, as a flexible platform, fiber textiles have made great progress in wearable and printed electronics and would enhance their wide adoption by combining other fabrication technologies, synthesizing new materials, improving performance, and so on.

### 11.3.5 Others

Besides commercially available polymers, temporary tattoo papers, and fiber textiles, nitrile finger cots also have been used as flexible substrates for wearable and printed devices due to high stretchability, durability, as well as convenience of detection. To satisfy the need for on-site detection of gunshot residue and explosives, Joseph Wang and coworkers designed a solid-state forensic finger sensor [72]. In short, three electrodes were screen-printed on nitrile index finger cots and the solid-state electrolyte was coated on thumb cots. The index finger that came into contact with the test analysts lightly abraded the thumb, which would form an integrated sensor. Based on the rapid square wave voltammetry (SWV) technology, this finger cots sensor with a field-portable electrochemical analyzer exhibited good sensing performance (selectivity, stability, sensitivity). Similarly, in 2016, Karolien De Wael and coworkers reported a finger cots-based electrochemical sensor for fast on-site detection of cocaine by using pre-prepared baseline-corrected response curves of cocaine and cutting agents as comparison [73]. In 2017, glove-based biosensors were reported by Joseph Wang's group [74]. As shown in Figure 11.8, the serpentine patterns were printed on the index finger glove by using stretchable inks to obtain high stretchability, and the enzyme organophosphorus hydrolase (OPH) was immobilized on the carbon ink of the thumb as the collector of organophosphate (OP). By connecting with wireless-based portable potentiostat, this biosensor achieved real-time detection of OP. In addition, because of the combination of highly sensitive detection technology (SWV) and high-specificity biocatalysts (OPH), the new glove-based biosensor exhibited desirable selectivity for the detection of OP.

## 11.4 Application

With economic development and ageing populations, precision medicine and health detection have been receiving growing attention. Owing to the capability of continuously monitoring the health status (vital physical signs or chemical parameters) of different wearers, wearable smart electronics have become the hot point in scientific research as well. Wearable smart electronics not only involve physical information technology such as data communication, acquisition, and processing, but also include materials chemistry such as the synthesis of new sensing materials, analytical procedure, and power devices. The application of wearable smart electronics (especially flexible and printed electronics)

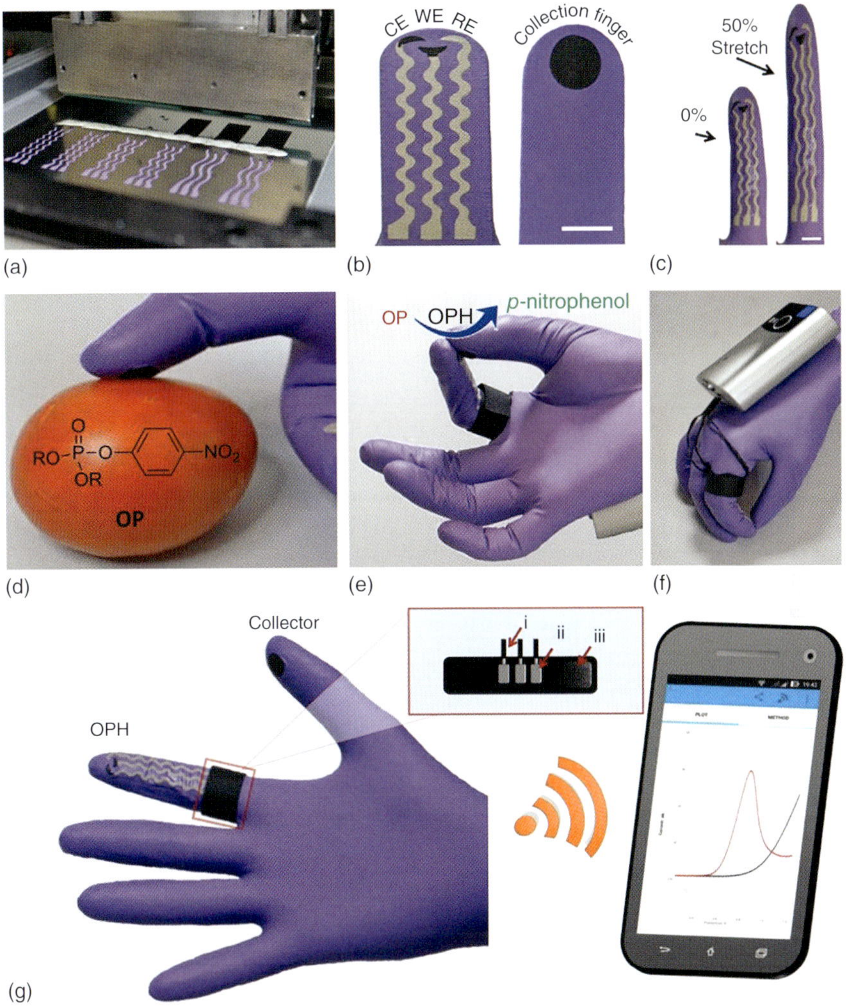


Figure 11.8 Flexible glove biosensor for detection of OP: fabrication, design, and performance. Source: Mishra et al. 2017 [74]. Copyright 2017, Reproduced with permission of Royal Society of Chemistry.

in wearable sensors/biosensors (focusing on non-electrochemical sensors, which have had little introduction in the previous Sections 11.2 and 11.3), BFC, supercapacitors, and batteries would be illustrated in this section.

### 11.4.1 Wearable Sensors/Biosensors

Organic electrochemical transistors (OECTs) (composed of organic semiconductor channel, source electrodes, drain electrodes, and gate electrodes) have an inherent amplification capability, which can sensitively convert biochemical signals into electronic signals. Therefore, combining OECTs with enzyme-based biosensors was an effective way of fabricating biosensors for different analytes with excellent sensitivity and selectivity. Based on the above information, Feng Yan and coworkers devised an OECTs-based biosensor on PET substrates with a different enzyme-based GO/PANI/Nafion-graphene/Pt gate electrode, which can be used for the detection of $H_2O_2$, uric acid, and glucose both in phosphate buffer solution (PBS) and saliva samples [75]. There are other highly selective wearable biosensors where the working electrode has been modified using various sensing materials. For example, Edilso Reguera and coworkers developed a biosensor for immunoglobulin M (IgM) antibodies by modifying screen-printed carbon electrodes with polydopamine (PDA) particles and lead sulfide (PbS) quantum dots; further tagging with anti-IgM antibodies, dengue-specific antigen, and streptavidin moieties exhibited good specificity, large detection range (0–0.5 mg/ml), and low limit of detection [76]. Recently, a novel screen-printed biosensor for detection of ethanol and nicotinamide adenine dinucleotide (NADH) was reported by Dalibor M. Stanković and coworkers [77]. On coating a graphene nanoribbon/ruthenium dioxide composite on the working electrode surface as the active component, this biosensor showed better response to ethanol and NADH in PBS at pH 7.5. Overall, the development of flexible electronics in the field of biosensors is based on both analytical methods and the synthesis of sensing materials.

A series of sensors that can be used for detection of the physical parameters also has been making great progress due to relatively simple sensing mechanisms, such as tactility, strain, pressure, and humidity sensors. For instance, in 2014, Xiaodong Chen's group designed a flexible resistance-based tactile sensor with ultrahigh sensitivity at low pressure range ($-5.5\,kPa^{-1}$, <100 Pa), fast response time (0.2 ms), and low limit of detection [78]. To be specific, graphene layers were deposited on a pyramid patterned PDMS film via layer-by-layer assembly methods at first. Secondly, hydrazine vapor was used as the reductant to get reduced graphene oxide (rGO). Finally, by assembling as-prepared rGO/PDMS film with ITO-coated PET film, a tactile sensor was successfully fabricated. After repeatedly applying and releasing pressure (100 Pa) 5000 times, this sensor revealed excellent stability (resistance only had little change). In 2015, Saleem Khan and co-workers reported two kinds of tactile sensors, piezoelectric-based (multiwall carbon nanotube [MWCNT]/poly(dimethyl-siloxane) [PDMS]) and piezoresistive-based (polyvinylidene fluoride-trifluoroethylene [PVDF-TrFE]) sensors, using a cost-effective screen-printing method [79]. A study of the response of the two kinds of sensors at different forces and frequencies revealed that piezoresistive-based sensors showed better response to static forces, whereas piezoelectric-based sensors showed good linear response to tactile sensing. In addition, by filling hollow elastomeric fibers with liquid metal (LM), a capacitive sensor was fabricated with large measurement ranges for torsion, strain, and touch sensing [80]. In 2017, Luyi Sun' group devised a moisture sensor with tunable dynamics [81]. The fabrication process was as follows: first, liquid PDMS was coated on polyvinyl alcohol (PVA) films and this as-prepared PDMS/PVA composite was peeled after curing. Second, the composite based on the PVA top thin film was cross-linked under a UV radiation chamber to get uniform films. With the formation of wrinkling surface, transmittance changed

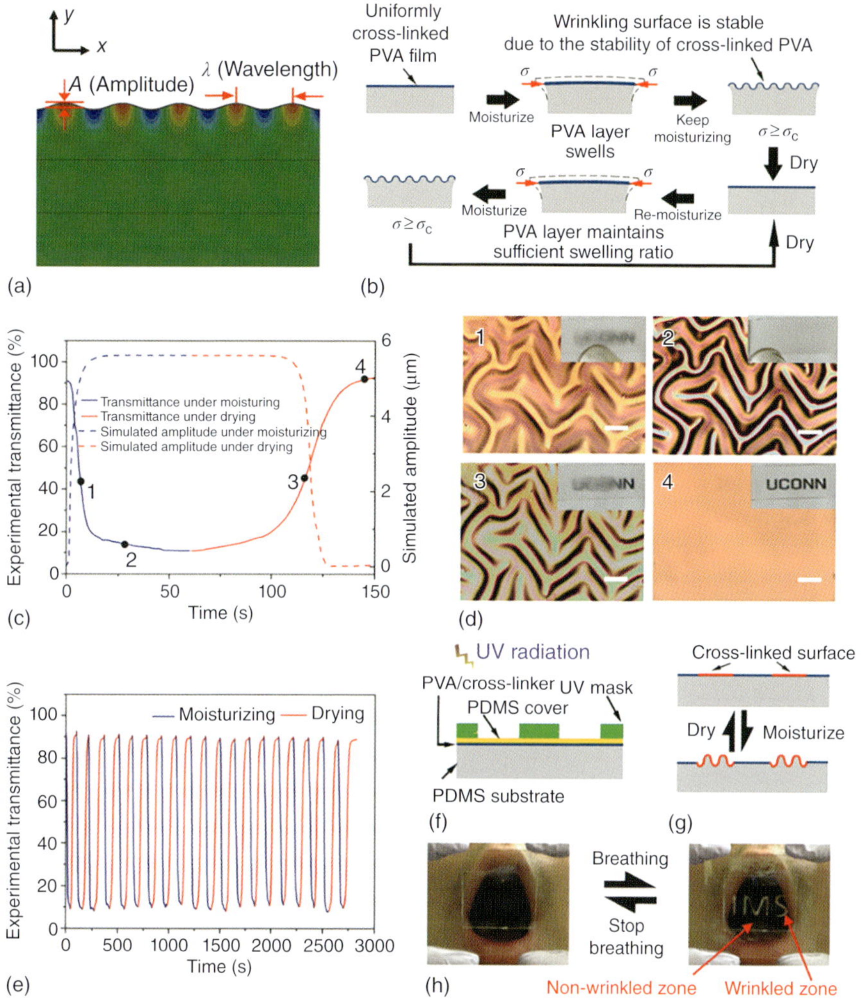


**Figure 11.9** (a) The first eigenmode of wrinkled shape obtained from linear buckling analysis; (b) schematic of the wrinkle dynamics; (c) evolution of the experimental UV–vis transmittance spectra and simulated wrinkle amplitude; (d) the relations between wrinkles formation and transparency change; (e) the UV–vis transmittance spectra; (f) the preparation approach for anticounterfeit tabs; (g) schematic of the responsive behavior of the anticounterfeit tab by creating wrinkles regionally; (h) the hidden pattern "IMS" can be reversibly revealed by breathing toward the anticounterfeit tab prepared based on sample A. Source: (a) Zeng et al. 2017 [81]. Copyright 2017, Reproduced with permission of John Wiley & Sons.

owing to light scattering of irregular surface (Figure 11.9). Consequently, according to UV–vis transmittance spectra under moisturizing or drying conditions, the wrinkles forming–disappearing dynamics can be expressly illustrated. Changhyun Pang' group designed a stretchable and skin-conformal sensor array with skin-like stretchability (<48%), high sensitivity (maximum gauge factor, GF ≈ 1697), and excellent bending performance by using assembled conductive graphene nanoplatelets to further transfer onto PDMS platforms [82]. As shown in Figure 11.10, this sensor can be used for monitoring spatial distribution and providing the amplification of E-blood vessel signals, which showed the potential for use as wearable strain sensors for ultrafast diagnosis of cardiovascular and cardiac illnesses.

### 11.4.2 Noninvasive Biofuel Cells

Noninvasive BFCs are a kind of energy-harvesting devices that can convert chemical energy (from metabolites present in sweat, tears or saliva) into electrical energy with the aid of biocatalytic redox enzymes (lactate oxidase (LOx), glucose oxidase (GOx)) [83]. Owing to widely sourced fuels and the potential for powering implantable devices, noninvasive BFCs have seen great development. In 2013, Joseph Wang's group first fabricated a lactate-based epidermal BFC via simple screen-printing technology on a temporary tattoos platform, and demonstrated the feasibility of harvesting energy from sweats [62]. The BFC was composed of anode "U" and cathode "C," and CNTs (for improving the conductivity) coated with tetrathiafulvalene (TTF) (electronic transfer mediator) and LOx/albumin layer (catalysts) and chitosan (for preventing the efflux of LOx) as anode electrocatalysts for the reaction of lactate oxidation; the cathode was prepared by immobilizing platinum black (catalysts) and Nafion layer (protective layer for skin) on the printed carbon electrode to catalyze the reduction of oxygen (Figure 11.11). In vitro studies indicated that this epidermal BFC can harvest energy from sweats, and the power density at various concentrations of lactate ranged from 5 to 70 $mW/cm^2$. Subsequently, BFCs-typed energy storage devices by integrating textile-based BFCs with PCB for powering electronic products [84], stretchable lactate-based BFCs by changing electrode array patterns and recipe of inks [44] were successively reported. At the same time, several studies about glucose BFCs also have shown great development. To obtain a high current density biocathode for glucose-based BFCs, S. Cosnier and coworkers prepared a freestanding redox buckypaper for the immobilization of horseradish peroxidase (HRP) and GOx as biocathodes, which exhibited excellent stability over weeks, as well as good electrocatalytic properties for oxygen reduction [85]. In 2016, Joseph Wang and coworkers designed a highly stretchable CNTs-based glucose BFC via fully printed methods, with the aid of the optimum serpentine structure (for improving the stretchable performance) and recipe of inks (stretchable CNT and Ag/AgCl inks) [42]. In 2017, to further improve the stretchable performance and power density, an island–bridge architecture skin-based BFC was fabricated by Joseph Wang's group [86]. The superiorities of island–bridge architecture were not only embodied in hexagonal close-packed structure, which can reduce internal resistance, but also showed in the circular geometry to mitigate non-form strain. Lithography was used for fabricating the patterned Au islands, and a layer of carbon ink was screen-printed onto the islands to bond the active materials. The active materials of anode were comprised of carbon nanotube-naphthoquinone (CNT-NQ), LOx, and chitosan, whereas CNT-$Ag_2O$ and Nafion were modified on the cathode islands. As shown in Figure 11.12, this skin-based BFC can generate stable power even after repeated stretching test,

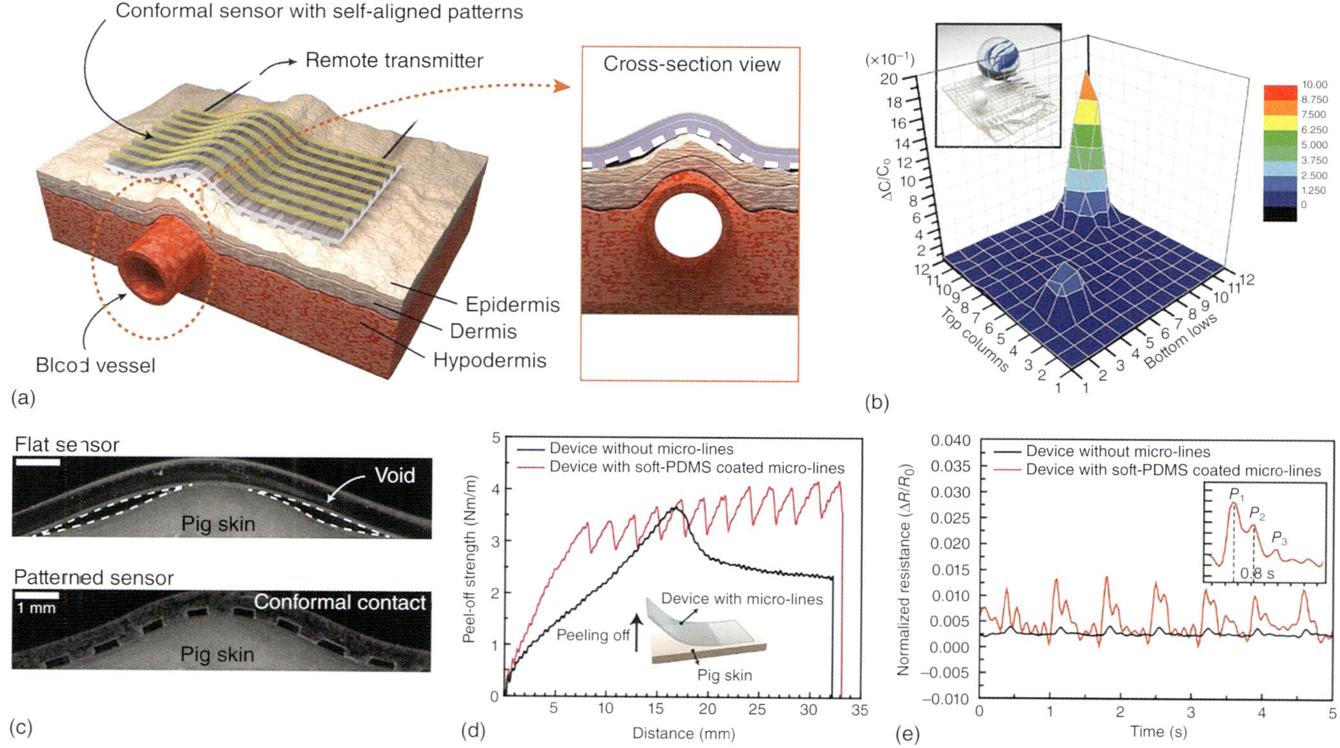


**Figure 11.10** Highly skin-conformal sensor array simply fabricated with graphene nanoplatelet network conductive patterns. (a) Schematic illustration of a sensor array to monitor biosignals. (b) The resulting intensity profile measured by mapping of the capacitive signals (144 pixels, $2 \times 2\ cm^2$). (c) Detailed cross-sectional images, (d) peel-off strengths of sensors, and e) pulse waves of the radial artery of sensors having flat geometry and microlines. (f) (i) Schematic illustration of biocompatible and waterproof GNN-embedded E-blood vessel sealed with a thin PDMS and (ii) change in resistance from monitoring electric signals of E-blood vessel applied by stepwise strains (<6.2%). (g) (i) Electric signals monitoring tiny strain-stimuli of tapping and grapping E-balloon and (ii) expansion or contraction of E-balloon by cool and hot streams (70 °C). Source: Park et al. 2017 [82]. Copyright 2017, Reproduced with permission of John Wiley & Sons.

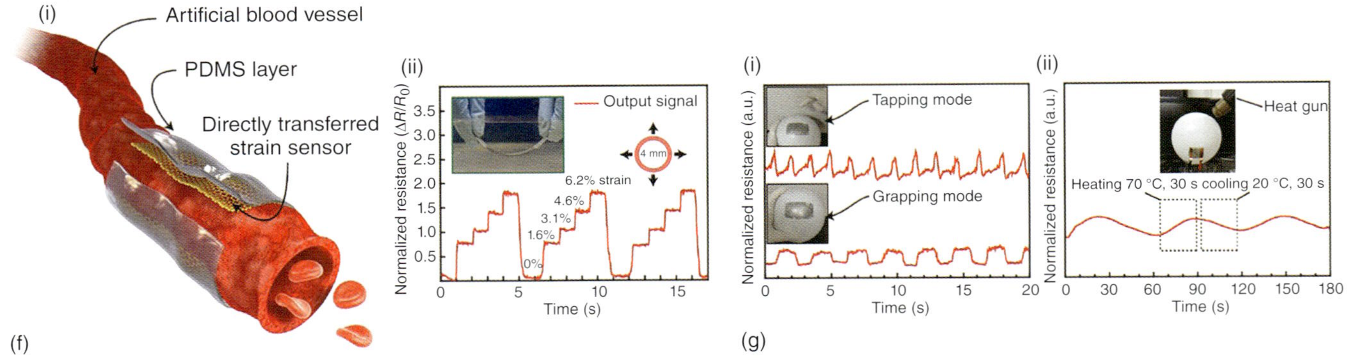


Figure 11.10 (*Continued*)

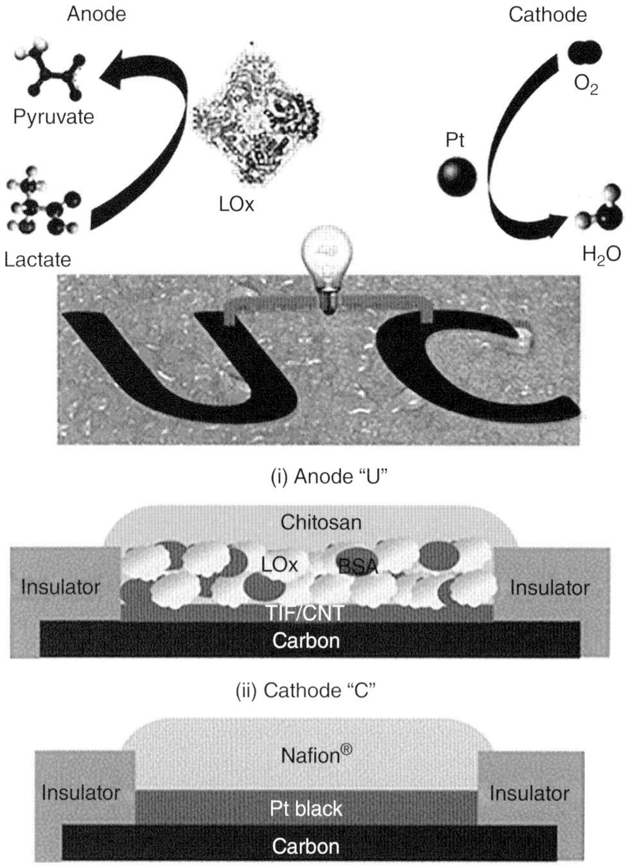


**Figure 11.11** Illustration of the epidermal BFCs and the constituents of the anode (i) and cathode (ii). Source: Jia et al. 2013 [62]. Copyright 2013, Reproduced with permission of John Wiley & Sons.

and the power increased linearly with increasing coverage from 20% to 100%. Recently, alcohol edible BFCs based on food materials were reported, which provided a new direction for developing smart self-powered biosensors [87].

### 11.4.3 Wearable Energy Storage Devices

Ziqiang Shao and coworkers selected polyaniline (PANI), poly(3,4-ethylenedioxythiophene: poly(styrene sulfonate)) (PEDOT:PSS), and graphene oxide (GO) as the active materials, and employed cellulose nanofibers (CNFs) paper as the substrate to fabricate two kinds of supercapacitors (one was prepared by CNFs/[PANI-GO]*n* called CPG-*n*, and another used CNFs/[PANI-PEDOT:PSS]*n*, CPP-*n* for short; *n* is the number of bilayers) by using layer-by-layer assembly methods [88], as illustrated in Figure 11.13. Results of a comparison of electrochemical performances between two kinds of supercapacitors demonstrated that S-PG-8 (5.86 $mF/cm^2$) showed higher

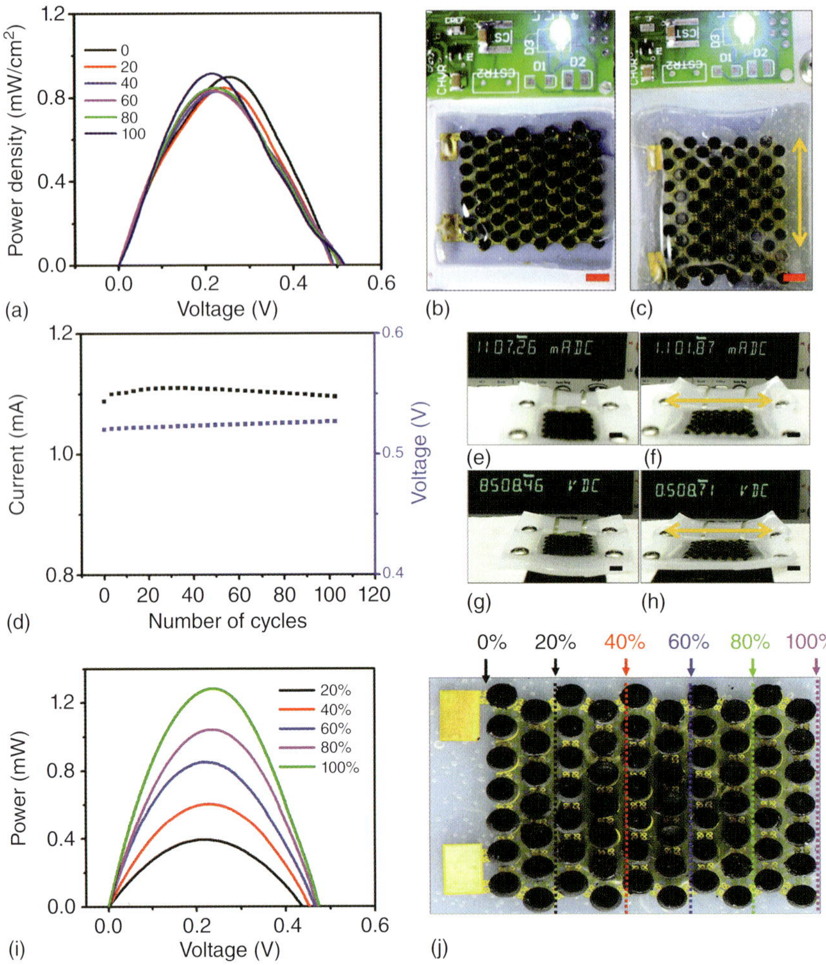


**Figure 11.12** Mechanical resilience studies and effect of fuel coverage on the E-BFC. (a) Power density versus voltage curves for E-BFC after repeated stretching iterations (e = 50%). Image of a blue LED powered by E-BFC when subjected to (b) e = 0% and (b) e = 50% strain. (d) Effect of repeated e = 50% strains on the short-circuit current and open-circuit voltage. Images showing the effect of strain on the (e and f) short-circuit current and (g and h) open-circuit voltage. (i) Effect of increasing the areal coverage of E-BFC by fuel on the power output. (j) Image of the E-BFC illustrating the extent of spatial coverage by fuel during the study conducted in (i). (a–j) 14 mM lactic acid was used as fuel. (b, c and e–h) Scale bar, 5 mm. Source: Bandodkar et al. 2017 [86]. Copyright 2017, Reproduced with permission of Royal Society of Chemistry.

area capacitance than S-PP-8 (4.22 $mF/cm^2$) at the same current density of 0.0043 $mA/cm^2$, as well as better cycling stability. Although this work was irrelevant to printing electronics, it offered another simple method to apply for flexible energy devices and can be used as the modification method for flexible and printing electronics. Owing to the high resolution and facile fabrication process, inkjet printing was widely used for printing supercapacitors. In 2016, Sang-Young

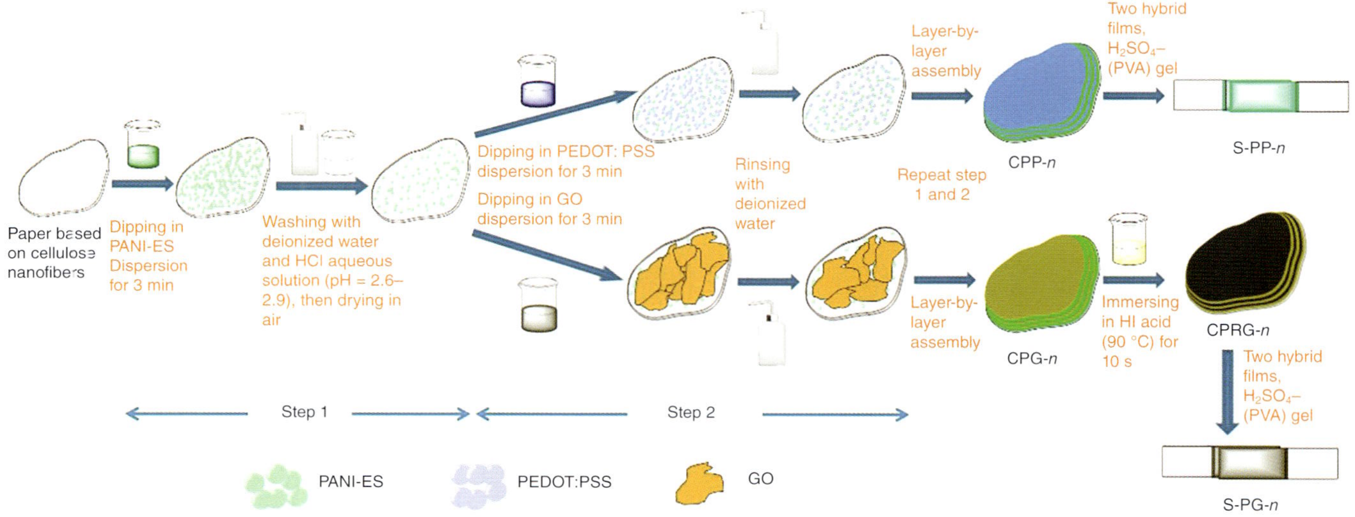


**Figure 11.13** Schematic illustration of the layer-by-layer assembly processes used to prepare the CPP-*n*, CPG-*n* multilayer films, and the preparation of the flexible supercapacitors. Source: Wang et al. 2014 [88]. Copyright 2014, Reproduced with permission of Elsevier.

Lee and coworkers proposed a full-solid-state supercapacitor employing conventional A4 paper as platform via inkjet-printing technology [51]. At first, CNF was printed onto A4 paper as a primer layer. Secondly, SWCNTs/AC ink and AgNWs were in turn printed on the CNF-based paper and further treated with UV irradiation to get highly interwelded AgNWs networks. Then, ionic liquid-based gel polymer electrolyte was inkjet-printed. Finally, a polypropylene-based adhesive film was used for sealing to produce a complete supercapacitor. As shown in Figure 11.14, the cyclic voltammetry curve of this supercapacitor was rectangular in shape. After 10 000 discharge–charge tests, capacitances had little change at a constant current density (0.2 $mA/cm^2$). Through 1000 repeated bending cycles, this all-solid-state supercapacitor displayed excellent bending stability. In addition, by connecting five cells in series or in parallel, this supercapacitor can show fivefold increase in voltage (in series) or capacitance (in parallel) than the mono cell. This printed supercapacitor not only can drive an LED lamp but also provide power for LED lamp even at high temperature (150 °C) conditions due to the high thermal stability of gel polymer electrolyte. In 2017, C. Daniel Frisbie' group devised a symmetrical graphene-based supercapacitor with the aid of photolithography (to fabricate Si master mold) and inkjet-printing methods (for preparing electrode arrays) [89]. The fabrication process of this symmetrical supercapacitor included the fabrication of Si master mold, PDMS stamp and imprinted substrate, the preparation of graphene inks, and the fabrication of supercapacitor. It is noteworthy that the graphene ink needed mixing of ethyl cellulose (EC) as the stabilizer with pristine graphene, and the electrolyte ink required poly(styrene-*b*-methylmethacrylate*b*-styrene) (PS-PMMA-PS), [EMIM][TFSI] ionic liquid (EMD Chemicals), and *n*-butyl acetate at a mixing ratio of 2 : 8 : 90 (w/w/w). By using these homemade graphene ink and electrolyte ink, this fully solid-state supercapacitor demonstrated excellent resolution, high specific capacitance, cycling stability, and bending performance, which established a facile approach for efficient fabrication of large-area energy devices with high resolution. Zhiyong Fan and coworkers also presented a nanocoral-structured supercapacitor with outstanding performance via inkjet-printing methods [90]. At first, Ag nanoparticle colloidal ink was inkjet-printed on PET. Then, Au layer, Ni nanocoral layer, and pseudocapacitive $MnO_2$ nanoflakes were covered on Ag-based PET via electrochemical deposition methods, respectively. Finally, lithium acetate/poly (vinyl acetate) gel was modified onto the top of as-prepared interdigitated electrodes. Significantly, by electrochemical deposition, Ni@$MnO_2$ hierarchical nanocoral structure was formed on the surface of electrodes, which would help improve pseudocapacitive performance. As illustrated in Figure 11.15, after 25 cycles of $MnO_2$ deposition, this supercapacitor showed the highest areal capacitance. The cyclic voltammetry (CV) profiles displayed quasi-rectangular shape and galvanostatic charging–discharging measurement and showed a high coulombic efficiency (up to 80%). In addition, this device also exhibited good rate capability and cycling stability.

Moreover, Ching-ping Wong and coworkers fabricated an asymmetric supercapacitor (Ni/$MnO_2$-based positive electrode, negative electrode was prepared with Ni/active carbon, PVA-$H_2SO_4$ as the solid electrolyte) by employing

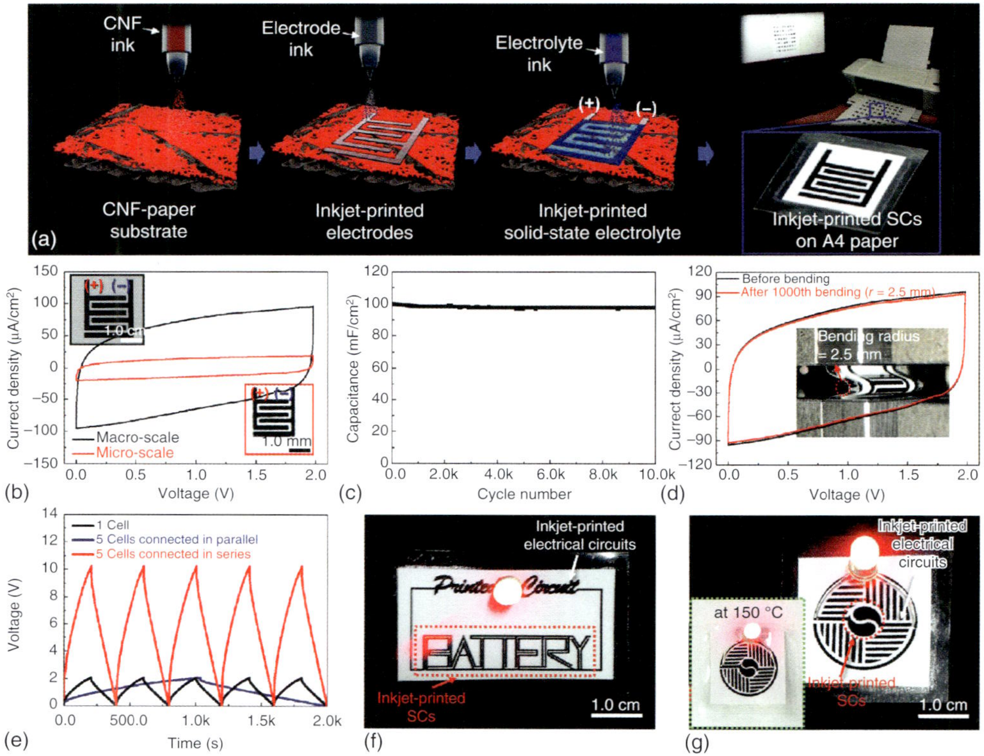


**Figure 11.14** Fabrication and characterization of the inkjet-printed Supercapacitors (SCs) with various form factors. (a) Schematic illustration of the stepwise fabrication procedure of the inkjet-printed SC. (b) CV profiles of the inkjet-printed SCs with different dimensions. (c) Cycling performance of the inkjet-printed SC at a constant charge/discharge current density. (d) CV profiles of the inkjet-printed SC after repeated bending deformation. (e) Galvanostatic charge/discharge profiles of the inkjet-printed SCs connected in series or parallel without the use of metallic interconnects. Source: Choi et al. 2016 [51]. Copyright 2016. Reproduced with permission of the Royal Society of Chemistry.

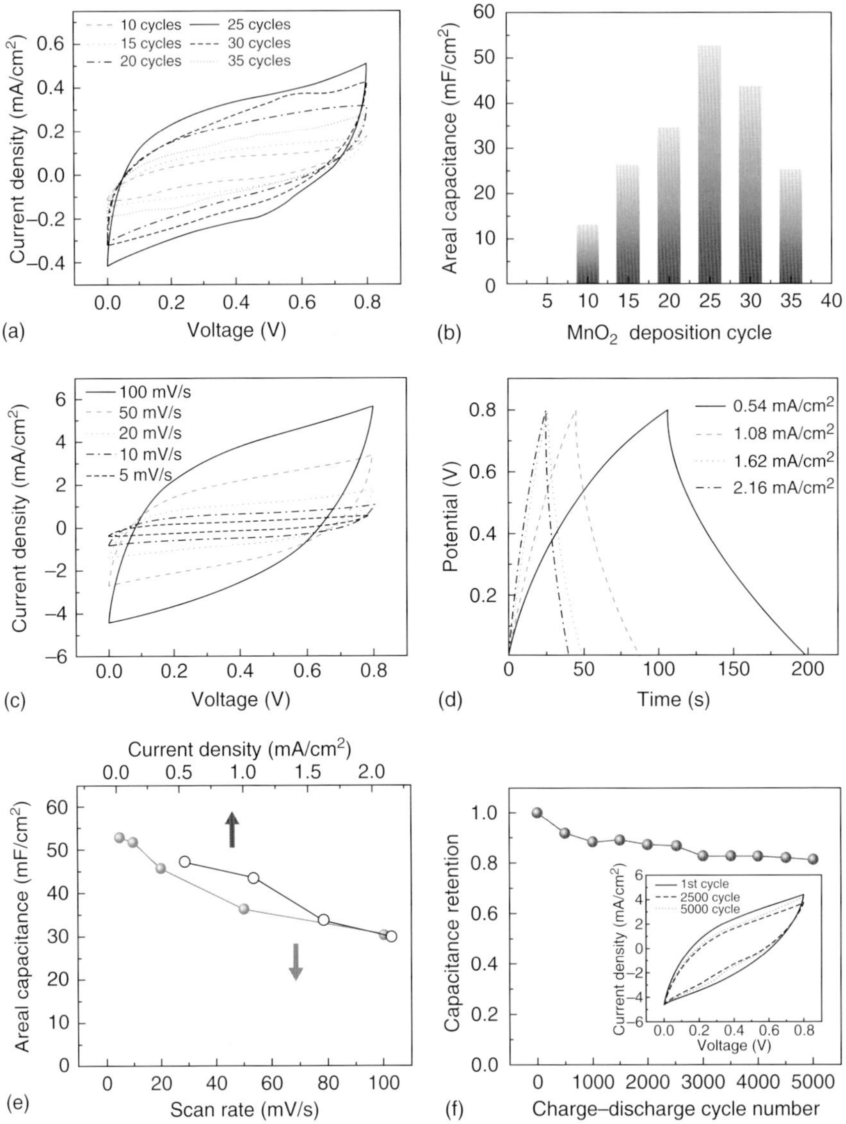


**Figure 11.15** Electrochemical measurements of Ni@$MnO_2$ nanocoral-structured planar supercapacitors. (a) CV curves of devices with different numbers of $MnO_2$ deposition cycles recorded at a constant scan rate. (b) Areal capacitance of devices with different numbers of $MnO_2$ deposition cycles. (c) CV curves of the device with optimized $MnO_2$ mass loading (25 deposition cycles) recorded at different scan rates. (d) GCD curves of the device with optimized $MnO_2$ mass loading at different discharge currents. (e) Areal capacitances of device with optimized $MnO_2$ mass loading calculated from CV and GCD measurements, respectively. (f) The cycling stability test of the device at a scan rate of 100 mV/s. Inset: CV curves at the first cycle, after 2500 and 5000 cycles. Source: Lin et al. 2017 [90]. Copyright 2017, Reproduced with permission of John Wiley & Sons.

laboratory filter paper by combining electroless plating and electrodeposition methods, which also displayed outstanding bending performance, large volume energy density, and high operating voltage [91]. Sac-like structured polypyrrole nanowires (PPy NWs), micro-supercapacitors on fluorine doped tin oxide (FTO) coated glass substrate, and Ag@PPy nanocomposites fully printed supercapacitors via screen-printing technology were all successfully reported [92]. Printed primary batteries based on $Ag_2O$ and Zn electrodes [93], printable solid-state lithium-ion batteries through facile stencil printing and ultraviolet cross-linking methods [94], 3D printed fiber-based Li-ion batteries [95], and printed batteries using Ni as current collector, graphite anode, and lithium cobalt oxide as cathode material [96] were successfully fabricated, which have potential for flexible electronics to serve as power source by further optimizing the synthesis of materials or the mechanical behavior of devices.

## 11.5 Prospects

In this chapter, we have summarized the fields of printing technology and flexible substrate, and the application of flexible and printed electronics in electrochemical sensors/biosensors, BFC, supercapacitors, and batteries. The rapid development of information technology (such as IoTs), printing technology, and new materials science has opened up new opportunities for flexible and printed electronics. Despite significant progress in these flexible and printing electronics, there are many key challenges to address before realizing the full potential of wearable sensors, noninvasive BFCs, as well as energy storage devices. First, we believe that with wearable sensors more attention should be paid to reduce the limit of detection of biomarkers in body fluids and establish the relationship with disease management. Second, noninvasive BFCs need to further improve current density for combining with other wearable sensors to indeed accomplish self-powered devices. Third, energy storage electronics need to better integrate with energy-harvesting devices and achieve the complete harvest–storage–power process for driving wearable sensors, not only LED lumps. Fourth, various printing technologies, lithographic methods, electroless deposition, as well as vacuum filtration should be flexibly used to bring their individual advantage for fabrication of flexible electronics.

## References

1 Kim, J., Kumar, R., Bandodkar, A.J., and Wang, J. (2017). *Adv. Electron. Mater.* 3: 1600260.
2 van Osch, T.H.J., Perelaer, J., de Laat, A.W.M., and Schubert, U.S. (2008). *Adv. Mater.* 20: 343.
3 Määttänen, A., Vanamo, U., Ihalainen, P. et al. (2013). *Sens. Actuators, B* 177: 153.
4 Delannoy, P.E., Riou, B., Lestriez, B. et al. (2015). *J. Power Sources* 274: 1085.

5 Zhou, Q., Park, J.G., Kim, K.N. et al. (2018). *Nano Energy* 48: 471.
6 Lim, J., Jung, H., Baek, C. et al. (2017). *Nano Energy* 41: 337.
7 Wang, Y., Zhang, Y.-Z., Dubbink, D., and ten Elshof, J.E. (2018). *Nano Energy* 49: 481.
8 Sun, K., Wei, T.S., Ahn, B.Y. et al. (2013). *Adv. Mater.* 25: 4539.
9 Duan, S., Yang, K., Wang, Z. et al. (2016). *ACS Appl. Mater. Interfaces* 8: 2187.
10 Braam, K. and Subramanian, V. (2015). *Adv. Mater.* 27: 689.
11 Liu, L., Tian, Q., Yao, W. et al. (2018). *J. Power Sources* 397: 59.
12 Chauhan, N., Gupta, S., Avasthi, D.K. et al. (2018). *ACS Appl. Mater. Interfaces* 10: 30631.
13 Lee, C.H., Kim, D.R., and Zheng, X. (2011). *Nano Lett.* 11: 3435.
14 Tiefenauer, R.F., Tybrandt, K., Aramesh, M., and Voros, J. (2018). *ACS Nano* 12: 2514.
15 Bariya, M., Shahpar, Z., Park, H. et al. (2018). *ACS Nano* 12: 6978.
16 Nathan, A., Ahnood, A., Cole, M.T. et al. (2012). *Proc. IEEE* 100: 1486.
17 Kolliopoulos, A.V., Metters, J.P., and Banks, C.E. (2013). *Anal. Methods* 5: 3490.
18 Kim, J., Valdes-Ramirez, G., Bandodkar, A.J. et al. (2014). *Analyst* 139: 1632.
19 Kim, J., Imani, S., de Araujo, W.R. et al. (2015). *Biosens. Bioelectron.* 74: 1061.
20 Sempionatto, J.R., Nakagawa, T., Pavinatto, A. et al. (2017). *Lab Chip* 17: 1834.
21 Gao, W., Nyein, H.Y.Y., Shahpar, Z. et al. (2016). *ACS Sensors* 1: 866.
22 Nyein, H.Y., Gao, W., Shahpar, Z. et al. (2016). *ACS Nano* 10: 7216.
23 Gao, W., Emaminejad, S., Nyein, H.Y.Y. et al. (2016). *Nature* 529: 509.
24 Emaminejad, S., Gao, W., Wu, E. et al. (2017). *Proc. Natl. Acad. Sci. U.S.A.* 114: 4625.
25 Gonzalo-Ruiz, J., Mas, R., de Haro, C. et al. (2009). *Biosens. Bioelectron.* 24: 1788.
26 Moyer, J., Wilson, D., Finkelshtein, I. et al. (2012). *Diabetes Technol. Ther.* 14: 398.
27 Matsuhisa, N., Kaltenbrunner, M., Yokota, T. et al. (2015). *Nat. Commun.* 6: 7461.
28 Choi, D.H., Kim, J.S., Cutting, G.R., and Searson, P.C. (2016). *Anal. Chem.* 88: 12241.
29 Su, M., Li, F., Chen, S. et al. (2016). *Adv. Mater.* 28: 1369.
30 Li, C.Y. and Liao, Y.C. (2016). *ACS Appl. Mater. Interfaces* 8: 11868.
31 Martin, A., Kim, J., Kurniawan, J.F. et al. (2017). *ACS Sensors* 2: 1860.
32 Zhang, B., Zhang, P., Zhang, H. et al. (2017). *Macromol. Rapid Commun.*: 38, 1700110.
33 Son, D., Kang, J., Vardoulis, O. et al. (2018). *Nat. Nanotechnol.* 13: 1057.
34 Kagie, A., Bishop, D.K., Burdick, J. et al. (2008). *Electroanalysis* 20: 1610.
35 Sekitani, T., Zschieschang, U., Klauk, H., and Someya, T. (2010). *Nat. Mater.* 9: 1015.
36 Kuribara, K., Wang, H., Uchiyama, N. et al. (2012). *Nat. Commun.* 3: 723.
37 Selvam, A.P., Muthukumar, S., Kamakoti, V., and Prasad, S. (2016). *Sci. Rep.* 6: 23111.

38 Munje, R.D., Muthukumar, S., and Prasad, S. (2017). *Sens. Actuators, B* 238: 482.

39 Nakata, S., Arie, T., Akita, S., and Takei, K. (2017). *ACS Sensors* 2: 443.

40 Li, Z., Zhang, R., Moon, K.-S. et al. (2013). *Adv. Funct. Mater.* 23: 1459.

41 Yao, H.B., Ge, J., Wang, C.F. et al. (2013). *Adv. Mater.* 25: 6692.

42 Bandodkar, A.J., Jeerapan, I., You, J.M. et al. (2016). *Nano Lett.* 16: 721.

43 Kaltenbrunner, M., Sekitani, T., Reeder, J. et al. (2013). *Nature* 499: 458.

44 Jeerapan, I., Sempionatto, J.R., Pavinatto, A. et al. (2016). *J. Mater. Chem. A* 4: 18342.

45 Hong, S.Y., Lee, Y.H., Park, H. et al. (2016). *Adv. Mater.* 28: 930.

46 Curto, V.F., Fay, C., Coyle, S. et al. (2012). *Sens. Actuators, B* 171–172: 1327.

47 Tobjork, D. and Osterbacka, R. (2011). *Adv. Mater.* 23: 1935.

48 Yuan, L., Xiao, X., Ding, T. et al. (2012). *Angew. Chem.* 51: 4934.

49 Shitanda, I., Kato, S., Hoshi, Y. et al. (2013). *Chem. Commun.* 49: 11110.

50 Zhang, T., Wang, X., Li, T. et al. (2014). *J. Mater. Chem. C* 2: 286.

51 Choi, K.-H., Yoo, J., Lee, C.K., and Lee, S.-Y. (2016). *Energy Environ. Sci.* 9: 2812.

52 Sundriyal, P. and Bhattacharya, S. (2017). *ACS Appl. Mater. Interfaces* 9: 38507.

53 Windmiller, J.R., Bandodkar, A.J., Valdes-Ramirez, G. et al. (2012). *Chem. Commun.* 48: 6794.

54 Guinovart, T., Bandodkar, A.J., Windmiller, J.R. et al. (2013). *Analyst* 138: 7031.

55 Bandodkar, A.J., Hung, V.W., Jia, W. et al. (2013). *Analyst* 138: 123.

56 Bandodkar, A.J., Molinnus, D., Mirza, O. et al. (2014). *Biosens. Bioelectron.* 54: 603.

57 Kim, J., de Araujo, W.R., Samek, I.A. et al. (2015). *Electrochem. Commun.* 51: 41.

58 Jia, W., Bandodkar, A.J., Valdes-Ramirez, G. et al. (2013). *Anal. Chem.* 85: 6553.

59 Bandodkar, A.J., Jia, W., Yardimci, C. et al. (2015). *Anal. Chem.* 87: 394.

60 Kim, J., Jeerapan, I., Imani, S. et al. (2016). *ACS Sensors* 1: 1011.

61 Soto, F., Mishra, R.K., Chrostowski, R. et al. (2017). *Adv. Mater. Technol.* 2: 1700210.

62 Jia, W., Valdes-Ramirez, G., Bandodkar, A.J. et al. (2013). *Angew. Chem.* 52: 7233.

63 Berchmans, S., Bandodkar, A.J., Jia, W. et al. (2014). *J. Mater. Chem. A* 2: 15788.

64 Mannoor, M.S., Tao, H., Clayton, J.D. et al. (2012). *Nat. Commun.* 3: 763.

65 Guinovart, T., Parrilla, M., Crespo, G.A. et al. (2013). *Analyst* 138: 5208.

66 Coyle, S., Lau, K.T., Moyna, N. et al. (2010). *IEEE Trans. Inf. Technol. Biomed.* 14: 364.

67 Jost, K., Perez, C.R., McDonough, J.K. et al. (2011). *Energy Environ. Sci.* 4: 5060.

68 Jost, K., Stenger, D., Perez, C.R. et al. (2013). *Energy Environ. Sci.* 6: 2698.

69 Sun, J., Huang, Y., Fu, C. et al. (2016). *Nano Energy* 27: 230.

70 Zhu, J., Tang, S., Wu, J. et al. (2017). *Adv. Energy Mater.* 7: 1601234.

71 Shin, S., Kumar, R., Roh, J.W. et al. (2017). *Sci. Rep.* 7: 7317.
72 Bandodkar, A.J., O'Mahony, A.M., Ramirez, J. et al. (2013). *Analyst* 138: 5288.
73 de Jong, M., Sleegers, N., Kim, J. et al. (2016). *Chem. Sci.* 7: 2364.
74 Mishra, R.K., Hubble, L.J., Martin, A. et al. (2017). *ACS Sensors* 2: 553.
75 Liao, C., Mak, C., Zhang, M. et al. (2015). *Adv. Mater.* 27: 676.
76 Ortega, G.A., Zuaznabar-Gardona, J.C., and Reguera, E. (2018). *Biosens. Bioelectron.* 116: 30.
77 Vukojevic, V., Djurdjic, S., Ognjanovic, M. et al. (2018). *Biosens. Bioelectron.* 117: 392.
78 Zhu, B., Niu, Z., Wang, H. et al. (2014). *Small* 10: 3625.
79 Khan, S., Tinku, S., Lorenzelli, L., and Dahiya, R.S. (2015). *IEEE Sens. J.* 15: 3146.
80 Cooper, C.B., Arutselvan, K., Liu, Y. et al. (2017). *Adv. Funct. Mater.* 27: 1605630.
81 Zeng, S., Li, R., Freire, S.G. et al. (2017). *Adv. Mater.* 29 https://doi.org/10.1002/adma.201700828.
82 Park, Y., Shim, J., Jeong, S. et al. (2017). *Adv. Mater.* 29 https://doi.org/10.1002/adma.201606453.
83 Bandodkar, A.J. and Wang, J. (2016). *Electroanalysis* 28: 1188.
84 Jia, W., Wang, X., Imani, S. et al. (2014). *J. Mater. Chem. A* 2: 18184.
85 Elouarzaki, K., Bourourou, M., Holzinger, M. et al. (2015). *Energy Environ. Sci.* 8: 2069.
86 Bandodkar, A.J., You, J.-M., Kim, N.-H. et al. (2017). *Energy Environ. Sci.* 10: 1581.
87 Jeerapan, I., Ciui, B., Martin, I. et al. (2018). *J. Mater. Chem. B* 6: 3571.
88 Wang, X., Gao, K., Shao, Z. et al. (2014). *J. Power Sources* 249: 148.
89 Hyun, W.J., Secor, E.B., Kim, C.-H. et al. (2017). *Adv. Energy Mater.* 7: 1700285.
90 Lin, Y., Gao, Y., and Fan, Z. (2017). *Adv. Mater.* 29: 1701736.
91 Zhang, L., Zhu, P., Zhou, F. et al. (2016). *ACS Nano* 10: 1273.
92 (a) Liu, L., Lu, Q., Yang, S. et al. (2018). *Adv. Mater. Technol.* 3: 1700206. (b) Zhu, M., Huang, Y., Huang, Y. et al. (2017). *Adv. Mater.* 29: 1605137.
93 Braam, K.T., Volkman, S.K., and Subramanian, V. (2012). *J. Power Sources* 199: 367.
94 Kim, S.H., Choi, K.H., Cho, S.J. et al. (2015). *Nano Lett.* 15: 5168.
95 Wang, Y., Chen, C., Xie, H. et al. (2017). *Adv. Funct. Mater.* 27: 1703140.
96 Ostfeld, A.E., Gaikwad, A.M., Khan, Y., and Arias, A.C. (2016). *Sci. Rep.* 6: 26122.

# 12

# Flexible and Wearable Electronics: from Lab to Fab

*Yuanyuan Bai#, Xianqing Yang#, Lianhui Li, Tie Li, and Ting Zhang**

*I-Lab, Suzhou Institute of Nano-Tech and Nano-Bionics (SINANO), Chinese Academy of Sciences (CAS), 398 Ruoshui Road, Suzhou 215123, China*

## 12.1 Introduction

The research on flexible and wearable electronics has increased in the past two decades [1–6]. Various flexible and wearable electronic devices with the functionalities of sensing [7–9], energy harvesting and storing [10–17], displaying [18–20], etc. have been demonstrated. To push forward the flexible and wearable electronic devices from lab to fab, several aspects ranging from the materials and manufacturing techniques to the cost and market need to be considered. Currently, printing is the most potential technique for mass production of flexible electronic devices [21]. It allows for low-temperature and highly efficient manufacturing of electronic devices on flexible substrates (e.g. metal foils, plastics, papers, and textiles) by creating electronic devices by printing inks of functional materials layer by layer in sequence. Several advanced printing equipment such as jet printers, screen printers, and gravure printers are commercially available for automatic production of flexible electronic devices. On the other side, various kinds of high-performance and environmentally stable inorganic nanomaterials (e.g. nanoparticles, nanowires, nanotubes) and highly flexible and processable organic materials are easily purchased for creating functional inks. All these achievements pave the way for flexible and wearable electronics toward industrialization, and some flexible and wearable electronic products such as flexible force sensors, paper batteries, flexible solar cells, and displays smart clothes are on sale now, which also promotes the development of smart clothing.

In this chapter, the materials for flexible and wearable electronics will be first introduced and discussed from the perspective of their compatibility with printing technologies. Then, the working principles and characteristics of several common printing techniques will be introduced, and their advantages and disadvantage will be discussed. The rest of the chapter will overview the industrialized status of typical state-of-the-art flexible electronic devices, and discuss the strategies to achieve smart clothing.

#Yuanyuan Bai and Xianqing Yang contributed equally to this work.
*Corresponding author: tzhang2009@sinano.ac.cn

*Flexible and Wearable Electronics for Smart Clothing*, First Edition.
Edited by Gang Wang, Chengyi Hou, and Hongzhi Wang.

## 12.2 Materials

### 12.2.1 Substrates

The characteristics of some widely used substrate materials in commercial flexible electronic devices are summarized in Table 12.1, focusing on the flexibility, transparency, thermal and chemical stability, electrical conductivity as well as the price.

Commercially available polyethylene terephthalate (PET), polyethylene naphthalate (PEN), and polyimide (PI) tapes are the most widely used flexible substrates. Among them, PET is the cheapest material, but its thermotolerance is lower than 120 °C [21], which is not compatible with the heat treatment process in the printing technique for some materials. PEN has better thermotolerance (<160 °C [22]) and chemical resistance than PET with medium price. PI has the best thermotolerance up to 450 °C [23], but its price is much higher, especially for transparent PI tapes. From the perspective of environmental friendliness and biodegradability, paper is also a good choice of flexible substrate [24–29]. Conducting metal foils with superior thermotolerance are often used as the substrates of flexible solar cells [30], with aluminum (Al) foil being the most commonly used due to its flexibility and low price. Highly stretchable substrates are currently not applied to commercial flexible and wearable electronic devices because the substrates are liable to deform under external forces, which may cause electrical instabilities, or interface sliding or deadhesion problems.

### 12.2.2 Functional Materials

Inorganic electronic materials with high performance and superior environmental stability are dominating in traditional electronic devices. For example,

**Table 12.1** Characteristics of some widely used substrate materials in commercial flexible electronic devices.

| Materials | Flexibility | Transparency | Thermotolerance (°C) | Chemical resistance | Electrical conductivity | Price ($/kg) |
|---|---|---|---|---|---|---|
| PI | Bendable | Yellow | <450 | Weak acids and alkali<br>Ethanol and acetone | Insulating | 20–150 |
| | | Yes | | | | 50–600 |
| PET | Bendable | Yes | <120 | Dissolvable in acetone | Insulating | 1–10 |
| PEN | Bendable | Yes | <160 | Weak acids and alkali<br>Ethanol and acetone | Insulating | 30–200 |
| Paper | Bendable | No | <100 | Poor | Insulating | 1–2 |
| Al foils | Bendable | No | <660 | Weak alkali<br>Ethanol and acetone | Conductive | 1–12 |

metals are widely used as interconnects, electrodes, inductors, antennas, etc. while silicon and germanium semiconductors are the key materials for fabricating transistors. To be compatible with the printing technique, these rigid inorganic materials require to be made into ink forms [21]. Generally, inorganic functional inks are made from aqueous or organic suspensions of inorganic micro/nano-sized particles [31], or some precursor solutions such as metallic salt solutions and organic metallic compound solutions [32, 33]. For those suspensions, a small amount of surfactants or lubricants are commonly added to achieve better dispersion and stability [34, 35]. Sometimes, polymers are added as well to enlarge the ink viscosity for screen printing, letterpress printing, or offset printing [36–40]. Post-treatments such as thermal annealing and sintering are generally required to obtain high film qualities [41]. The required temperature values vary with different materials, but are limited by the thermotolerance of the flexible substrates. Lowering the processing temperature of inorganic functional materials is critical for creating high-performance printed electronic devices.

Compared with inorganic electronic materials, organic materials have the advantages of high processability and natural compatibility with flexible polymeric substrates. A wide range of organic conducting, semiconducting, and insulating materials are emerging and playing key roles in creating flexible electronic devices. For example, breakthroughs in conductive polymers have pushed forward the development of organic solar cells [35, 42, 43] and organic thin-film transistors [44–46]. Superior film-forming ability of the organic materials and proper tuning of the ink formulation are of vital importance for creating high-performance printed electronic devices [47].

## 12.3 Printing Technologies

Printing technologies for manufacturing flexible and wearable electronic devices can be divided into two categories according to whether a template is employed or not. Jet printing is the only templateless printing method available, while printing relying on a template can be further divided into screen printing, gravure printing, letterpress printing, and offset printing, according to the difference in the template.

### 12.3.1 Jet Printing

Jet printing is a kind of non-contact digital printing technique that takes advantage of a computer-controlled ejector to propel ink drops out of one or more nozzles and to spray them onto a substrate to form specific patterns. It is widely used in the laboratories and factories for fundamental research and small-scale productions. Electronic components including batteries, antennas, sensors, field-effect transistors (FETs), solar cells, flexible displays, and so forth, can be feasibly jet printed [48–54]. At present, commercial jet printers mainly include inkjet printer, aerosol jet printer, and electrohydrodynamic jet printer. A high-resolution printing ability down to 100 nm has been achieved, and the printing can be so fast and fully automatic that large-scale fabrication is also possible.

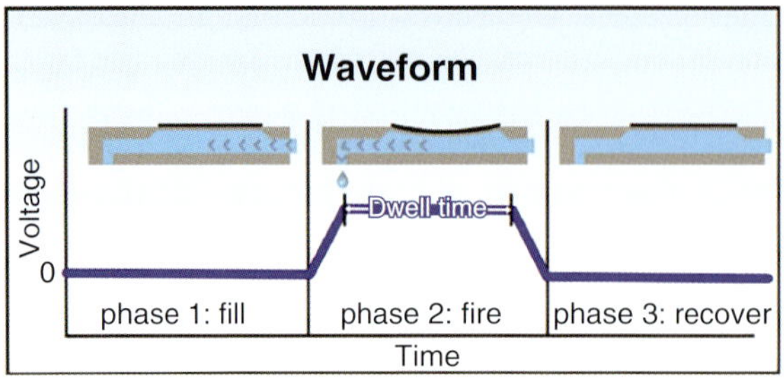


**Figure 12.1** Schematic illustration of the working process of piezoelectric driven printing, showing the repelling and retraction of the ink. An open access article. Source: Hill et al. 2015 [55]. Copyright 2019, Reproduced with permission of The Electrochemical Society.

#### 12.3.1.1 Inkjet Printing

A commercial inkjet printer commonly consists of an ink cartridge, a printer head, a computer, an X, Y, Z operating platform, and a charge coupled device (CCD) camera for observation. The ink cartridge is used to store electronic inks, and is connected to the printer head through tubules. The printer head mainly comprising one or more jet nozzles assembled together with piezoelectric transducers or heating units is used to repel ink droplets. For the former piezoelectric driven printer head, the piezoelectric transducer deforms when a voltage is applied, which repels the ink out of the nozzle (Figure 12.1). For the latter thermal driven printer head, the heating unit generates heat when an electric current passes through to vaporize the ink to form a bubble. The rapid thermal expansion of the bubble then propels a droplet of ink out of the nozzle. The major deficiencies of thermal driven printing are that the chemical property of the ink is prone to change at high temperature and that the volume of the ink is hard to control, making the piezoelectric driven printing more popular in printed electronics.

A stepping motor is connected to the printer head to make it movable in plane and vertically. Parameters such as the dot spacing, lines, arrays, and layers can be set manually on the computer using the processing software supplied by the manufacturer to create a drop matrix. A desired pattern can also be obtained by creating a digital image, which is converted into a monochrome bitmap digital file by the processing software, of which the black and white pixels represent the printed and non-printed area respectively. The printer control program then converts the bitmap digital file into electric signals, driving the printer head to propel ink droplets.

Various materials in the form of liquids, solutions, colloids, or suspension liquids are inkjet printable, but the ink should be of low viscosity (typically in the range of 10–30 cP) and low volatility to avoid nozzle clogging. Fine size of particles and superior stability are also required for colloids and suspension liquids. The resolution of inkjet printers is typically ~20 μm, which can be further degraded due to the spreading and placement errors of droplets. Some strategies have been utilized to improve the resolutions, such as modifying the wetting properties of the substrates [56], using lithographically defined chemical or topographical patterns [46], and enabling self-alignment process [57].

#### 12.3.1.2 Aerosol Jet Printing

Different from inkjet printer, the working principle of an aerosol jet printer, provided by Optomec, Inc., is shown in Figure 12.2. The ink is stored in an atomizer

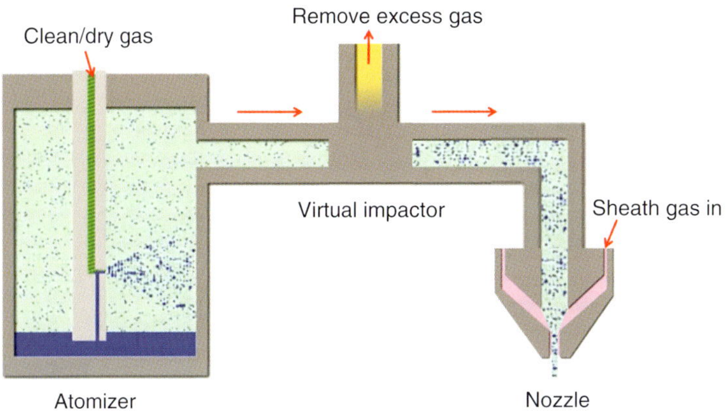


**Figure 12.2** Schematic illustration of the working principle of an aerosol jet printer provided by Optomec, Inc. The ink is stored in an atomizer and ultrasonically atomized into tiny droplets. Then excess carrier gas is removed to create a dense aerosol mist of droplets between 1 and 5 μm in diameter, which is then delivered to the jet nozzle and is focused by a sheath gas to form a high-velocity tight stream of droplets. The gas used in the system is typically clean, dry nitrogen or compressed air. Source: Figure is drawn based on an animation from Optomec with permission [58]. https://www.optomec.com/printed-electronics/aerosol-jet-technology/.

and ultrasonically atomized into tiny droplets, which is subsequently concentrated into a dense aerosol mist of droplets between 1 and 5 μm in diameter. The aerosol mist is then delivered to the jet nozzle and is focused by a sheath gas to form a high-velocity tight stream of droplets spraying onto a substrate. To create a desired pattern, a standard computer-aided design (CAD) data is first created and converted to make a vector based tool path to drive a 2D or 3D motion control system connected with the sample stage. The tool path can also be set using the system's built-in program.

The required ink viscosity for aerosol jet printing is typically in the range of 1–1000 cP; thus the aerosol jet printer has a much wider tolerance to different ink materials than the inkjet printers. Besides, since the stream of droplets is insulated from the nozzle by the sheath gas, the aerosol jet printer is less likely to have clogging problem. A resolution of <10 μm can be reached using the aerosol jet printer since the focused ink stream has a diameter as small as 1–5 μm.

#### 12.3.1.3 Electrohydrodynamic Jet (e-Jet) Printing

e-Jet printing was first used by J.A. Rogers' group for high-resolution fabrication of electronics [59]. The schematic diagram of an e-jet printer is shown in Figure 12.3. A metal (e.g. Au) coated glass capillary with an inner diameter well within the submicrometer range serves as a nozzle. A pressure controller is connected to the microcapillary to deliver fluid inks to the tip of the nozzle. A voltage is applied between the nozzle and a conductive support, generating an electric field in between and causing mobile ions in the ink to accumulate near the surface of the meniscus in the nozzle. The ink surface is then subjected to both electrostatic stresses and surface tension. When the electric field is high enough, the former overcomes the latter, and droplets eject from the nozzle. The

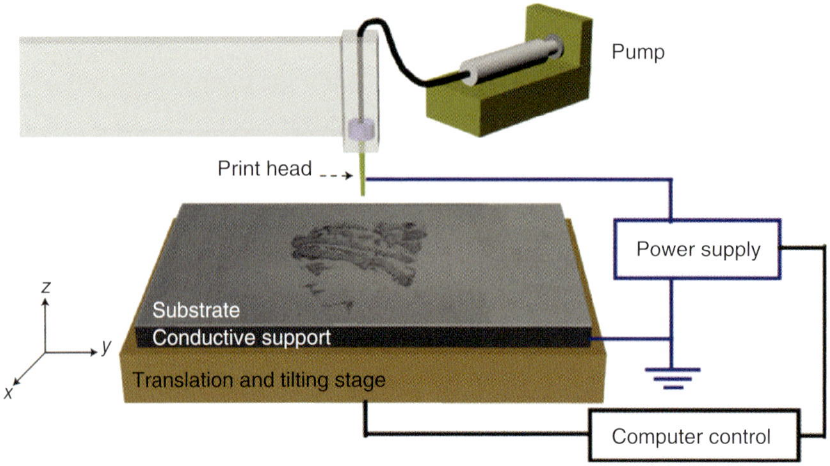


**Figure 12.3** Schematic diagram of an e-jet printer. Source: Park et al. 2007 [59]. Copyright 2007, Reproduced with permission of Springer Nature.

ion concentrations to enable this ejecting process can be very small, making a wide range of organic and inorganic inks; even liquids with low conductivities of $10^{-13}$ to $10^{-3}$ S/m are e-jet printable. The conductive support usually rests on a computer-controlled translation and tilting stage to enable patterned deposition of the inks.

A prominent advantage of e-jet printing over inkjet and aerosol jet printing is its high-resolution printing ability down to 100 nm to date, attributed to three characteristics: (i) the inner diameter of the nozzle is as small as ~100 nm; (ii) the size of the droplet generated under a high electric field can be significantly smaller than the diameter of the nozzle; (iii) the electric field suppresses the displacement errors of the droplets. Recently, J.A. Rogers' group summarized the development e-jet printing technique in terms of the mechanisms, capabilities, and applications [60].

### 12.3.2 Screen Printing

Screen printing is the simplest printing method widely used in industry. Five critical components constitute a screen printer, which are screen masks, a blade, electronic inks (or pastes), substrates, and a supporting stage. The screen mask is a frame supported fabric mesh with a specific stencil deposited on it. The stencil made of patterned light-sensitive materials divides the screen into the transparent and opaque areas respectively. By covering the screen with electronic ink of high viscosity, which is usually called paste, pressing the blade onto the past-covered screen, and moving it across the screen, the paste is able to penetrate through the patterned area and create a replication of the pattern of the screen on the substrate.

There are mainly three kinds of operating modes for commercial screen printers, as shown in Figure 12.4. The sheet-to-sheet mode adopts flat screen and

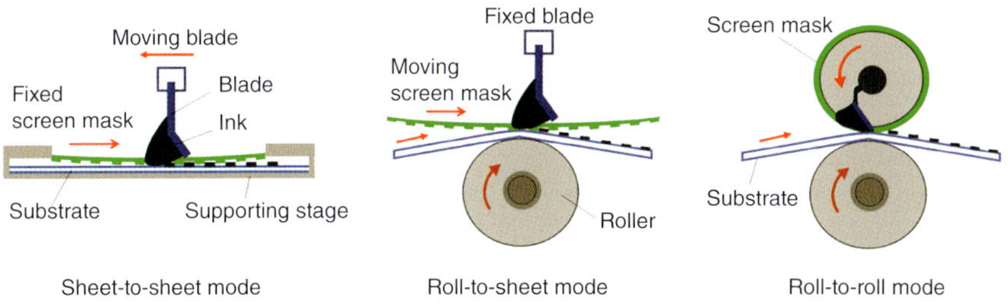


**Figure 12.4** Schematic diagrams of three typical operating modes for commercial screen printers, which are sheet-to-sheet mode, roll-to-sheet mode, and roll-to-roll mode.

substrate holder, and relies on the movement of the squeegee to transfer the printing ink onto the substrate. It has a simple structure and is convenient for manual operation, but the efficiency is low. The roll-to-sheet mode also adopts a flat screen, but the substrate is transmitted through a rotary holder. The synchronous movement of the screen and the substrate is critical for accurate printing. To enable high throughput printing, the screen can also be rotary, with the screen and the squeegee integrated inside the roller. A roll of flexible substrate is fed in synchrony with the screen, which is a true roll-to-roll working mode.

In terms of the printing inks, micro-sized materials, which are easier to obtain and use than nanoparticles, can be used. High viscosity of the ink (500–10 000 cP) is required. The benefit is that it is much easier to control the ink movements during the printing process and to obtain thick films (typically more than 10 μm). However, the drawback is that the high viscosity also results in rough pattern surface and that additives are usually added to the ink to increase the viscosity, which will possibly decrease the electrical properties. Compared with jet printing, screen printing has lower resolution (~50 μm), but its overall cost is lower and its efficiency is much higher, so it is adequate and quite suitable for manufacturing relatively simple and low-resolution structures.

### 12.3.3 Other Printing Techniques

Other commonly used printing techniques including gravure printing, flexographic printing, and gravure offset printing have similar working mode as screen printing, but differ in the template and the way of transfer of the printing inks.

Gravure printing adopts a gravure plate having engraved patterns in the form of dented dots [61, 62]. The gravure is immersed in the ink before printing. During printing, the ink contained in the engraved area is transferred to the substrate surface under the pressure of an impression roller in contact with the gravure, while any excess ink not in the dented dots is removed by a blade. Compared with screen printing, low viscosity ink (50–500 cP) can be used, and it is possible to obtain different thicknesses in a pattern by employing gravure cells of various sizes and depths. The printed layer thickness is typically in the range of 0.8–8 μm. The disadvantages of gravure printing are that the cost of making a gravure plate is high and that the printing plate is rigid, which is only suitable for flexible substrates since the required impression pressure for printing on rigid substrate is too high to achieve.

Flexographic printing [63] is a kind of letterpress printing that uses an elastic printing plate with pattern areas as surface reliefs protruding above non-pattern areas. An anilox roller with evenly distributed engraved cells covering the entire surface of the roller is first immersed in the ink before printing. During printing, the ink in the cells is firstly transferred onto the flexographic printing plate, and then transferred to the substrate. More rollers are often used for accurate tension control. A prominent advantage of using elastic plate is that it is possible to print ink on a wider range of substrates, ranging from rigid and flexible substrates to substrates with rough surface. Besides, a lower printing pressure is required for flexography printing than gravure printing. However, the soft nature of the elastic plate also has a short service life due to the abrasion under printing pressure.

Gravure offset printing [64] is developed from gravure printing to expand the range of applicable substrates. It uses an elastic intermediate carrier (blanket) to transfer the printing ink from a printing plate to the substrate, so it is a kind of indirect printing since the substrate is not in direct contact with the printing plate. Gravure offset printing inherits the merits of gravure printing; moreover, it can print on more substrates, even rigid, fragile, rough, and irregularly shaped substrates. The deficiency is that the device is more complex than a direct gravure machine.

## 12.4 Flexible and Wearable Electronic Products

### 12.4.1 Flexible Force Sensors

Flexible force sensors that transduce force signals caused by tension, pressure, weight, torque, and internal stress/strain into electrical signals have great potential applications in the fields of human physiological signal and motion detection [65–67], medical diagnosis [68], human–machine interactions [69–71], and so forth. A great many high-performance flexible force sensors with different configurations have been reported in the literature [72–81]. Nevertheless, only a few have achieved commercialization, which are available from corporations such as Interlink Electronics and Tekscan Inc., USA; Suzhou Leanstar Tech and Shenzhen Salens Technology Co., Ltd., China, etc. A typical construction of commercial standard flexible force sensors is shown in Figure 12.5a.

Two layers of flexible polymers (substrates) are coated with conductive electrodes and sensitive materials (typically are piezoresistive materials) on the inside surfaces. The electrode is typically circular or rectangular, with a long strip of tail for external electrical connection. A layer of spacer adhesive around the perimeter of the sensor is used to separate the substrates. The sensor commonly has a high resistance due to the existence of insulating air gaps between the electrodes. When an external force is applied to one substrate of the sensor, the two substrates touch each other, causing electrical connections of the electrodes through the piezoresistive materials and a decrease of the sensor resistance proportional to the applied force.

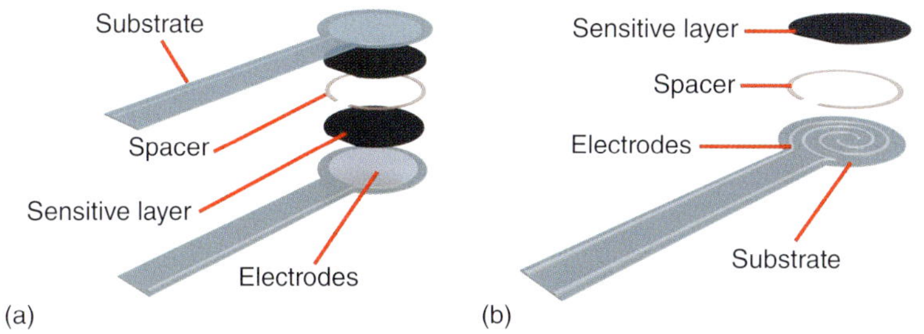


**Figure 12.5** Structural schematic diagram of two typical kinds of commercial standard flexible force sensors: (a) sensors with parallel face-to-face electrodes; (b) sensors with interdigitated electrodes.

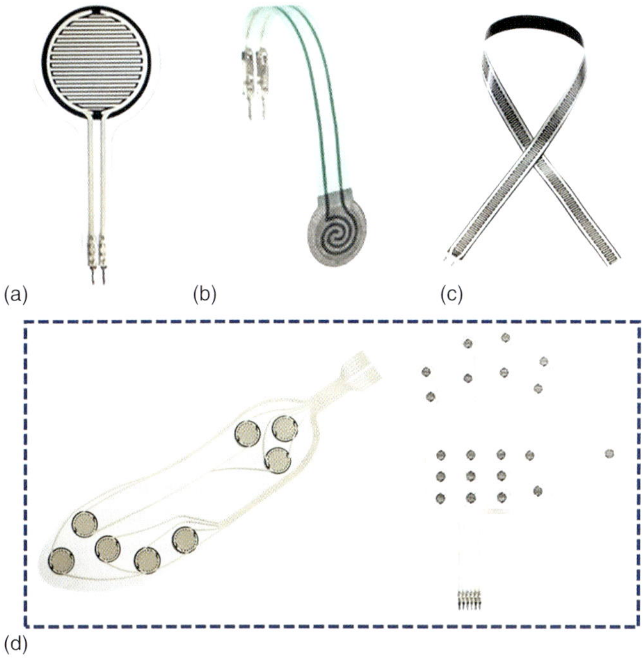


**Figure 12.6** (a) Structural schematic diagram of a typical kind of commercial standard flexible force sensors; (b–c) standard flexible sensors, and (d) customized flexible sensors. Source: Courtesy of Suzhou Leanstar Electronic Technology Co,. Ltd. http://www.lssensor.com/products/.

Another typical construction of commercial standard flexible force sensors is shown in Figure 12.5b. Instead of being coated on both substrates, the electrodes are coated on the inside surface of one substrate in an interdigitated form. The sensitive material is coated on the inside surface of the other substrate. When the two substrates are pressed together, the sensitive material shorts across the interdigitated fingers of the facing surface, resulting in a decrease of the resistance between the conducting fingers. For commercial flexible force sensors, the interdigitated electrodes are versatile in structure (Figure 12.6a–c). Owing to the merits of material and time saving, this kind of sensor is becoming more and more popular.

The substrates used in commercial flexible force sensors include PET, PI, polyesters, and other flexible polymers. The temperature sensitivity of the polymer substrate is directly related to the sensor's workable temperature range, while its mechanical modulus and film thickness have great influences on the sensor's response time and sensing range.

The electrodes and the sensitive layers are mostly made from silver-polymer inks, and carbon based semiconductive inks respectively using the screen printing technique. By simply varying the template patterns, customized sensors composed of multiple discrete force sensors or strip-like sensors for applications such as smart insoles, smart gloves, or pressure mapping are easily achieved (Figure 12.6d).

For most commercial flexible force sensors, broad sensing range on the order of $10^{-3}$ to $10^4$ N, fast response time (<100 ms), and superior durability up to hundreds of thousands of times have been achieved, while the sensitivity and linearity of the sensors need to be further improved. Besides, current commercial sensors are only bendable and cannot fit soft surfaces well. To realize the commercialization of laboratory highly flexible and stretchable force sensors, much effort is required to solve the problems of electrical instabilities, interface sliding, or deadhesion, etc. caused by the deformation of the sensors.

### 12.4.2 Paper Battery

Paper battery is a kind of printable micro-power source cell based on paper or other flexible films, which is manufactured using simple, industry-standard printing, drying, and laminating processes. With the characteristics of flexibility, ultrathinness (<1 mm), lightweight, environmental friendliness, low cost, etc. [82] paper batteries are quite suitable for a wide range of applications such as cosmetic patches, medical devices, smart cards, electronic labels, smart toys, and other occasions where portable, small-scale and highly integrated low-power supplies are needed.

According to whether it can be recharged or not, paper battery is divided into primary paper battery and storage paper battery. The former has been industrialized, and the representative companies include the Power Paper in Israel and Enfucell Flexible Electronics, Ltd. in China. The storage paper battery, mainly lithium-ion paper battery, is still restricted to the laboratory [83–85].

As shown in Figure 12.7a, a piece of paper battery typically has a multilayered structure composed of the current collectors (typically are bendable and foldable thin fibrous materials), cathode and anode, electrolyte, and separator (a piece of thin soft diaphragm paper). All layers are printable from proprietary inks onto flexible film substrates. The entire battery can be extremely thin (<0.7 mm) (Figure 12.7b).

Currently, an annual output of 100 million pieces of paper batteries can be achieved from a fully automatic roll-to-roll production line (data from Enfucell).

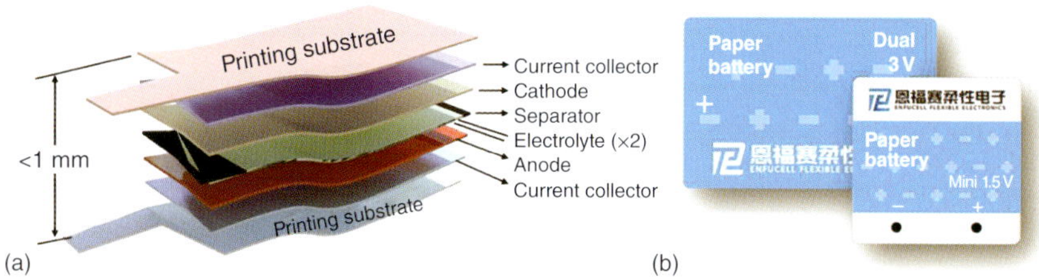


**Figure 12.7** (a) Structural schematic diagram of a piece of paper battery; (b) A typical paper battery. Source: Courtesy of China Enfucell Flexible Electronics.

**Table 12.2** Specifications for paper batteries from Enfucell.

| Product specifications | | DS3535WD | 3654 |
|---|---|---|---|
| Performance | Nominal voltage (V) | 1.5 | 3.0 |
| | Nominal capacity (mAh) | 30 | 20 |
| | Nominal pulse peak current | 10 mA (5 ms) | 10 mA (5 ms) |
| | Nominal internal resistance (Ω) | 60 | 90 |
| Size | Maximum thickness (mm) | 0.7 | 0.7 |
| | Outer dimensions | 36 mm × 46 mm | 36 mm × 54 mm |
| | Weight (g) | 1–2 | 1–2 |

Source: Courtesy of China Enfucell Flexible Electronics.

The specifications for the commonly used paper batteries from Enfucell are listed in Table 12.2.

For a paper battery with an area of approximately 20 cm2 and a weight of only 1–2 g, a nominal voltage (3 V) equivalent to the voltage of two alkaline batteries that are commonly used in daily life can be generated. The nominal capacity and continuous current density is 20 mAh and 10 mA, respectively. In real applications, multiple batteries can be stacked to obtain larger voltage or current.

A huge disadvantage of current commercial paper batteries is that they are primary batteries that cannot be recharged and reused, so their service lives are limited by their capacities, which restricts their wide applications. For example, for some highly integrated long-term service devices such as active radio frequency identification (RFID) tags, smart cards, and microsensors, it is inconvenient to replace their batteries, so long-term stable power supplies are required.

### 12.4.3 Flexible Solar Cell

Solar energy is a kind of green and renewable energy, which is occupying an important place in the world's energy consumption and has the potential to become the main part of the world's energy supply. Solar cells convert light energy directly into electrical energy through photoelectric or photochemical effects [86–91]. The solar cell industry has made great progress since 2000, with three representative technological innovations (Table 12.3).

**Table 12.3** Three representative technological innovations during the development of solar cell industry.

| Technology parameters | I. Wafer cells | II. Vacuum-based thin film on glass | III. Roll-printed thin film on foil |
|---|---|---|---|
| Process | Silicon wafer processing | Sputtering, evaporation in a vacuum chamber | Printing in plain air |
| Process control | Fragile wafers | Expensive metrology | Built-in bottom-up reproducibility |
| Process yield | Robust | Fragile | Robust |
| Materials utilization | 30% | 30–50% | Over 95% |
| Substrate | Wafer | Glass | Conductive foil |
| Continuous | No–wafer | No–glass | Yes |
| Processing | Handling | Handling | / |
| Cell matching | Yes | No | Yes |
| Panel current | High | Low | High |
| Energy payback | 3 years | 1.7 years | <1 month |
| Throughput/capital expenditure (Cap. Ex.) | 1 | 2–5 | 10–25 |

Source: Pagliaro et al. 2008 [92]. Copyright 2008, Reproduced with permission of John Wiley & Sons.

Flexible substrate based thin-film solar cells based on roll-to-roll manufacturing processes have the advantages of more flexibility, less thickness and lighter weight, lower cost, easier mass production, and reduced payback time compared with silicon-wafer based solar cells and glass based solar cells [90]. Besides, they are able to be integrated with existing buildings, fabrics, tents, sails, glass, and all sorts of surfaces, which significantly broadens the applications of solar cells.

Commercial inorganic flexible solar cells are mainly amorphous silicon (a-Si) based or copper indium gallium selenide (CIGS) based solar cells, and their performances are listed in Table 12.4.

A-Si based solar cells are currently the most mature type of commercialized flexible batteries. They are commonly manufactured by vapor depositing a-Si at low temperatures, together with sputtering depositing conducting and reflecting layers on rolls of stainless steel or metal coated plastic substrates. The whole manufacturing process is fully automatic with high throughput. Commercial flexible a-Si cell modules have efficiencies of ~8%, while for hybrid cell modules with tandem junctions of a-Si and microcrystalline silicon (μc-Si), their efficiencies are higher (~12%).

More and more companies are engaged in developing and producing CIGS based flexible solar cells due to their high efficiencies. Cell efficiencies in the range of 12–15% that optimally trade off cost and performance have been successfully achieved. The manufacturing processes of CIGS based solar cells can be the same with those of a-Si based solar cells; nevertheless, to further reduce the amount of raw materials, especially In with small reserve and high price, the

**Table 12.4** The properties of typical commercial solar cell modules.

| Specifications company | Type | Area ($cm^2$) | Thickness (mm) | Efficiency | Power (W) | Date |
|---|---|---|---|---|---|---|
| HyET solar | a-Si | 19272 | 0.4 | 8% | 116 | 2017 |
| | a-Si/μc-Si | 19272 | 0.4 | 12% | 161 | |
| Solar voltaic | a-Si | 1336 | 1.5 | 8% | 81 | 2017 |
| Nano solar | CIGS | 223 | 0.4 | 17.1% | 2.6 | 2017 |
| Ascent solar | CIGS | 7243 | 1.9 | — | 50 | 2018 |
| BIPV | CIGS | 18140 | 2 ± 1.5 | 15.5% | 240 | 2018 |
| MiPV | CIGS | 8741 | 2.7 | — | 110 | 2018 |
| Solar J | CIGS | 7906 | 1.2 | — | 50 | 2010 |
| Global solar | CIGS | 9964 | 3 | 12.7% | 100 | 2018 |
| Solar motion electronics | CIGS | 9964 | 3 | 12.7% | 100 | 2015 |

Source: Official website of the company.

material-saving and high-throughput roll-to-roll printing technology based on functional nano-inks have been utilized for production since the end of 2007.

In recent years, organic solar cells using organic compounds as semiconductors are rising in the market of solar cells due to the advantages of high flexibility, extremely thin and lightweight, ultra-low-cost manufacturing, and so forth. Heliatek in Germany is the first company in the world to produce large-scale commercialized flexible organic solar cells using efficient roll-to-toll processes in vacuum and inert atmospheres. It holds the world record of 13.2% cell efficiency for non-transparent organic solar cells. Organic solar cells with transmissions up to 50% and efficiencies of 6–7% are also achieved. It is possible to adapt the color and the transparency of organic solar cells by modifying the electrical properties of organic semiconductors and changing the thickness of the active and conductive layers.

Another new type of solar cell on the rise is dye-sensitized solar cell (DSSC), which uses nano titanium dioxide and photosensitive dyes as the main raw materials to convert solar energy into electric energy, similarly to the photosynthesis of plants using solar energy in nature [93]. An outstanding advantage of DSSC is the abundance and low cost of the raw materials for mass production. G24 Power in the United Kingdom is recognized as a global leader in the design and manufacture of flexible DSSC. It started roll-to-roll printing of DSSC based solar modules in 2007. Their portable solar charger products have thickness less than 1 mm, working indoors and outdoors, charging for mobile phones, laptops, audio players, and digital cameras. The efficiency of DSSC has reached 14.1% in the laboratory, while that measured under standard test conditions is 10% (https://gcell.com/dye-sensitized-solar-cells/advantages-of-dscc/efficiency). There is no doubt that the efficiency of DSSC needs to be further improved for better commercialization. Apart from efficiency, improving the long-term stability and lifetime during outdoor use are also big challenges in the commercialization and wide use of DSSCs [94, 95].

### 12.4.4 Flexible Display

Displays enable the visualization of information, and are widely used in computers, mobile phones, watches, and many other electronic products. Developing flexible display is an essential part of flexible electronics, which has been attempted by many scholars and companies. It will make the display of information more convenient and diverse. Currently, the technologies to realize flexible display mainly include electrophoresis display (EPD), liquid crystal display (LCD), and organic light-emitting diode (OLED) display [96].

EPD is mostly used to create electronic paper (e-paper), which is firstly commercialized by e-ink (purchased by Prime View International, Taiwan in 2009 and renamed e-ink Houdings Inc in 2010). It is fabricated by depositing liquid electronic inks on thin flexible substrates (e.g. ultrathin glass, plastics, or metal foils), and relies on the alternate movement of colored charged matters in the electronic ink under applied electric field to show two or more different colors [97]. Other representative companies engaged in developing e-paper display include SiPix, USA; Hitachi and Fujitsu, Japan; Samsung, Korea; Guangzhou open explore dream (OED) Technologies, China, etc. EPD display has great potentials in manufacturing flexible indicators, digital signatures, and electronic display cards and readers. However, a huge disadvantage is that full-color display and video display are still challenging. Besides, it cannot be used on information media devices such as mobile phones because it is a passive display.

Thin-film transistor-liquid crystalline display (TFT-LCD) is a kind of non-emissive display technology widely used in notebook computer, and animation/image processing products. Flexibility can be achieved by replacing the glass substrate with thin plastic substrate, and further replacing silicon based thin-film transistor (TFT) with organic TFT (organic LCD, organic liquid crystal display [OLCD]) [96]. Many companies used to devote to developing OLCD; however, deficiencies in the performance of OLCD such as slow response speed, narrow operating temperature range, and short service life have not been solved, and the manufacturing technology is still not mature enough for industrialization. Now, the FlexEnable, UK is providing small quantities of curved OLCD modules for early prototyping, concepts, and demonstrators; nevertheless, the mass production and fully flexible display have not been realized.

OLED, first proposed by C.W. Tang and S.A. Vanslyke [98], is a kind of emissive display with the advantages of wide viewing angle, high color contrast, fast response time, ultrathin profile, etc. [99] It is the most potential technology to realize flexible display presently [18]. Many companies such as Samsung and LG Display, Korea; and Royole, BOE, and Visionox, China, etc. are able to produce foldable, rollable, and even fully flexible OLED displays, and claim to have set up mass production lines for OLED flexible displays. For example, in 2014, Royole released the world's thinnest flexible full-color OLED display with a thickness down to 0.01 mm (Figure 12.8a). The bending radius can be as small as 1 mm, with a high-quality display after being bended for 50–100 thousand times. In

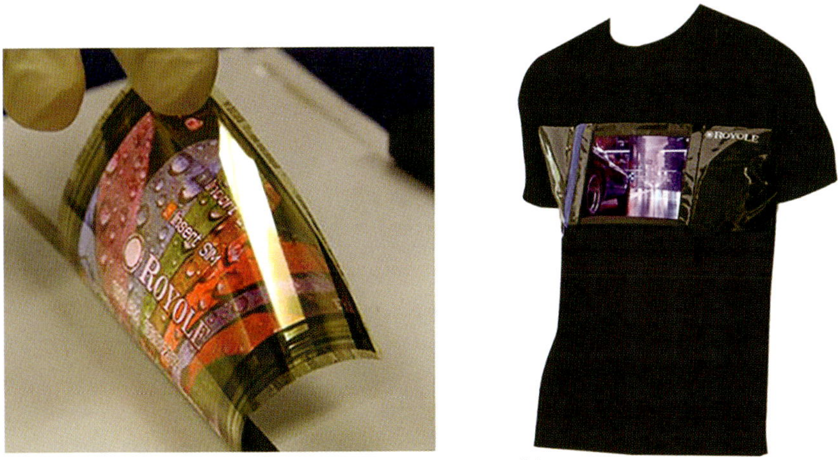


(a) (b)

**Figure 12.8** OLED flexible displays from Royole and BOE, China (a, b). Source: Courtesy of China Royole, Inc. http://www.royole.com/.

2017, BOE achieved the mass production of the sixth-generation OLED flexible display production line, which can annually produce 70 million pieces of flexible display screens with thickness of only 0.03 mm. These flexible displays have been used to create wearable electronics (Figure 12.8b).

## 12.5 Strategy Toward Smart Clothing

Smart clothing is a kind of clothes with new functions such as sensing, energy harvesting, and display [100–102], which are different from traditional clothes for protection and decoration. The most traditional and direct way to achieve smart clothing is to integrate lightweight and small size rigid silicon based electronic devices on/in textiles. Owing to the maturity of the development of silicon based electronics, diverse high-performance devices are available, and it is easy to achieve mass production. However, restricted by the inherent rigidity, this kind of clothes are slightly inconvenient and uncomfortable when worn on soft and curved human body.

Mechanically flexible electronic devices, nevertheless, make up for the aforementioned deficiency. Smart clothes integrated with flexible electronic devices have the potential to meet the requirements of many occasions such as monitoring heartbeat, muscle tension, and other body motions where flexibility is needed [66, 68]. Some smart clothes integrated with flexible force sensors, solar cells, displays, etc. are already on sale presently; nevertheless, it is still challenging for flexible electronics to replace traditional rigid electronics in terms of performance parameters such as the accuracy, reliability, consistency, and so forth. In particular, chips for transmitting and processing signals are still limited to traditional silicon based electronics. Therefore, in practical applications, it is usually necessary to combine flexible and rigid silicon devices to create the real sense of smart clothing. For example, Ralph Lauren Corporation, USA Corporation is selling a PoloTech™ smart shirt that integrates both flexible silver fiber sensors woven into the fabric for detecting the wearer's heart rate, breathing depth, calories burned, etc. and rigid electronics such as a detachable Bluetooth-enabled "black box" composed of silicon based devices to transmit collected data to a mobile terminal device such as an iphone or ipad. A big drawback of this smart shirt is that its price (~US$295) is a little too high.

## 12.6 Summary and Perspective

In this chapter, the requirements for materials and the printing technologies for achieving low temperature mass production of flexible electronics toward industrialization are discussed. In general, the creation of high-performance electronic inks, i.e. inks with high stability, appropriate viscosity, superior volatility, etc., is critical, while flexible electronics with resolution typically higher than 20 μm can be created at low temperature using screen printing, gravure printing, letterpress printing, and offset printing techniques, typically in a roll-to-roll working mode to achieve high throughout. Inkjet printing and aerosol jet printing are available for printing flexible electronics with resolution down to 10 μm; however, the yield is lower and the cost is a little higher. For flexible electronic devices with smaller size, the newly developed e-jet printing can be used; however, large-scale production has not been achieved.

The industrialized status of some flexible electronic devices including flexible sensors, paper batteries, flexible solar cells displays are summarized, with emphasis on introducing the structure, performance, characteristics, limitations, and challenges of typical commercial products. Also, the strategy to achieve smart clothing is discussed. A combination of flexible and rigid silicon devices is necessary to create real sense of smart clothing. Despite this, the popularization of smart clothing still has a long way to go since problems still exist in terms of security, effectiveness, reliability, and so on. Besides, most of the existing smart clothes have very high prices, which restricts their wide applications.

## References

1 Kim, D.H., Lu, N.S., Ma, R. et al. (2011). *Science* 333: 838.
2 Bauer, S. (2013). *Nat. Mater.* 12: 871.
3 Choi, S., Lee, H., Ghaffari, R. et al. (2016). *Adv. Mater.* 28: 4203.
4 Someya, T., Bao, Z.N., and Malliaras, G.G. (2016). *Nature* 540: 379.
5 Wang, X.W., Liu, Z., and Zhang, T. (2017). *Small* 13: 1602790.
6 Hammock, M.L., Chortos, A., Tee, B.C.K. et al. (2013). *Adv. Mater.* 25: 5997.
7 Amjadi, M., Kyung, K.U., Park, I., and Sitti, M. (2016). *Adv. Funct. Mater.* 26: 1678.
8 Yang, T.T., Xie, D., Li, Z.H., and Zhu, H.W. (2017). *Mat. Sci. Eng., R* 115: 1.
9 Trung, T.Q. and Lee, N.E. (2016). *Adv. Mater.* 28: 4338.

10 Huang, Q.Y., Wang, D.R., and Zheng, Z.J. (2016). *Adv. Energy Mater.* 6: 1600783.
11 Wu, H., Huang, Y.A., Xu, F. et al. (2016). *Adv. Mater.* 28: 9881.
12 Shi, J.J., Guo, X., Chen, R.J., and Wu, F. (2016). *Progr. Chem.* 28: 577.
13 Wang, X.F., Lu, X.H., Liu, B. et al. (2014). *Adv. Mater.* 26: 4763.
14 Wang, X.Y., Li, Z., Xu, W.J. et al. (2015). *Nano Energy* 11: 728.
15 Meng, C.Z., Liu, C.H., Chen, L.Z. et al. (2010). *Nano Lett.* 10: 4025.
16 Chai, Z.S., Gu, J.W., Khan, J. et al. (2015). *RSC Adv.* 5: 88052.
17 Ma, M.M., Tang, Q.W., Yang, P.Z., and He, B.L. (2016). *RSC Adv.* 6: 82933.
18 Xu, R.P., Li, Y.Q., and Tang, J.X. (2016). *J. Mater. Chem. C* 4: 9116.
19 Wang, X.L. and Liu, J. (2016). *Micromachines* 7: 206.
20 Kim, W., Kwon, S., Lee, S.M. et al. (2013). *Org. Electron.* 14: 3007.
21 Rim, Y.S., Bae, S.H., Chen, H.J. et al. (2016). *Adv. Mater.* 28: 4415.
22 Nathan, A., Ahnood, A., Cole, M.T. et al. (2012). *Proc. IEEE* 100: 1486.
23 Liaw, D.J., Hsu, P.N., Chen, W.H., and Lin, S.L. (2002). *Macromolecules* 35: 4669.
24 Russo, A., Ahn, B.Y., Adams, J.J. et al. (2011). *Adv. Mater.* 23: 3426.
25 Lien, D.H., Kao, Z.K., Huang, T.H. et al. (2014). *ACS Nano* 8: 7613.
26 Zhong, Q.Z., Zhong, J.W., Cheng, X.F. et al. (2015). *Adv. Mater.* 27: 7130.
27 Liao, X.Q., Zhang, Z., Liao, Q.L. et al. (2016). *Nanoscale* 8: 13025.
28 Liu, H., Qing, H.B., Li, Z.D. et al. (2017). *Mat. Sci. Eng., R*: 112.
29 Guo, H.Y., Yeh, M.H., Zi, Y.L. et al. (2017). *ACS Nano* 11: 4475.
30 Yim, T.H., Koo, S.H., and Lee, H. (2010). *Curr. Appl. Phys.* 10: S241.
31 Meier, H., Loffelmann, U., Mager, D. et al. (2009). *Phys. Status Solidi A* 206: 1626.
32 Perelaer, J., Smith, P.J., Mager, D. et al. (2010). *J. Mater. Chem.* 20: 8446.
33 Dearden, A.L., Smith, P.J., Shin, D.Y. et al. (2005). *Macromol. Rapid Commun.* 26: 315.
34 Yang, C., Wong, C.P., and Yuen, M.M.F. (2013). *J. Mater. Chem. C* 1: 4052.
35 Kommeren, S., Coenen, M.J.J., Eggenhuisen, T.M. et al. (2018). *Org. Electron.* 61: 282.
36 Baudry, H. and Franconville, F. (1978). *Acta Electron.* 21: 283.
37 Chu, Z.Y., Peng, J.M., and Jin, W.Q. (2017). *Sens. Actuators, B* 243: 919.
38 Menon, H., Aiswarya, R., and Surendran, K.P. (2017). *RSC Adv.* 7: 44076.
39 Somalu, M.R., Muchtar, A., Daud, W.R.W., and Brandon, N.P. (2017). *Renew. Sustain. Energy Rev.* 75: 426.
40 Sun, J.Y., Cheng, C.M., and Liao, Y.C. (2015). *Anal. Sci.* 31: 145.
41 Gong, S.C., Jang, S.K., Ryu, S.O. et al. (2010). *Curr. Appl. Phys.* 10: E192.
42 Hoth, C.N., Schilinsky, P., Choulis, S.A., and Brabec, C.J. (2008). *Nano Lett.* 8: 2806.
43 Cho, C.K., Hwang, W.J., Eun, K. et al. (2011). *Sol. Energy Mater. Sol. Cells* 95: 3269.
44 Lin, Y., Liu, C.F., Song, Y.J. et al. (2016). *RSC Adv.* 6: 40970.
45 Fukuda, K., Takeda, Y., Yoshimura, Y. et al. (2014). *Nat. Commun.* 5: 4147.
46 Sirringhaus, H., Kawase, T., Friend, R.H. et al. (2000). *Science* 290: 2123.
47 Calvi, S., Maita, F., Rapisarda, M. et al. (2018). *Org. Electron.* 61: 104.
48 Huang, L., Huang, Y., Liang, J.J. et al. (2011). *Nano Res.* 4: 675.

49 Matyas, J., Munster, L., Olejnik, R. et al. (2016). *Jpn. J. Appl. Phys.* 55: 02BB13.
50 Song, J.Z. and Zeng, H.B. (2015). *Angew. Chem. Int. Ed.* 54: 9760.
51 Dankoco, M.D., Tesfay, G.Y., Benevent, E., and Bendahan, M. (2016). *Mater. Sci. Eng., B* 205: 1.
52 Gao, M., Li, L.H., and Song, Y.L. (2017). *J. Mater. Chem. C* 5: 2971.
53 Deiner, L.J. and Reitz, T.L. (2017). *Adv. Eng. Mater.* 19: 1600878.
54 Kim, K., Kim, G., Lee, B.R. et al. (2015). *Nanoscale* 7: 13410.
55 Hill, T.Y., Reitz, T.L., Rottmayer, M.A., and Huang, H. (2015). *ECS J. Solid State Sci. Technol.* 4: P3015.
56 Kuang, M.X., Wang, L.B., and Song, Y.L. (2014). *Adv. Mater.* 26: 6950.
57 Noh, Y.Y., Zhao, N., Caironi, M., and Sirringhaus, H. (2007). *Nat. Nanotechnol.* 2: 784.
58 Optomec. Optomec Aerosol Jet process schematic for 3D printed electronics.
59 Park, J.U., Hardy, M., Kang, S.J. et al. (2007). *Nat. Mater.* 6: 782.
60 Onses, M.S., Sutanto, E., Ferreira, P.M. et al. (2015). *Small* 11: 4237.
61 Jung, M., Kim, J., Koo, H. et al. (2014). *J. Nanosci. Nanotechnol.* 14: 1303.
62 Pudas, M., Halonen, N., Granat, P., and Vahakangas, J. (2005). *Prog. Org. Coat.* 54: 310.
63 Liu, X. and Guthrie, J.T. (2003). *Surf. Coat. Int. Part B: Coat.* 86: 91.
64 Pudas, M., Hagberg, J., and Leppavuori, S. (2004). *Prog. Org. Coat.* 49: 324.
65 Gu, Y., Wang, X.W., Gu, W. et al. (2017). *Nano Res.* 10: 2683.
66 Wang, X.W., Gu, Y., Xiong, Z.P. et al. (2014). *Adv. Mater.* 26: 1336.
67 Li, L.H., Bai, Y.Y., Li, L.L. et al. (2017). *Adv. Mater.* 29: 1702517. https://doi.org/10.1002/adma.201770310.
68 Li, L.H., Xiang, H.Y., Xiong, Y. et al. (2018). *Adv. Sci.* 5: 1800558.
69 Roh, E., Hwang, B.U., Kim, D. et al. (2015). *ACS Nano* 9: 6252.
70 Amjadi, M., Pichitpajongkit, A., Lee, S. et al. (2014). *ACS Nano* 8: 5154.
71 Kim, J., Lee, M., Shim, H.J. et al. (2014). *Nat. Commun.* 5: 5747.
72 Park, J., Lee, Y., Hong, J. et al. (2014). *ACS Nano* 8: 4689.
73 Choong, C.L., Shim, M.B., Lee, B.S. et al. (2014). *Adv. Mater.* 26: 3451.
74 Trung, T.Q., Tien, N.T., Seol, Y.G., and Lee, N.E. (2012). *Org. Electron.* 13: 533.
75 Tien, N.T., Jeon, S., Kim, D.I. et al. (2014). *Adv. Mater.* 26: 796.
76 Kim, D.I., Trung, T.Q., Hwang, B.U. et al. (2015). *Sci. Rep.-Uk* 5: 12705.
77 Zirkl, M., Sawatdee, A., Helbig, U. et al. (2011). *Adv. Mater.* 23: 2069.
78 Lipomi, D.J., Vosgueritchian, M., Tee, B.C.K. et al. (2011). *Nat. Nanotechnol.* 6: 788.
79 Yao, S.S. and Zhu, Y. (2014). *Nanoscale* 6: 2345.
80 Yamada, T., Hayamizu, Y., Yamamoto, Y. et al. (2011). *Nat. Nanotechnol.* 6: 296.
81 Wang, H.Y., Liu, Z.F., Ding, J.N. et al. (2016). *Adv. Mater.* 28: 4998.
82 Nguyen, T.H., Fraiwan, A., and Choi, S. (2014). *Biosens. Bioelectron.* 54: 640.
83 Berchmans, S., Bandodkar, A.J., Jia, W.Z. et al. (2014). *J. Mater. Chem. A* 2: 15788.
84 Zhou, G.M., Li, F., and Cheng, H.M. (2014). *Energy Environ. Sci.* 7: 1307.

85 Hu, L.B., Wu, H., La Mantia, F. et al. (2010). *ACS Nano* 4: 5843.
86 Yu, Z.B., Li, L., Zhang, Q.W. et al. (2011). *Adv. Mater.* 23: 4453.
87 Yoon, J., Baca, A.J., Park, S.I. et al. (2008). *Nat. Mater.* 7: 907.
88 Simoes, R. and Neto, V. (2016). *J. Mater. Res.* 31: 1633.
89 Tilley, S.D. (2019). *Adv. Energy Mater.* 9: 1802877.
90 Genene, Z., Mammo, W., Wang, E., and Andersson, M.R. (2019). *Adv. Mater. (Deerfield Beach, Fla.)* 31: e1807275.
91 Xu, J.-Y., Tong, X., Yu, P. et al. (2018). *Adv. Sci.* 5: 1800221.
92 Pagliaro, M., Palmisano, G., and Ciriminna, R. (2008). *Flexible Solar Cells.* Weinheim: Wiley-VCH.
93 Miettunen, K., Vapaavuori, J., Poskela, A. et al. (2018). *Wires Energy Environ.*: 7.
94 Berginc, M., Krasovec, U.O., and Topic, M. (2014). *Sol. Energy Mater. Sol. Cells* 120: 491.
95 Mozaffari, S., Nateghi, M.R., and Zarandi, M.B. (2017). *Renew. Sustain. Energy Rev.* 71: 675.
96 Jang, J. (2006). *Mater. Today* 9: 46.
97 Comiskey, B., Albert, J.D., Yoshizawa, H., and Jacobson, J. (1998). *Nature* 394: 253.
98 Tang, C.W. and Vanslyke, S.A. (1987). *Appl. Phys. Lett.* 51: 913.
99 Chen, H.W., Lee, J.H., Lin, B.Y. et al. (2018). *Light-Sci. Appl.* 7: 17168.
100 Park, S. and Jayaraman, S. (2003). *MRS Bull.* 28: 585.
101 Shim, B.S., Chen, W., Doty, C. et al. (2008). *Nano Lett.* 8: 4151.
102 Stoppa, M. and Chiolerio, A. (2014). *Sensors-Basel* 14: 11957.

# 13

# Materials and Processes for Stretchable and Wearable e-Textile Devices

*Binghao Wang*[1] *and Antonio Facchetti*[1,2]

[1] *Northwestern University, Department of Chemistry and the Materials Research Center, 2145 Sheridan Road, Evanston, IL 60208, USA*
[2] *Flexterra Corporation, 8025 Lamon Avenue, Skokie, IL 60077, USA*

## 13.1 Introduction

Recent developments in materials and device design, specifically for mechanically flexible/stretchable, wearable electronics has broadened the applicability of "Internet of Things" (IoTs) concepts to a myriad of new fields and commercial products [1–8]. Specifically, when addressing the needs of human beings, such as monitoring of body motions, detection of health data, and integration with human tissues, the optoelectronic devices (capacitors, transistors, sensor, energy generator, light emitter/absorber, battery), the interconnect/circuits enabling their functions, and the core passive components (such as the substrate, adhesion promoters, planarization, and passivation materials) from which the whole device is fabricated must sustain different degrees of mechanical stress [9–14]. All these materials and components are critical for proper device function; however, the electrical conductor is required in all electronic building blocks and, equally important, to connect them within the device and, for active power supply, with the external power source. Consequently, as a vital component of device assembly, conductors in the electronic system mainly determine the reliability, and dominate the bulk mechanical flexibility and stretchability to mechanical stresses, of the entire device.

The most common electronic devices are based on mechanically flexible but poorly deformable metal lines enabling rigid circuits with excellent performance parameters but very limited mechanical compliance to several types of stresses [15, 16]. This results in considerable limitations for their integration with/at mechanically stretchable points of contact and biological interfaces due to severe mismatch between mechanical characteristics. The Young's moduli of common textiles and the human skin or organs are typically $<$100 to 1000 times smaller than those of bulk metals [17, 18]. Consequently, to overcome this limitation, approaches involved the development of new conducting materials, the design of conducting element having mechanically insensitive geometries, and the utilization of conductors having a micro/nanostructure exhibiting greater tolerability to mechanical stresses as well as combination thereof.

*Flexible and Wearable Electronics for Smart Clothing*, First Edition.
Edited by Gang Wang, Chengyi Hou, and Hongzhi Wang.

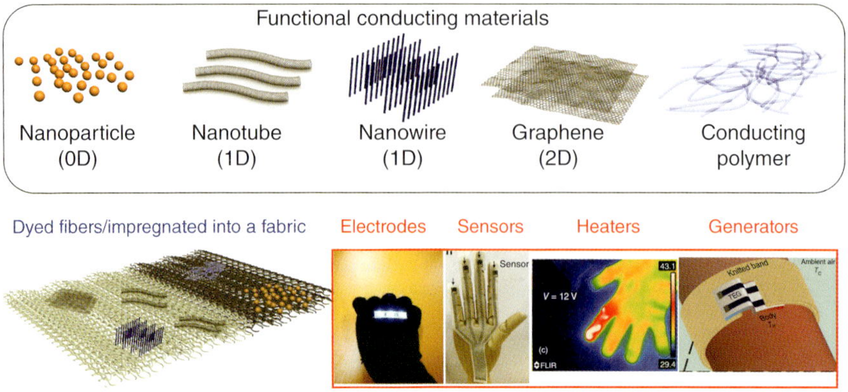


**Figure 13.1** Schematic of the materials, structures, and applications of e-textiles. Source: Zhao et al. 2019 [19]. Copyright 2019, Reproduced with permission of Wiley-VCH Verlag GmbH & Co. KGaA; Liu et al. 2017 [20]. Copyright 2017, Reproduced with permission of Wiley-VCH Verlag GmbH & Co. KGaA; Lima et al. 2018 [21]. Copyright 2018, Reproduced with permission of American Chemical Society; Allison and Andrew 2019 [22]. Copyright 2019, Reproduced with permission of Wiley-VCH Verlag GmbH & Co. KGaA; Wang and Facchetti 2019 [23]. Copyright 2019, Reproduced with permission of Wiley-VCH Verlag GmbH & Co. KGaA.

In this chapter, we summarize the recent advances in developing flexible and stretchable conducting elements for e-textile device applications, mostly based on conducting nanomaterials (Figure 13.1). Since these materials must be integrated with a textile to achieve the desired mechanical properties, in the first part we introduce several material classes for use as textile layer and/or support fiber. In the following sections, recent studies addressing specific devices, including interconnects, sensors, heaters, supercapacitors, and energy generators are summarized by using metal nanomaterials (e.g. Ag nanoparticles [AgNPs], Ag flakes, Au nanowires [AuNWs]), carbon nanomaterials (amorphous carbon, carbon nanotubes/nanofibers, graphene) and conducting polymers (e.g. poly(3,4-ethylenedioxythiophene)–poly(styrenesulfonate) [PEDOT:PSS], polyaniline [PANI]), and supporting fibers (e.g. cotton, polyacrylates) [24]. Finally, we conclude with a summary, outlining the major challenges and opportunities that this research field is experiencing.

## 13.2 Materials for e-Textiles

In this section, we summarize the basic characteristics and major classes of conducting and passive textile materials employed to blend with or be the support of the conducting element as well as the substrate to fabricate e-textile devices.

### 13.2.1 Conducting Materials

#### 13.2.1.1 Metal Nanomaterials

Silver and gold are the most investigated metals for e-textile applications. Silver exhibits the highest bulk electrical conductivity ($6.3 \times 10^5$ S/cm) of all metals and it is quite resistant against oxidation, making it commonly used in electrical circuits and as metal contacts [25]. Silver-based nanomaterials, including 0D silver nanoparticles (AgNPs), 1D silver nanowires (AgNWs), and 2D silver nanoflakes have been synthesized using different methodologies depending on precursor availability and specific applications [26–35]. Among them, the polyol method remains the most common procedure due to the facile implementation into mass production, low cost, and simplicity [36, 37].

Gold, as a precious metal, is widely used throughout history for its beauty, ductility, and resistance to corrosion as well as an excellent bulk conductor ($\sigma = 4.10 \times 10^5$ S/cm) [38]. Similar to silver, gold nanostructures include 0D gold nanoparticles (AuNPs), 1D gold nanorods (AuNRs), nanobelts, nanohelix, and AuNWs as well as 2D Au nanoflakes. 0D and 1D gold nanostructures are usually synthesized by reduction of chloroauric acid ($H[AuCl_4]$) with citrates, tannic acid, sodium borohydride, or triisopropylsilane in the presence of a stabilizer (e.g. oleylamine) [39–42].

#### 13.2.1.2 Carbon Nanomaterials

Conducting carbon nanotubes (CNTs) and graphene are certainly the most important carbon nanomaterials. CNTs are molecules with a cylindrical nanostructure consisting of a hexagonal arrangement of $sp^2$-hybridized carbon atoms [43]. Metallic CNTs offer several advantages for soft and stretchable electronics, including high intrinsic carrier mobility ($\sim$10 000 $cm^2/(V\,s)$), high current density ($4 \times 10^9$ $A/cm^2$), superior mechanical performance, and a high chemical stability [44–46]. The chemical vapor deposition (CVD) growth method is the most popular for CNT synthesis since it offers a certain degree of control over the tube diameter, length, and morphology, and it can be scaled [47]. The CNT length depends on the temperatures, growth time, gas concentration, etc. [48] To date, half-meter long CNTs with superhigh mechanical strength ($\sim$120 GPa) and breaking strain ($\sim$16%) have been reported [49, 50]. Carbon nanofibers (CNFs) are cylindrical nanostructures having graphene sheets arranged as stacked cones, cups, or plates, thus some of them resemble multiwall carbon nanotubes (MWCNTs) with a larger diameter (50–200 nm) [51].

Graphene, an all-carbon 2D material, is the most widely investigated since its emergence [52, 53]. It is a semimetal with zero bandgap and exhibits excellent electrical properties (electron mobility up to 200 000 $cm^2/(V\,s)$, conductivity up to $\sim 10^6$ S/cm), thermal conductivity, mechanical strength, and chemical stability [44]. Except high-quality CVD growth method, graphene can be produced by chemical exfoliation of graphite or CVP processes [10], Graphene can also be synthesized from the reduction of exfoliated graphene oxide (GO) sheets and is compatible with solution-processing techniques [54, 55].

#### 13.2.1.3 Conducting Polymers

Conducting polymers (CPs) are a class of organic polymers conducting electricity where the chain comprises a $\pi$-conjugated backbone mostly/entirely consisting of $sp^2$-hybridized carbon atoms. CPs may have several advantages over other (semi)conductors including solution processability, lightweight, and mechanical properties fundamentally more compatible to flexible and wearable electronic applications [56]. CPs can be synthesized by several methods including electrochemical polymerization, oxidative coupling, and polycondensation reactions. PEDOT:PSS, PANI, and polypyrrole (PPy) are the most widely investigated CP for flexible electronics due to the excellent solution processability from aqueous solutions, commercial availability, and tuned conductivity based on formulation additives and chemical functionalization [57].

### 13.2.2 Passive Textile Materials

Humans use textiles for a broad range of applications from clothes, towels, and furniture upholstery as well as medical aids. Furthermore, textiles employed in the transportation and construction industries are essential for human and commercial activities [58]. Microscopically, textiles are composed of fibers having a high aspect ratio, typically greater than 1000. These fibers can be composed of filaments, which are also called continuous fibers, or staple fiber of a certain length, which are integrated into a yarn to produce a continuous structure. These fibrotic structures are typically between 15 and 150 mm in length with a thickness ranging between 10 and 50 μm [59].

The fiber of textiles is divided into natural and manufactured, with the latter produced from both natural and synthetic polymers. Typical natural fibers can be cellulose based such as cotton or protein based such as wool and silk. In natural fibers the polymers composing them are highly aligned along the fiber axes and exhibit a high molecular mass, with cotton exceeding 1500 kDa. The silk fiber is the only natural filament and is among the toughest known fibers thanks to its high molecular weight of about 600 kDa and a hydrogen bond network with uniaxial alignment. Manufactured fibers, such as polyamide (PA), poly(ethyleneterephthalate) (PET), polyacrylonitrile (PAN), and polypropylene (PP), are widely used for textiles. Other synthetic fibers include carbon fiber, glass fiber, and mixed fibers.

To produce wearable textiles such as clothes or a piece of fabric, the fibers must be arranged into a two-dimensional pattern via a manufacturing process such as weaving, knitting, braiding, embroidering, felting, and crocheting [58]. Weaving consists of interlacing a set of longer threads (called the warp) with a set of crossing threads (called the weft), which is carried out on a loom. Knitting and crocheting involve interlacing loops of yarn, which are formed either on a knitting needle, or on a crochet hook, together in a line. These processes are different since knitting has several active loops at one time on the knitting needle waiting to interlock with another loop, while crocheting never has more than one active loop on the needle. Felting involves pressing a mat of fibers together, and working them together until they become entangled. A liquid, such as soapy water, is usually added to lubricate the fibers, and to open up the microscopic scales on strands of wool. Braiding or plaiting involves twisting threads together into a cloth. Knotting involves tying threads together and is used in making macrame. Nonwoven textiles are manufactured by the bonding of fibers to make a fabric. Finally, bonding may be thermal or mechanical, and adhesives can be used.

Table 13.1 Performance parameters for various wearable e-textile-based devices.

| Device | Performance parameters |
|---|---|
| Interconnects and electrodes | Conductivity, stretchability, conductivity–strain relationship |
| Strain sensor | GF and its relationship with strain, strain range, response/recovery time, conductivity |
| Strain gauge based pressure sensor | Sensitivity and its relationship with pressure, pressure range, response/recovery time, conductivity |
| Heater | Conductivity, temperature range, response/recovery time, stretchability, conductivity |
| Supercapacitors | Specific capacitance, energy density, stretchability, conductivity |
| Thermoelectric generators | Thermoelectric power, working temperature, stretchability, Seebeck coefficient, conductivity, thermal conductivity |
| Triboelectric generators | Open-circuit voltage, short-circuit current, power density, stretchability, conductivity |

## 13.3 Device Applications

Wearable and stretchable electronic functions and devices need the concomitant achievement of outstanding opto-electronic functions associated with superb mechanical robustness and performance compliance [60]. The fabrication of e-textile can be achieved in two ways: (i) via direct penetration/printing/in situ growth of the conducting materials into the fabric; (ii) by coating the conducting materials on fibers/yarns followed by fabric manufacturing. This section describes representative applications of wearable e-textiles, including interconnects, electrodes, strain sensors, temperature sensors and heaters, supercapacitors, and electric generators. Particularly, we provide a brief introduction of the device working mechanism and critical requirements (Table 13.1) before reporting recent advances in their utilization in e-textiles.

### 13.3.1 Interconnects and Electrodes

For stretchable and wearable electronic applications, internal wiring (or interconnects) having a high conductivity under considerable mechanical stress is required to fully integrate electronic devices [61–64]. For most applications, interconnects should have a conductivity $>10^3$ S/cm when stretched at the operation mode of the device of interest. Oh et al. pioneered stretchable conductive fabrics based on PPy and nylon-spandex fabricated by the in situ polymerization of pyrrole [65]. The resulting textile had a conductivity of ~1 S/cm with negligible change upon stretching up to 80% strain. In 2005, a seminal work by Liu et al. achieved highly stretchable sheath–core conducting fibers created by wrapping CNTs on stretched rubber fiber cores with a diameter ($d$) of ~2 mm [66]. The resulting structure exhibited a conductivity of ~28 S/cm with a resistance change of less than 5% for a 1000% stretch.

Recently, Zhao et al. reported a moss-inspired Au-coating strategy to achieve strong metal/fiber adhesion under large mechanical strains [19]. To achieve this, first the authors fabricated a dry-spun gold nanowires (AuNWs)/poly(styrene-ethylene-butylene-styrene) (SEBS) elastic fiber ($r \sim 100\,\mu m$) longer than 10 m by solution extrusion from a needle directly into the air at room temperature (Figure 13.2a,b). Next, electroless deposition of an Au film on the outmost surface of a prestrained AuNWs/SEBS fiber was carried out to form a thick Au coating. The presence of AuNWs served as seeds to initiate both the nucleation and growth processes. As expected, wrinkled structures and microcracks formed upon releasing the prestrained fibers. The initial conductivity ($\sigma$) values of the fibers resulting from different prestrains (100–600%) were 60–100 S/cm, but their resistance under higher prestrains are more strain-insensitive. For example, the resistance of the 600% and 100% prestrained fibers increased by approximately two and eight times, respectively, when a 380% strain was applied (Figure 13.2c). As a wearable e-textile, the stretchable conductive fiber was woven in a glove as part of a wearable light-emitting diode (LED) circuit. When the LED was powered by a DC voltage of 12 V, a constant illumination was achieved also when the glove was severely stretched and bent (Figure 13.2d).

More recently, a continuous dyeing process was reported by Lund et al. to convert a commercial silk sewing thread into a wash- and wear-resistant functional thread having a $\sigma$ of ~70 S/cm [67]. A laboratory-scale continuous dip-coating process was built to coat the silk yarn with PEDOT:PSS, which was achieved by continuously passes of the silk through the conductor bath (here, aqueous PEDOT:PSS dispersion containing 5% EG), followed by drying and collection onto a take-up roller. The resulting PEDOT:PSS-dyed silk yarn had a length of more than 100 m (Figure 13.2e). These yarns exhibited excellent flexibility by bending them at a radius of 5 mm for 1000 cycles and the electrical resistance ($\sigma = 74$ S/cm) increased only by 1.3 times after storage in ambient for 100 days (Figure 13.2f). The authors also used an embroidery machine to design a variety of conducting designs (Figure 13.2g). Furthermore, they demonstrated that even after this abrasive process, the resistance of these yarns remained sufficiently low (13 kΩ) to function as interconnect between a battery and an LED (Figure 13.2h). Importantly, the yarns survived machine laundering with the electrical resistance remaining stable even after 15 machine washing cycles.

Wu et al. demonstrated a solution-based Au metallization methodology (electroless nickel (Ni) immersion gold [ENIG] plating) to cover the fibers of a weft-knitted polyester textile [68]. The ENIG process involved electroless deposition of Ni on the surface of the textile and an immersion Au process. The resulting fibers have a core–shell architecture with a ~80 nm-thick Au shell. The Au-coated polyester textile exhibited sheet resistances of 1.0–3.5 Ω/sq. Next, the authors investigated the resistance variation with strain in both $X$ and $Y$ directions. For both directions, stretching the system to a 15% strain reduced the resistance by ~80%; beyond that the resistance remained constant. Also, the resistance of Au-coated textile did not change considerably after one hour

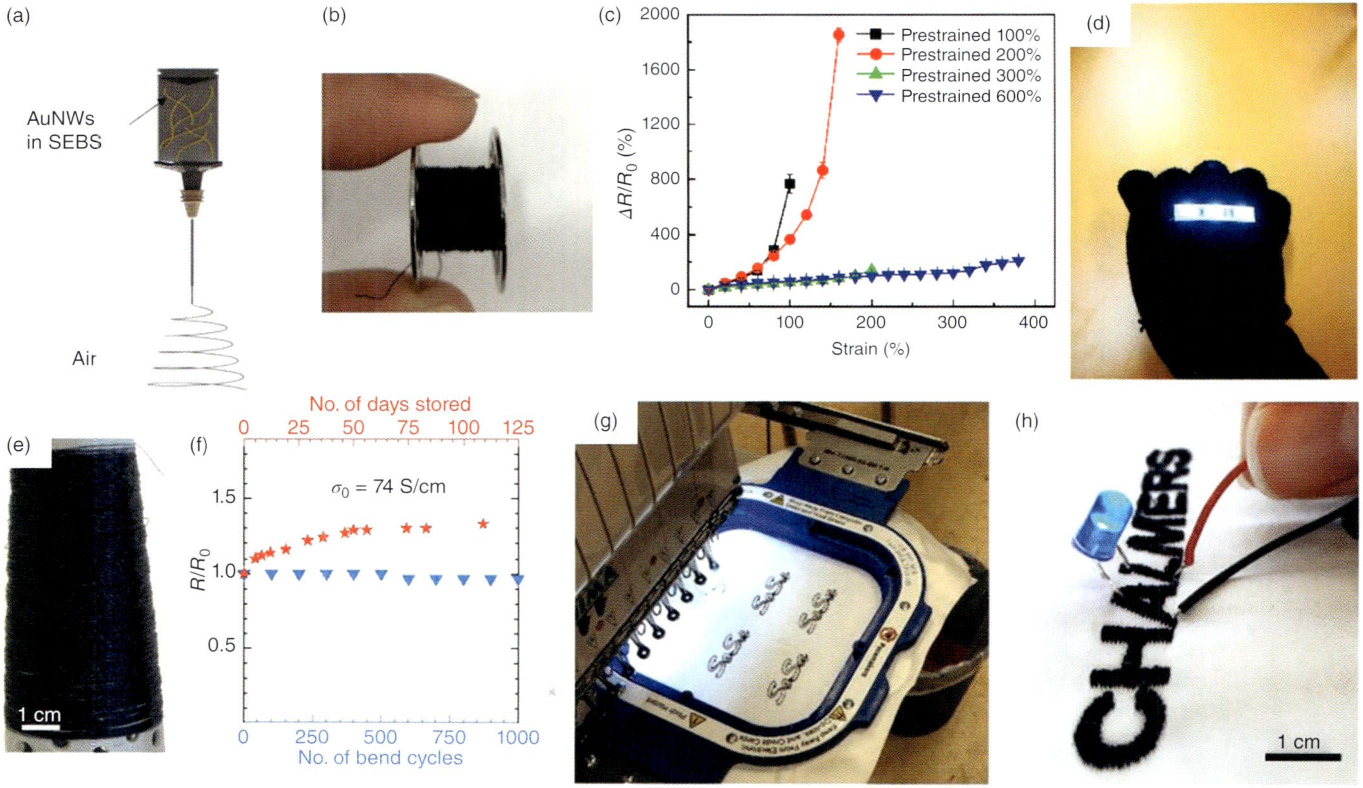


**Figure 13.2** (a) Schematic illustration of the dry spinning process to fabricate the AuNWs/SEBS fiber. (b) Photograph of a long AuNWs/SEBS fiber (>10 m) collected on a metal bobbin. (c) Normalized resistance change–strain curves on the effect of prestrain level. (d) Digital photographs of a conductive stretchable fiber woven in the finger area of a black glove as a part of a wearable LED circuit. (e) Photograph of a PEDOT:PSS-coated silk yarn more than 100 m long. (f) The change in resistance as a function of bending and storage under ambient conditions. (g) Photographs of a machine embroidery using dyed yarns. (h) A machine embroidered text that functions as an electrical connector between an LED and a battery. Source: (a–d) Zhao et al. 2019 [19]. Copyright 2019, Reproduced with permission of John Wiley & Sons; (e–h) Lund et al. 2018 [67]. Copyright 2018, Reproduced with permission of John Wiley & Sons.

washing in an aqueous solution with a laundry detergent. Furthermore, the authors applied the conductive textile as the bottom electrode for an electroluminescent fabric. When applying a 165 V AC, the device exhibited a uniform blue emission and persistent functionality without degradation up to 40% strain.

Electromyography (EMG) is a diagnostic procedure that records the electrical activity of muscles [69]. Jin et al. used knitted fabric/Ag e-textiles as conformal electrodes for EMG test [70]. The Ag flake (average particle size = 2–3 μm) ink was screen-printed on the FR3600 fabric (76% nylon and 24% polyurethane [PU]) for five passes, followed by hot pressing at 160 °C/30 kPa for 30 seconds. As shown in the scanning electron microscopic (SEM) images of Figure 13.3a,b, the Ag flakes penetrated well into the yarns. The initial sheet resistance of the printed traces was 0.06 Ω/sq, and it increased only by ~70× after stretching it to 450% (Figure 13.3c). As a proof of the good quality of this conductor, an LED connected to the printed electrode remained functional while stretched up to 400%. Next, four-channel fabric/Ag electrodes were mounted on an arm for EMG monitoring (Figure 13.3d). The results showed that the noise levels of the textile electrodes were only 0.07 mV and different activations of muscle during hand opening and closing motion were observed (Figure 13.3e). Similarly, Chung's group used screen-printed PU/Ag e-textile electrodes for EMG and electroencephalography (EEG) monitoring of several muscle activities and brain waves [73]. The EEG signal from these e-textile electrodes exhibited fewer movement artifacts than that based on commercially available gel electrodes.

Since screen printing requires high-viscosity (500–5000 mPa s) inks, conductive pastes for screen printing were developed using metal micro/nanoparticles with binding materials, which need to be eliminated upon high-temperature annealing (200–300 °C). Kim et al. reported e-textiles by inkjet printing particle-free organometallic silver inks, which requires relatively low viscosity (1–25 mPa s) [72]. The organometallic solution is formed by mixing silver carboxylates ($AgO_2C\text{-}R_1$) with aqueous proprietary ammonia ligands ($-NH_2/-NR_2H$) (Figure 13.3f). Differential scanning calorimetry and thermogravimetric analysis spectra of the ink showed an endothermic process at 150 °C, due to the solvent evaporation and volatilization of organic components. Thus, an annealing temperature of 150 °C was chosen after inkjet printing the solution on PET knit fabrics. SEM image showed that 300–2000 nm thick Ag conductive layers uniformly coated the PET fibers (Figure 13.3g). The $R/R_0$–$\varepsilon$ curve showed that $R/R_0$ first decreased with increasing the strain and reached ~0.7 for a strain of 20%, and then it remained constant when the strain increased from 30% to 115% (Figure 13.3h). Next, a commercial LED was connected with two printed line trace and mounted on a glove, which remained functional up to a ~35% strain or upon bending of fingers (Figure 13.3i,j).

### 13.3.2 Strain Sensors

A resistive strain gauge is a sensor converting a mechanical deformation into a change in electrical resistance that can then be measured. The gauge factor (GF), an important performance parameter of a strain sensor, is defined as the relative change in electrical resistance ($R$) to the mechanical strain ($\varepsilon$), given by the

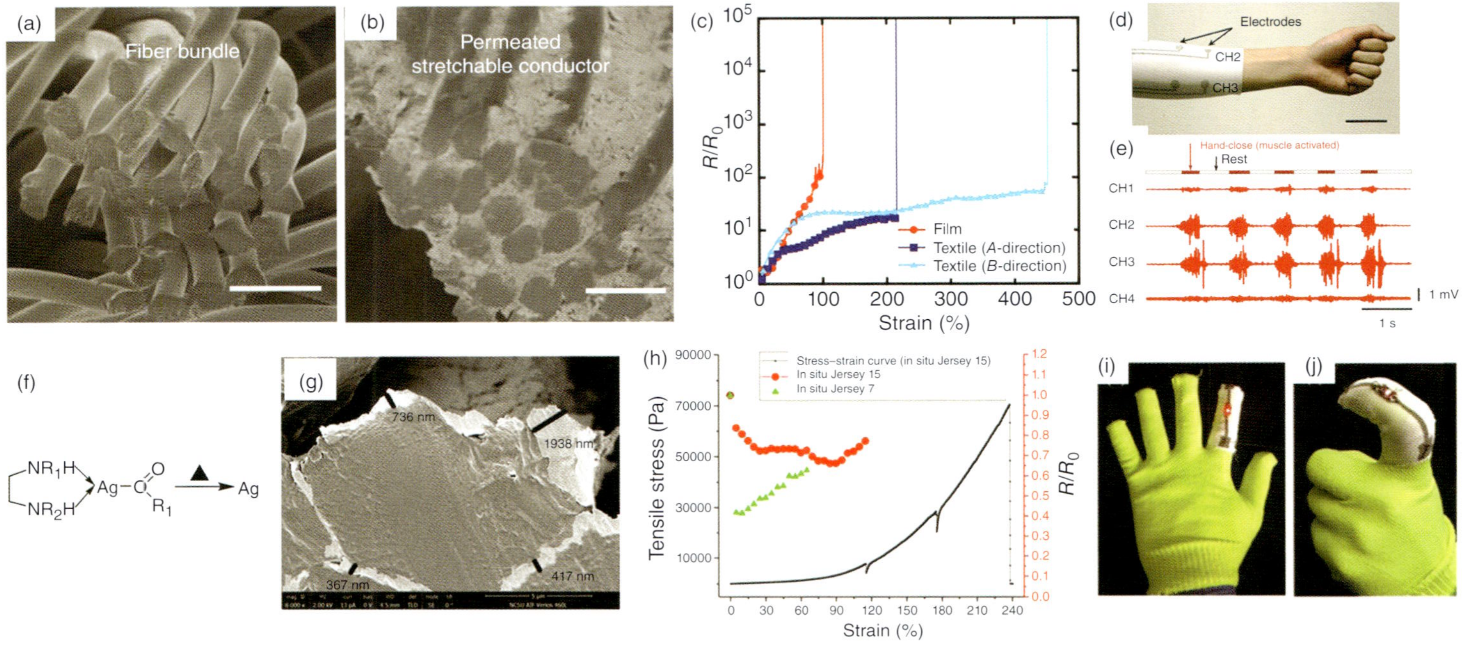


**Figure 13.3** Cross-section SEM image of the bundle (a) before and (b) after screen printing, respectively. The scale bars are 30 μm. (c) Normalized $R/R_0$ vs. strain of the silver flucroelastomer printed on a PU film substrate, on a textile in the *A*-direction (parallel to fiber bundles), and on a textile in the *B*-direction (perpendicular to fiber bundles). (d) An image of a printed four-channel EMG measurement sleeve before wearing and after wearing it. The scale bars are 5 cm. (e) EMG signals of forearm muscles measured by a four-channel sleeve during hand-close (five times) and rest. (f) Formation of silver from the organometallic amine compound when exposed to heat. (g) A SEM image of the fiber coated with Ag layer. (h) Electromechanical properties of the inkjet-printed e-textiles. (i) An image of an LED connected to conductive patterns on a textile glove. (j) LED illumination during finger joint flexion. Source: (a–e) Jin et al. 2017 [70]. Copyright 2017. Reproduced with permission of John Wiley & Sons; (f) Shahariar et al. 2019 [71]. Copyright 2019, Reproduced with permission of American Chemical Society; (g–j) Kim et al. 2019 [72]. Copyright 2019, Reproduced with permission of John Wiley & Sons.

equation $GF = \frac{\Delta R/R_0}{\varepsilon}$, where $\Delta R$ equals the difference between the resistance at a specific stain and the initial resistance ($R_0$) [74]. Other parameters include the operating temperature, the range of the strain to be detected, and the stability. When measuring the $\sigma$ variation with pressure-induced strain, the device is called a strain gauge based pressure sensor. The sensitivity ($S$) is typically employed as the performance parameter instead of the GF, and it is given by the relationship $S = \frac{\Delta R/R_0}{P}$, where $P$ is the applied pressure. High $S$ is needed to enhance detectivity while achieving a linear relationship between $\Delta R/R_0$, and $P$ is more important for quantifying the applied pressure level. With regard to stretchable strain (or pressure) sensors, they may find applications not only to detect minimal strains (<2%), subtle vibrations (e.g. heart beating), and small pressure variations (<1 Pa) from fluid or gas (e.g. blood flow), but also to monitor mechanical deformations due to large strains (e.g. >50%) or large pressure change (e.g. >100 kPa) [75]. Furthermore, fast response and recovery times are needed especially for monitoring high-frequency periodic stresses.

De Rossi et al. were the first to utilize conducting fabrics for strain and temperature sensors in 1999 [76]. As a strain sensor, the PPy/Lycra® fabric exhibited a negative GF of −13 when stretched up to 1.2%; also, it had a negative temperature coefficient of resistance (TCR) of −2%/°C when measured over a temperature range spanning from 20 to 60 °C. Later, Li et al. fabricated PPy-coated Tactal (83%)/Lycra (17%) fabrics by using a CVD method, with a GF as high as 80 when stretched to 50% [77]. In 2010, Ajayan's group in situ synthesized zinc oxide (ZnO) nanorods (NRs) on cellulose fibers, which demonstrated a GF of ~21 [78]. Two years later, a graphene woven fabric (GWF)-based strain sensor with an ultrahigh GF of $\sim 10^3$ under 2–6% strains was reported by the group of Zhu and coworkers [79]. Lee et al. reported a seminal work for detecting pressure with the use of AgNPs/SBS coated Kevlar® fibers [80]. The conducting fiber exhibited an electrical conductivity of 0.15 Ω/cm and the e-textile device based on it exhibited a $S_p$ of 0.21 $kPa^{-1}$ for pressures <1 kPa, a fast response time, and high operation stability for $10^4$ loading/unloading cycles.

The most recent attractive applications of e-textiles are based on motion monitoring, for instance, detecting the degree of bending of a knee, extension of an arm, and hand motions. To this end, silver/polyester textiles were fabricated by Song's group using a low-temperature in situ reduction of a silver precursor into AgNPs [81]. The plasma-treated polyester textile with an interlock-stitch structure was first dyed with the Ag precursor, followed by a 90 °C-annealing with formic acid. Then, the two ends of the resulting fabric/AgNP conductive composite were connected with copper wires as external electrodes. The $R/R_0$–$\varepsilon$ curve of the AgNP-based e-textile strain sensor showed that $R/R_0$ first decreased for strains less than 6% and then increased when the strain was increased between 6% and 50%. Next, multiple e-textile strain sensors were mounted on a heel, upper knee, and lower knee for real-time monitoring. The sensors mounted at different joints exhibited different resistance response for different body movements such as walk, squat, and jump.

Liu et al. reported large-area all-textile-based highly sensitive pressure sensor arrays [20]. For the fabrication, a CNT-coated cotton fabric was prepared via dip-coating, which exhibited a sheet resistance of 75.4 kΩ/sq. SEM images demonstrated that the cotton fibers were uniformly coated with a layer of a CNT network (Figure 13.4a). After laminating the CNTs/cotton on a polyester textile with interdigitated Ni electrodes, the authors measured the current response of this e-textile to pressure (0–24 kPa). As shown in Figure 13.4b, the e-textile exhibited a high $S$ of 14.4 $kPa^{-1}$ for pressures below 3.5 kPa, which decreased to 7.8 $kPa^{-1}$ for pressures between 3.5 and 15 kPa. The great mechanical stability and durability were confirmed by cycling pressing (2.9 kPa, 1000 cycles) and bending ($r = 25$ mm, 900 cycles) test. Also, when a leaf (pressure ~2 Pa) was placed on the e-textile, the current changed with a rapid response/recovery time of ~20 ms (Figure 13.4c,d). Furthermore, the e-textile was used to detect finger movements, acoustic vibrations, and pulse beat by attaching it to fingers, neck, and wrist, respectively. For instance, the e-textiles could sense the motion of each finger precisely, and the output of each action (holding all fingers straight, stretching the index and middle finger into a "V" sign, shaking hands, and holding a pencil or a beaker) could be monitored and distinguished (Figure 13.4e,f).

Song et al. reported a MWCNTs/polypropylene (PP) fabric e-textile as a sensitive pressure sensor [82]. The roll-to-roll synthetic process for the production of the MWNTs@fabric is schematically illustrated in Figure 13.5a. During the process, a piece of PP fabric was unraveled from the roller I, passed through a MWNTs/polydimethylsiloxane (PDMS) dispersion, dried by the lamp, and then rolled up by roller II, resulting in the final large-scale MWNTs/PP fabric products as shown in Figure 13.5b. The use of PDMS enhances the adhesion of the MWNT with the fabric, which maintains its original resistance after washing in water for several times. Figure 13.5c shows the relative resistance change with pressure. Pressing in the normal direction led to a resistance decrease with a sensing GF of −1.4 $kPa^{-1}$ for a pressure <610 Pa and −0.114 $kPa^{-1}$ for pressures in the range of 610–1500 Pa. Also, the MWNTs@fabric exhibited high durability and the resistance change is reversible even after 5000 times of cyclic pressing. Furthermore, one piece of the wearable e-textile was attached to the fingertip and the force loading details during the pick-up process were recorded in the resistance change waveforms (Figure 13.5d,e). Picking up different objects (a foam bulk, an egg, and a bottle of water) led to similar $\Delta R/R_0$ waveforms but different amplitudes. Figure 13.5f shows the enlarged $\Delta R/R_0$-time curves for picking up an egg, which exhibited three distinct stages (grasping, holding, and releasing).

Wang et al. reported e-textile strain sensors based on reduced graphene oxide (rGO)-decorated thermoplastic PU electrospun fibrous mats [83]. The functionalization of the PU textile with rGO was achieved by ultrasonication (380 W for 20 minutes) of an rGO solution (0.6 mg/ml in water) containing the PU mat. SEM images show that the unfunctionalized PU fiber exhibited a smooth surface with an average diameter of 1.3 μm, while it became rough like fish scales after ultrasonication, indicating that an rGO coating (~200 nm thick) was successfully anchored onto the fiber surface. The initial $R$ of the rGO/PU textile was ~100 kΩ. Next, the e-textile was used as pressure sensor, which exhibited a fast response time (0.2 seconds). The GF of this device was 15.2 when the strain was <30%, and then it gradually increased to 79 when the strain increased to 100%. Also,

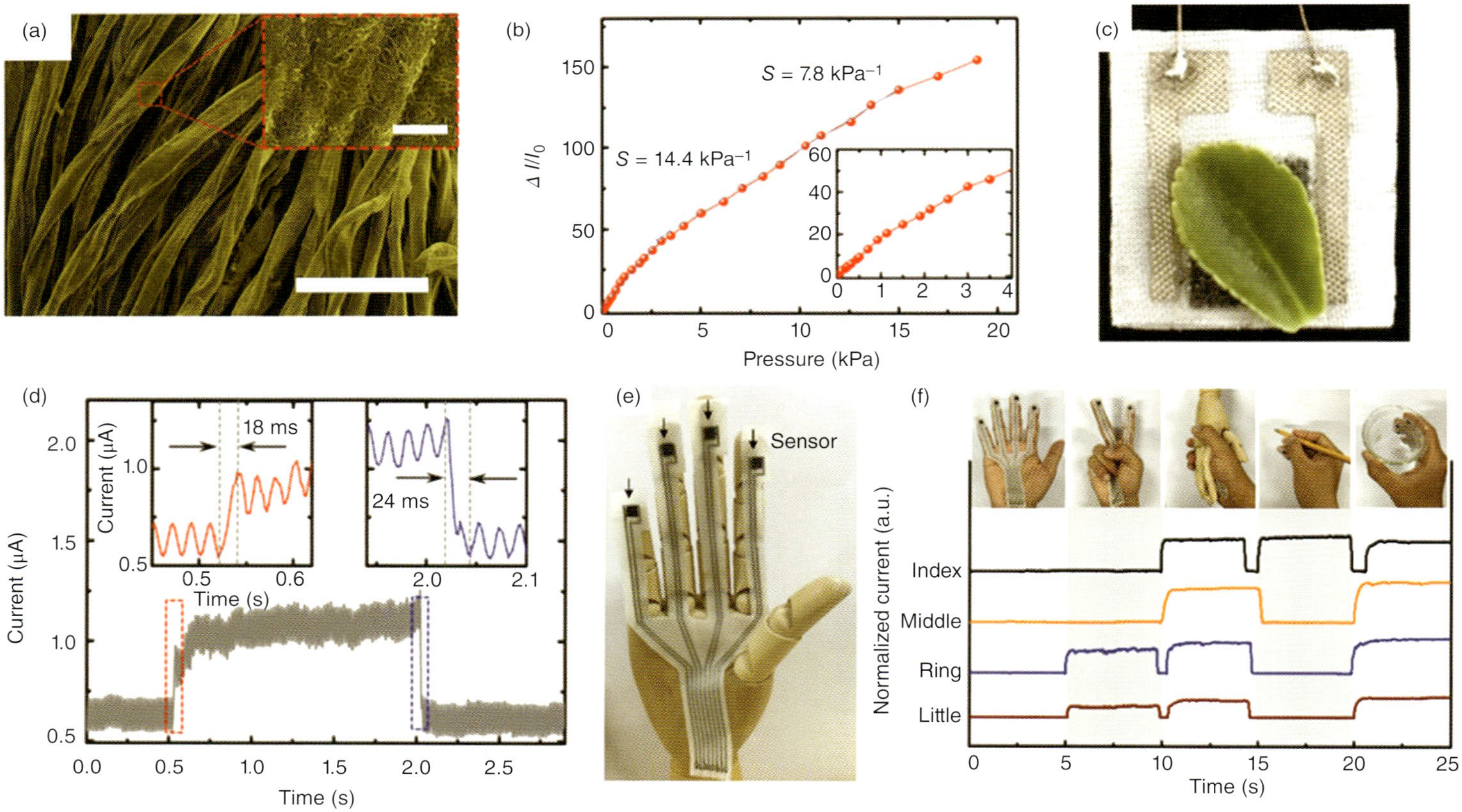


**Figure 13.4** (a) SEM image of a CNT coated cotton fabric. Scale bar is 100 μm. (b) Relative current change with pressure for the all-textile pressure sensor. (c) Optical image and (d) current curve of the e-textile loaded/unloaded with a leaf. (e) Photograph of an wood hand covered by textile pressure sensors. (f) Real-time current signal in response to various hand motions. Source: Liu et al. 2017 [20]. Copyright 2017, Reproduced with permission of John Wiley & Sons.

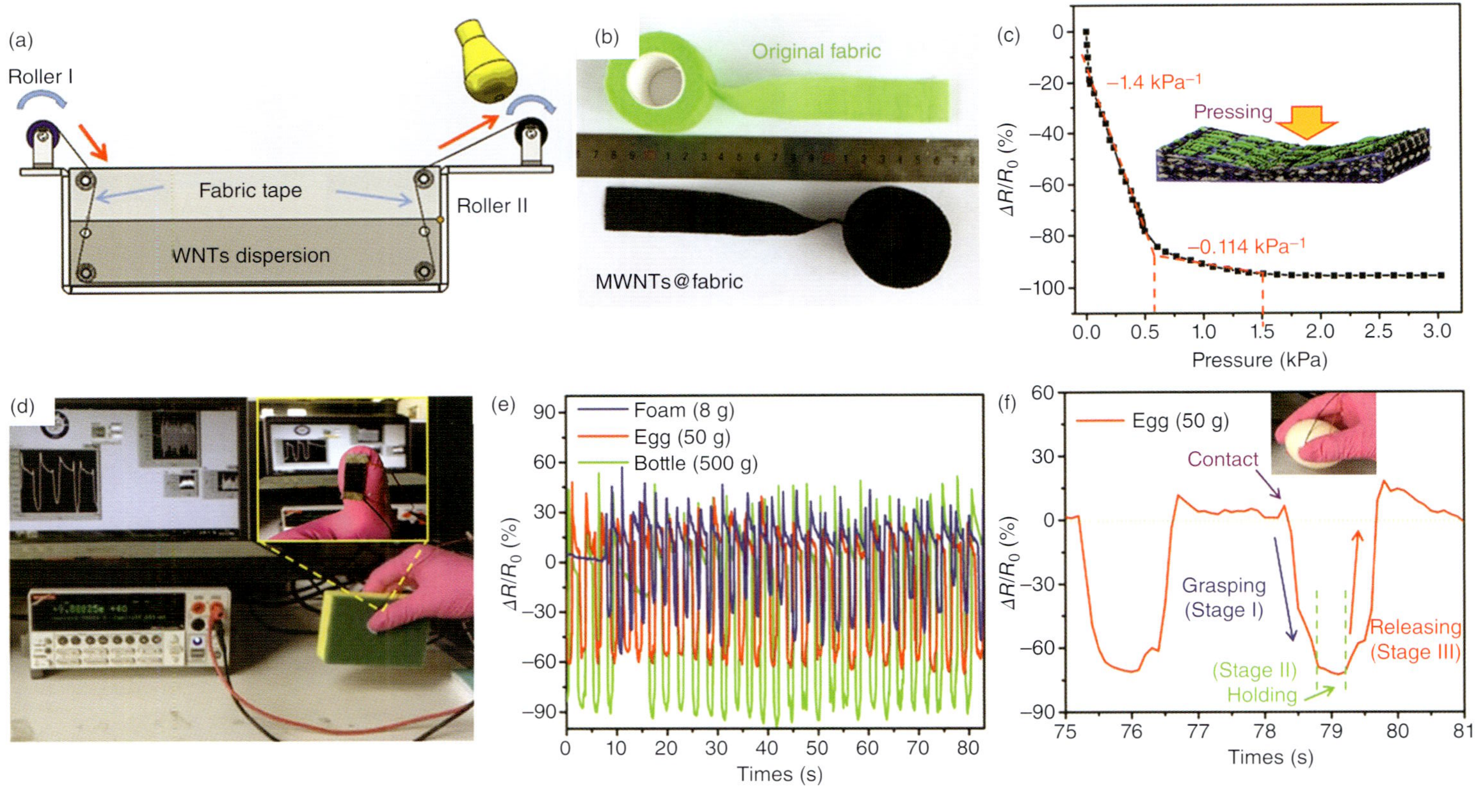


**Figure 13.5** (a) Schematic illustration of the roll-to-roll synthetic process for the production of a MWNTs@fabric e-skin. (b) The adopted original fabric (top) and the final MWNTs@fabric product (bottom). (c) Relative resistance changing plot of the e-fabric with normal pressure. (d) Configuration of the setup used for loading forces monitoring during grasping a body with a piece of the e-fabric attached on a person's fingertip (the inset). (e) Relative resistance–time plot for the e-fabric when grasping up a foam, an egg, and a bottle of water having the indicated weights. (f) The enlarged view of a selected area of the resistance change plot when picking up an egg. Source: Song et al. 2019 [82]. Copyright 2019, Reproduced with permission of John Wiley & Sons.

this e-textile exhibited remarkable stability and repeatability during 6000 loading/unloading cycles with a strain of 50%. Furthermore, various large (knee, wrist, finger bending) and subtle (phonation, coughing, muscle movements) human motions were monitored with this e-textile sensor. For instance, the $\Delta R/R_0$ of the e-textile mounted on knee changed during leg bending; thus it could distinguish the motions of squatting, walking, jumping, and running.

### 13.3.3 Heaters

A resistive heater is a device that produces heat through the passage of an electric current. This device has been utilized extensively in a wide range of medical, consumer, industrial, and automotive applications [84]. The operating voltage, maximum operating temperature, temperature uniformity, and temperature saturation time are critical parameters for a resistive heater. With regard to wearable heaters, they are projected to be very valuable for personal thermal management and health-care purposes, such as heat preservation and thermotherapy [85, 86]. Desirable wearable heaters should be conformal to the human body, exhibit >20% stretchability, ~40 °C operating temperature, and a low driving voltage (<5 V) [87, 88].

Wang et al. reported the use of $Ti_3AlC_2$ nanosheets/PPy-coated polyester textile as a heater [86]. First, the 2D $Ti_3AlC_2$ sheets were modified with in situ polymerized PPy to generate an electrically conductive and stable $Ti_3AlC_2$/PPy ink. Next, a PET textile was immersed in the $Ti_3AlC_2$/PPy ink to afford the conducting element, which was laminated on a PDMS layer. The resulting textile exhibits an electrical conductivity of ~10 S/cm. Figure 13.6a compares the Joule heating performances of this e-textile under different input voltages, and the insets present the related temperature distribution images recorded by an infrared camera. A voltage of 2 V yields a saturated temperature of ~40 °C, which is elevated to 79 °C at 4 V.

Zhang's group reported a highly stretchable electrically driven heater based on electrically conductive weft-knitted fabrics [87]. The fabrication process is given in Figure 13.6b. First, a piece of modal textile was treated under an inert atmosphere at 1050 °C, leading to the formation of highly conductive carbonized modal textile (CMT). After connecting with external Cu electrodes, the CMT was placed on an Ecoflex® substrate and encapsulated with a layer of liquid Ecoflex precursor. The resistance of CMT under different strains was measured and the results showed that negligible change in resistance (~70 Ω) was observed under large strains up to 80%. Importantly, operation stability was accessed by cyclic loading/unloading for a 60% strain over 2000 cycles. Figure 13.6c,d reports the Joule heating performances of the CMT heater under static and dynamic states. The CMT exhibited a fast thermal response (<4 seconds) under various voltages and the temperature increased to 140 °C when operated at 3.5 V. Figure 13.6e shows the temperature of a CMT heater at a constant input voltage of 4 V and under dynamic stepwise strains from 0% to 70%. The Joule heating performance of the CMT heater with a maximum temperature of ~135 °C did not deteriorate even under a large strain of 70%.

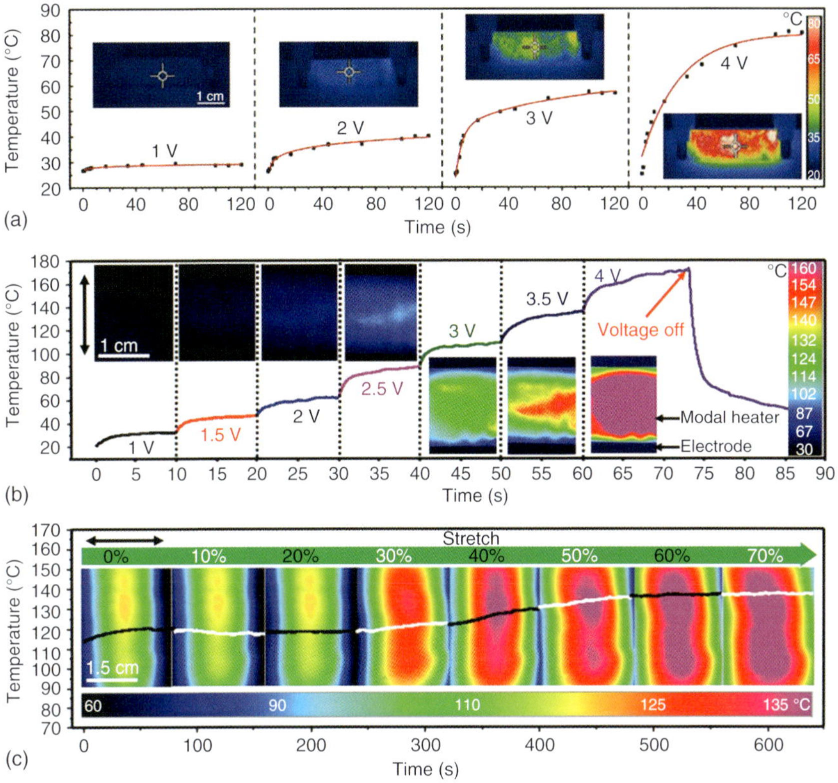


**Figure 13.6** (a) Joule heating performances of a piece of $Ti_3AlC_2$/PPy-coated PET textile. (b) Transient temperature evolution of the CMT heater without strain at various DC voltages. Insets show the distribution of the temperature in the CMT heater at various DC voltages recorded by an IR camera. (c) Transient temperature evolution of the CMT heater under stepwise strains at a DC voltage of 4 V. Insets are the corresponding images showing the distribution of temperature. Source: (a) Wang et al. 2019 [86]. Copyright 2019, Reproduced with permission of John Wiley & Sons; (b, c) Zhang et al. 2017 [87]. Copyright 2017, Reproduced with permission of John Wiley & Sons.

### 13.3.4 Supercapacitors

A supercapacitor is a high-capacity capacitor, which is used to store electrical energy that bridges the gap between electrolytic capacitors and rechargeable batteries. A supercapacitor is typically composed of two active electrodes separated by an ion-permeable membrane (separator) and an electrolyte medium connecting the electrodes. Currently, supercapacitors are used for vehicles, cranes, and elevators, where temporary storage/release of energy is needed [89]. Critical parameters of supercapacitor are specific capacitance ($C$), specific energy, charging/discharging time, working voltage, and cycle life. In earlier studies, Cui's group demonstrated stretchable textile-based supercapacitors using SWCNTs/cotton as the electrodes [90]. A high specific capacitance of 80 F/g at 20 μA/cm$^2$ was achieved and it did not change after being stretched up to a 120% strain for 100 times. Peng's group developed a superelastic fiber-shaped supercapacitor based on CNT/PANI composite electrodes, which could be stretched by >400% [91]. The reported specific capacitance was ~80 F/g and remained unaltered after stretching at a strain of 300% for 5000 cycles.

Recently, Lima et al. fabricated a wearable multifunctional electrically conducting cotton yarn by incorporating CNTs and PPy [21]. First, CNTs were anchored onto the cotton yarns by sonicating the yarns into a CNT/water solution, followed by in situ interfacial polymerization of pyrrole on the CNT-coated yarns (named CNT-I-PPy). Electrical measurements on the CNT-I-PPy yarns indicated a conductivity of ~10.5 S/cm, which was insensitive to textile bending and twisting. In addition the durability was confirmed by investigating the electrical performance of CNT-I-PPy yarns after several cycles of laundering. The conductivity exhibited a small decrease after the first washing cycle, due to the removal of weekly bounded conducting filaments; then it remained unchanged even after five successive washing procedures. The CNT-I-PPy cotton yarn could be used as the electrodes of a supercapacitor, composed by a symmetric assembly of parallel textile electrodes separated by a thin PVA/$H_3PO_4$ electrolyte layer. The resulting device exhibited a maximum specific capacitance and power density of 30 F/g and 11.33 mW/g, respectively. The device was cycled under a current of 1.5 mA for up to 2000 cycles to verify the cyclability. The results demonstrated that the capacitive retention increased in the first 500 cycles, possibly due to a self-activation process, and then decreased slightly and remained at ~80% of the initial value after 2000 cycles of use. Furthermore, the authors also demonstrated that the CNT-I-PPy cotton yarn can function as a wearable heater when it was sewed on a cotton knitted hand glove. When the device was connected to a 12 V DC voltage source, IR camera images revealed that the wearable heater could achieve a temperature of ~ 43 °C.

Naguib's group reported electrochemical capacitors and strain sensors based on polyaniline nanorod (PAniNR)-decorated PAN nanofibers [92]. For device fabrication, first PAN fibers with an average diameter of 72.2 nm were fabricated by electrospinning (Figure 13.7a). Then, the in situ growth of PAniNR on the PAN fiber was conducted in an aqueous solution of aniline monomer/HCl/ammonium persulfate, affording PAniNR@PAN fibers. The fiber average diameter increased to 173.6 nm after PAniNR coating as shown in Figure 13.7b. When using two pieces of PAniNR@PAN fabrics as parallel electrodes, 1 M $H_2SO_4$ aqueous solution as electrolyte, the gravimetric capacitance for the PAniNR@PAN electrode is 629.6 F/g, corresponding to an energy density of 63.0 Wh/kg when measured at a 10 mV/s scan rate (Figure 13.7c). With the utilization of the PVA–$H_2SO_4$ gel electrolyte between two PAniNR@PAN fabric electrodes, this device exhibited a good sensitivity to pressure. Figure 13.7d demonstrates a linear relationship between $\Delta R/R_0$ and pressure with a $S_p = 0.95\ \text{MPa}^{-1}$, where the resistance decreases at higher pressure due to the formation of a more dense conductive network. Furthermore, cyclic tests (Figure 13.7d) were conducted by applying a pressure onto the fabric cell by a fingertip revealing excellent stability after ~200 cycles (Figure 13.7e).

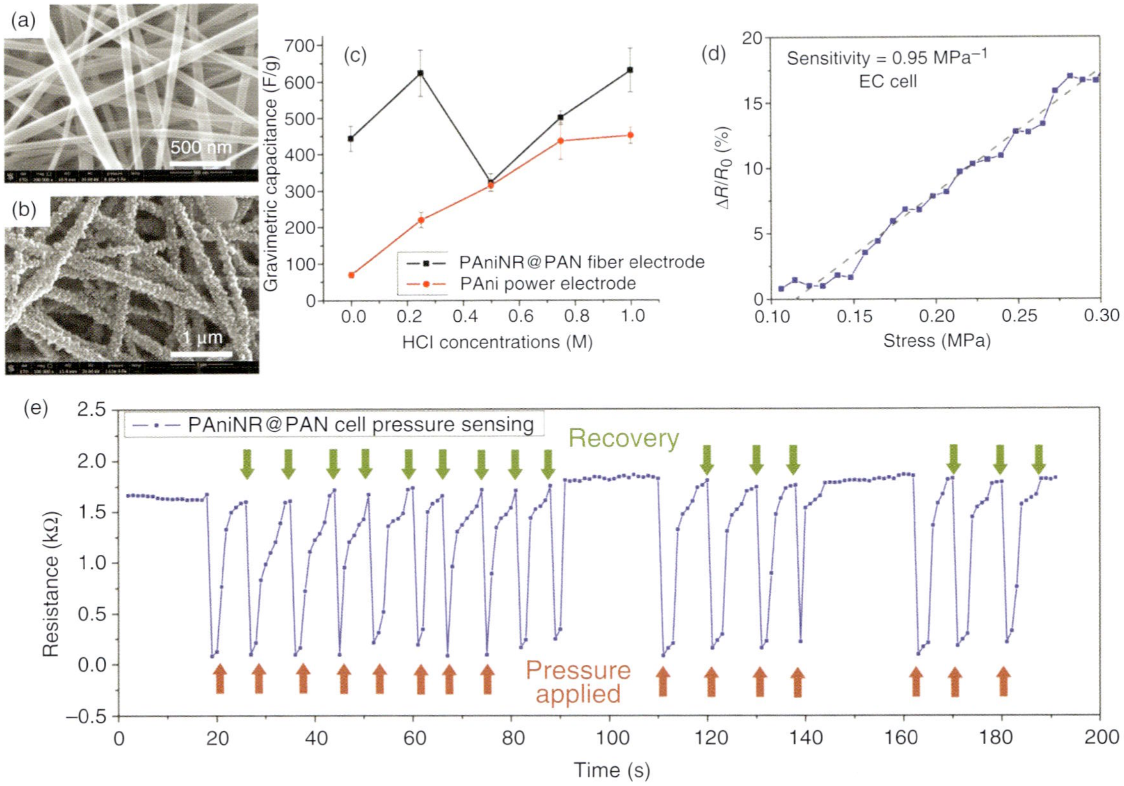


**Figure 13.7** SEM images of (a) pristine electrospun PAN fibers and (b) PAniNR@PAN fibers. (c) Comparison of gravimetric specific capacitance values of a PAniNR@PAN fiber electrode to that of the PAni powder electrode for different HCl concentrations. (d) Piezoresistive response of an electrochemical capacitor cell to compressive stress. (e) Piezoresistive response of the PAniNR@PAN cell during finger tapping of the electrochemical capacitors cell. Source: Shi et al. 2018 [92]. Copyright 2018, Reproduced with permission of American Chemical Society.

### 13.3.5 Energy Generators

A generator is a technology converting mechanical and thermal energy resulting from a small physical change into electricity [93, 94]. Integrating these self-powered systems in wearable electronic devices could generate power for their operation in an efficient way [95, 96]. In fact, the human body-produced biomechanical and thermal energies could be an effective power source for wearable electronics. For example, daily activities, such as the footsteps and arm movements of a 68-kg adult person, can deliver approximately 67 and 60 W of kinetic power, respectively [97]. There are four major types of generators: piezoelectric, triboelectric, pyroelectric, and thermoelectric. Here we focus mainly on the thermoelectric and triboelectric as they are more promising in our daily life.

#### 13.3.5.1 Thermoelectric Generators

The device structure of a thermoelectric generator is based on an array of n-type (electron transporting) and p-type (hole transporting) thermoelectric couples connected in series. This device functions by placing the two sides at different temperature, thus resulting in a temperature gradient ($\Delta T$) across it and producing a permanent output voltage ($V_{oc}$). A key dimensionless figure of merit of a thermoelectric material providing it the capability to produce power from a temperature gradient is $zT$, given by $zT = \frac{S_c^2 \sigma T}{\kappa}$, where $z$ is a measure of a material's thermoelectric properties, $T$ is the absolute temperature, $S_c$ is Seebeck coefficient, $\sigma$ is the electrical conductivity, and $\kappa$ the thermal conductivity [98]. Thus, high performing thermoelectric materials should have a large $\sigma$ and a large $S_c$ along with a small $\kappa$.

In the wearable/stretchable area, Cho's group first reported ~500 μm thick, lightweight (~0.13 g/cm$^2$), and wearable thermoelectric generators with printed n-type $Bi_2Te_3$ ($S_c$ = 141 μV/K) and p-type $Sb_2Te_3$ ($S_c$ = 98 μV/K) films on glass-fabric substrates, which exhibited a high output power density of 3.8 mW/cm$^2$ (corresponding to 28 mW/g$^1$) for a $\Delta T = 50$ K [99]. In 2015, Du et al. reported thermoelectric generators based on PEDOT:PSS-coated polyester fabrics [100]. The fabric device exhibited stable thermoelectric properties from 300 to 390 K and could generate a voltage output ($V_{oc}$) of 4.3 mV at a $\Delta T$ of 75.2 K.

Recently, Andrew's group reported an all-textile thermopile fabricated by vapor printing a chloride-doped poly(3,4-ethylenedioxythiophene) (PEDOT-Cl) film onto a commercial cotton fabric [22]. The coating of the fabric was accomplished by in situ polymerization of EDOT monomer and $FeCl_3$ (oxidant) vapors at 80 °C (Figure 13.8a,b). The resulting PEDOT-Cl/cotton exhibited a $\sigma$ of ~9 S/cm, which did not change after rubbing or laundering. Thermoelectric test revealed that the conductivity increased from ~15.4 S/cm for a $\Delta T = 1$ °C to ~16.2 S/cm for a $\Delta T = 20$ °C (Figure 13.8c). For a $\Delta T = 10$ °C the Seebeck coefficient was found to be 16 μV/K and the power factor calculated as 0.48 μW/(m K$^2$). Next, the authors demonstrated an all-fabric thermopile for wearable e-textile applications. As shown in Figure 13.8d, two rectangular PEDOT-Cl patterns (8 × 42 mm) were first vapor printed on a piece of cotton, which served as *p*-type elements; then one 50 mm-long carbon fiber thread (*n*-type) was sewed to connect them, finished by sewing silver-coated nylon as electrical contacts.

Figure 13.8e shows that the cotton thermopile was sewn into a band so that one surface was exposed to air while the other surface was buried under the band, enabling a heat gradient across the device. The resulting band displayed a clear thermoelectric behavior and generated ~5 and ~10 mV voltage outputs when the person was indoor ($\Delta T$~12 °C) and outdoor ($\Delta T$~35 °C), respectively (Figure 13.8f). Interestingly, the voltage output could increase to 24 mV after the person performed intense physical activities because of the perspiration.

Ryan et al. reported on conducting PET yarns prepared by dip-coating a composite of CNTs and polyvinylpyrrolidone (PVPP) [101]. The resulting electron-transporting (n-type) yarns had a conductivity of 1 S/cm and a $S_c$ of −14 μV/K, corresponding to a thermoelectric power factor of ~$10^{-2}$ μW/(m $K^2$). Furthermore, the authors combined the n-type yarns with PEDOT:PSS dyed silk yarns (hole conducting component) to fabricate an all-organic textile thermoelectric module with 38 n/p elements. One side of the textile module was placed on a hot plate and the other side on a cold plate. The textile module was capable of producing a $V_{oc}$ of 143 mV when exposed to a $\Delta T$ of 116 °C and a maximum power output of 7.1 nW at a $\Delta T$ of 80 °C.

#### 13.3.5.2 Triboelectric Generators

A triboelectric generator, based on a Maxwell displacement current, can convert mechanical energy into electricity via triboelectric effect [102–105]. This device was first demonstrated as an energy-harvesting architecture by Wang's group in 2012 [106]. The triboelectric effect is a particular contact electrification, in which electrical charges are separated and transferred from one material to the other material when they are brought into frictional contact [95]. Almost all known materials exhibit the triboelectrification effect, but the extent and type of charges produced during operation depends on the type of material, surface roughness, temperature, as well as other parameters. An AC power output is produced from a triboelectric device since the free electrons will continuously flow back and forth between the two electrodes. For this type of device, the open-circuit voltage ($V_{oc}$), the short-circuit current ($I_{sc}$), and the power density are critical parameters. In 2015, Kim et al. reported a fiber-shaped triboelectric nanogenerator (TENG) based on ZnO NR-coated Al wires and PDMS [107]. After weaving it into a fabric, the output current reached 210 μA, corresponding to a power output of 4 mW, under a cycled compressive force of 50 N. This fabric was also found to be quite stretchable to >25%.

Recently, Cao et al. reported a washable CNT-coated nylon textile as TENG [108]. For device fabrication, a ~20 μm thick CNTs electrode array was first screen-printed on a nylon fabric, followed by annealing at 80 °C for 10 minutes. Next, a silk fabric serving as a frictional material was laminated on the CNT electrodes by hot pressing. The resistance between the two ends of the electrode was 0.2 kΩ/sq, which remained stable during folding or after 15 hours of immersion in water. The TENG is capable of generating electrical signals when pressed with a finger. When applying a ~100 kPa pressure, the $V_{oc}$ of the device remained stable (~7 V) under various frequencies (1–2 Hz), whereas the $I_{sc}$ increased from 50 to 180 nA when the frequency increased from 1 to 2 Hz. However, the $V_{oc}$ was found to be dependent on applied pressure. Pressure-dependent sensitivity was

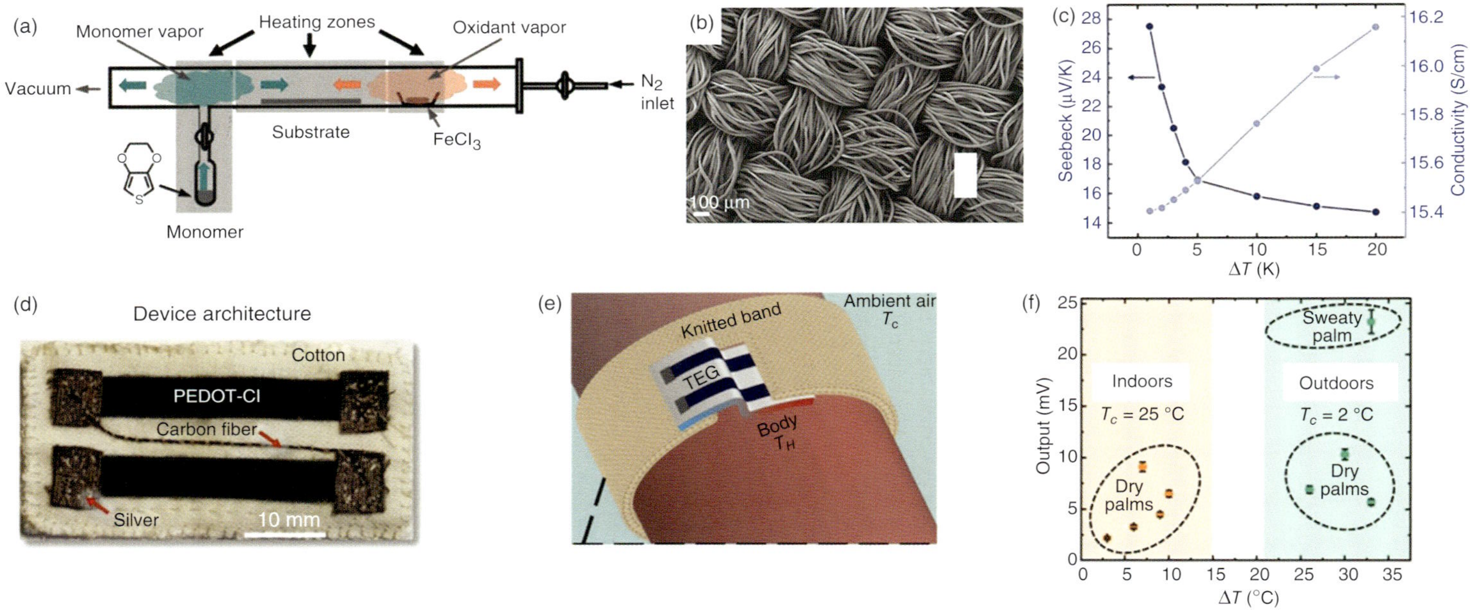


**Figure 13.8** (a) Schematic of the vapor printing process for PEDOT deposition. (b) An SEM image of PEDOT-Cl/cotton. (c) Seebeck coefficient and conductivity of PEDOT-Cl/cotton at different temperature gradients. (d) A photograph of an all-fabric thermopile. (e) Schematic of a wearable thermoelectric generator. (f) Thermovoltage outputs obtained when multiple devices wore the knit TEG on their palm. Source: Allison and Andrew 2019 [22]. Copyright 2019, Reproduced with permission of John Wiley & Sons.

found to exhibit linear behaviors in three different response regions. The $S$ was $0.048\,kPa^{-1}$ when the pressure was 0–100 kPa, which decreased to $0.019\,kPa^{-1}$ for pressures of 100–350 kPa, then further lowered to $0.003\,kPa^{-1}$ when the pressure was between 350 and 650 kPa. Last, the authors fabricated a $2\times3$ TENG-based pressure sensor array to detect fingertip motion paths.

Zhang's group reported the direct printing of e-textile composed of core–sheath fibers for use as triboelectric generators and supercapacitors [109]. Specifically, as shown in Figure 13.9a,b, the CNTs@silk fibroin (SF) core–sheath fiber was extruded from a coaxial spinneret and printed into designed patterns on a textile using a 3D printer. The optical image verified the core–sheath structure, and the core–sheath fibers printed on the textile have a diameter of ~500 mm with ~200 mm width conductive CNT core (Figure 13.9c). To fabricate the triboelectric generator, parallel CNTs@SF core–sheath fiber patterns and PET film were chosen as the triboelectric pair. The device can generate a $I_{sc}$ peak of 1.4 mA and an open-circuit voltage ($V_{oc}$) peak of 15 V at a displacement speed of 10 cm/s (Figure 13.9d). In addition, the power density peak could reach a maximum value of $18\,mW/m^2$ at an external resistance load of $4\,M\Omega$ (Figure 13.9e). With regard to the supercapacitor, the sheath of the printed fiber was sodium carboxymethyl cellulose, which serves as a solid-state electrolyte. The gap between two individual core–sheath fiber electrodes was then filled with PVA/$H_3PO_4$ gel to work as the solid-state electrolyte for the supercapacitor. Lastly, PDMS was used to encapsulate the printed supercapacitor. The areal capacitance of the supercapacitors was calculated to be $26.4\,mF/cm^2$ at a current density of $0.42\,mA/cm^2$, which corresponded to an areal energy density/power density of $0.33\,mWh/cm^2/31.9\,mW/cm^2$. Furthermore, the authors printed the CNTs@SF core–sheath fibers pattern near an underarm as shown in Figure 13.9f. The touching/separating of the smart pattern with an opposite PET film on the underarm sleeve induced by moving the arms could generate an alternating current. The inset of Figure 13.9f shows the typical alternating current with a maximal $I_{sc}$ peak of $\sim1.8\,mA/cm^2$, after employing a bridge rectifier to convert the alternating current to direct current; the collected energy can be stored in capacitors for powering LEDs.

## 13.4 Summary and Perspectives

In this chapter, we have collected the latest major achievements of utilizing conducting (nano)materials for fabricating conducting elements and optoelectronic devices in the areas of stretchable and wearable e-textile applications. This research field has greatly expanded during the past few years in order to incorporate a broader array of flexible/stretchable materials, more complex device architectures, and greater system integration.

The electrical conductor is a critical element for enabling flexible and stretchable electronic devices. Thus, several conducting materials were reviewed here including metals, carbon nanomaterials, and conducting polymers and provided the reader with examples on how the mechanical properties of these conductors, but intrinsically poorly stretchable, structures are enhanced [110–115].

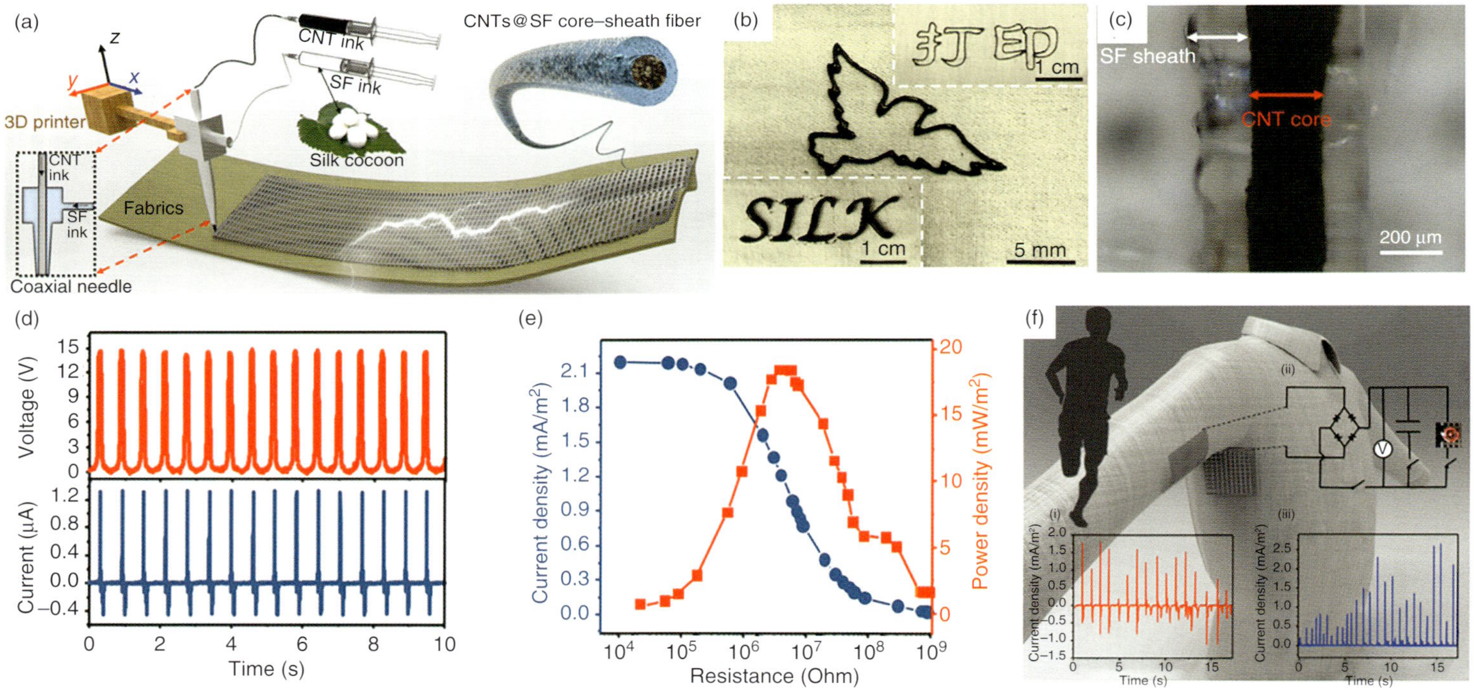


**Figure 13.9** (a) Schematic illustration showing the 3D printing process using a coaxial spinneret. (b) Photographs of customer-designed patterns on textile. (c) Optical image of top-view of a core–sheath fiber on textile. (d) Typical $V_{oc}$ and $I_{sc}$ of a parallel-line pattern on textile. (e) $I_{sc}$ density and power density as a function of external connected resistance. (f) Schematic illustration showing smart clothes for energy management and their performance. Source: Zhang et al. 2019 [109]. Copyright 2019, Reproduced with permission of Elsevier.

Furthermore, we provided several examples where the resulting conducting elements were used as interconnects for devices or to fabricate a number of devices including different type of sensors, power generators, and for power storage.

In addition, we have shown that the most widely utilized approach to confer flexibility/stretchability to a material exhibiting sufficient conductivity as a bulk is to incorporate their (preformed) nanostructures with passive and electrical insulating fibers. If the conductor loading is not excessive, the resulting composites can typically maintain most of the mechanical properties of the passive component, and gain considerable charge-transport conducting characteristics. It is important to stress that depending on the final use of the conductor, a different balance between conductivity and mechanical properties is required. For instance, for applications as interconnects or as electrodes in a device, the stretchable conductor must exhibit a conductivity as high as possible and, equally important, it should change minimally with strain. However, when the conductor is used as the sensor element, e.g. in a strain sensor, then the absolute conductivity value may be less important than achieving large and reliable sensitivities in a broad range of strains. Either approach relies on nanostructured conductors and thus morphologies where the role of the material surface greatly impacts not only charge transport and other physical properties but also environmental and chemical stability, both of which remain of great concern. Furthermore, the reliable synthesis at scale of some of these nanostructured materials remains problematic and expensive.

Although most of this chapter focuses on conducting materials and proof-of-principle prototypes, other essential requirements for technology readiness and commercialization at both material and device levels must be met in the future. Particularly, (i) a scalable and mass producible technology which thus has to obey the rigid standard of the microelectronic industry. (ii) Achieving human safety requirements as for any other wearable systems. (iii) Reliable supply of both passive and active materials. Therefore, addressing the fundamental materials and integration issues discussed in the previous sections as well as the unsolved questions in device engineering and commercialization will generate several opportunities for interesting research and development.

## References

1 Wang, B., Huang, W., Chi, L. et al. (2018). High-*k* gate dielectrics for emerging flexible and stretchable electronics. *Chem. Rev.* 118 (11): 5690–5754.

2 Bao, Z. and Chen, X. (2016). Flexible and stretchable devices. *Adv. Mater.* 28 (22): 4177–4179.

3 Liu, Y., Pharr, M., and Salvatore, G.A. (2017). Lab-on-skin: a review of flexible and stretchable electronics for wearable health monitoring. *ACS Nano* 11 (10): 9614–9635.

4 Zhan, Y., Mei, Y., and Zheng, L. (2014). Materials capability and device performance in flexible electronics for the internet of things. *J. Mater. Chem. C* 2 (7): 1220–1232.

5 Noh, Y.Y., Zhao, N., Caironi, M., and Sirringhaus, H. (2007). Downscaling of self-aligned, all-printed polymer thin-film transistors. *Nat. Nanotechnol.* 2 (12): 784–789.

6 Bucella, S.G., Luzio, A., Gann, E. et al. (2015). Macroscopic and high-throughput printing of aligned nanostructured polymer semiconductors for MHz large-area electronics. *Nat. Commun.* 6: 8394.

7 Nketia-Yawson, B., Kang, S.J., Tabi, G.D. et al. (2017). Ultrahigh mobility in solution-processed solid-state electrolyte-gated transistors. *Adv. Mater.* 29 (16): 1605685.

8 Sim, K., Rao, Z., Kim, H.J. et al. (2019). Fully rubbery integrated electronics from high effective mobility intrinsically stretchable semiconductors. *Sci. Adv.* 5 (2): eaav5749.

9 Hiremath, S., Yang, G., and Mankodiya, K. (2014). Wearable internet of things: concept, architectural components and promises for person-centered healthcare. *4th International Conference on Wireless Mobile Communication and Healthcare -Transforming Healthcare through Innovations in Mobile and Wireless Technologies (MOBIHEALTH).* Athens, Greece: IEEE. https://doi.org/10.4108/icst.mobihealth.2014.257440pp 304-307.

10 Hammock, M.L., Chortos, A., Tee, B.C. et al. (2013). 25th anniversary article: the evolution of electronic skin (e-skin): a brief history, design considerations, and recent progress. *Adv. Mater.* 25 (42): 5997–6038.

11 Castro, H.F., Correia, V., Sowade, E. et al. (2016). All-inkjet-printed low-pass filters with adjustable cutoff frequency consisting of resistors, inductors and transistors for sensor applications. *Org. Electron.* 38: 205–212.

12 Chang, J., Zhang, X., Ge, T., and Zhou, J. (2014). Fully printed electronics on flexible substrates: high gain amplifiers and DAC. *Org. Electron.* 15 (3): 701–710.

13 Choi, B.D., Park, J., Baeg, K.-J. et al. (2018). Optimized activation of solution-processed amorphous oxide semiconductors for flexible transparent conductive electrodes. *Adv. Electron. Mater.* 4 (1): 1700386.

14 Choi, S., Jo, J.W., Kim, J. et al. (2017). Static and dynamic water motion-induced instability in oxide thin-film transistors and its suppression by using low-*k* fluoropolymer passivation. *ACS Appl. Mater. Interfaces* 9 (31): 26161–26168.

15 Yang, T., Xu, Z., Wen, J., and Yang, L. (2009). Factors influencing bioleaching copper from waste printed circuit boards by acidithiobacillus ferrooxidans. *Hydrometallurgy* 97 (1–2): 29–32.

16 Merkel, T., Graeber, M., and Pagel, L. (1999). A new technology for fluidic microsystems based on PCB technology. *Sens. Actuators, A* 7: 98–105.

17 Pailler-Mattei, C., Bec, S., and Zahouani, H. (2008). *In vivo* measurements of the elastic mechanical properties of human skin by indentation tests. *Med. Eng. Phys.* 30 (5): 599–606.

18 Petersen, K.E. and Guarnieri, C.R. (1979). Young's modulus measurements of thin films using micromechanics. *J. Appl. Phys.* 50 (11): 6761–6766.

19 Zhao, Y., Dong, D., Gong, S. et al. (2019). A moss-inspired electroless gold-coating strategy toward stretchable fiber conductors by dry spinning. *Adv. Electron. Mater.* 5 (1): 1800462.

20 Liu, M., Pu, X., Jiang, C. et al. (2017). Large-area all-textile pressure sensors for monitoring human motion and physiological signals. *Adv. Mater.* 29 (41): 1703700.

21 Lima, R., Alcaraz-Espinoza, J.J., da Silva, F.A.G. Jr., and de Oliveira, H.P. (2018). Multifunctional wearable electronic textiles using cotton fibers with polypyrrole and carbon nanotubes. *ACS Appl. Mater. Interfaces* 10 (16): 13783–13795.

22 Allison, L.K. and Andrew, T.L. (2019). A wearable all-fabric thermoelectric generator. *Adv. Mater. Technol.* 4 (5): 1800615.

23 Wang, B. and Facchetti, A. (2019). Mechanically flexible conductors for stretchable and wearable e-skin and e-textile devices. *Adv. Mater.* 31: 1901408. https://doi.org/10.1002/adma.201901408.

24 Trung, T.Q. and Lee, N.E. (2016). Flexible and stretchable physical sensor integrated platforms for wearable human-activity monitoring and personal healthcare. *Adv. Mater.* 28 (22): 4338–4372.

25 Liu, C.H. and Yu, X. (2011). Silver nanowire-based transparent, flexible, and conductive thin film. *Nanoscale Res. Lett.* 6 (1): 75.

26 Wiley, B., Sun, Y.G., and Xia, Y.N. (2007). Synthesis of silver nanostructures with controlled shapes and properties. *Acc. Chem. Res.* 40 (10): 1067–1076.

27 Rycenga, M., Cobley, C.M., Zeng, J. et al. (2011). Controlling the synthesis and assembly of silver nanostructures for plasmonic applications. *Chem. Rev.* 111 (6): 3669–3712.

28 Yang, C., Gu, H., Lin, W. et al. (2011). Silver nanowires: from scalable synthesis to recyclable foldable electronics. *Adv. Mater.* 23 (27): 3052–3056.

29 Liang, J., Tong, K., and Pei, Q. (2016). A water-based silver-nanowire screen-print ink for the fabrication of stretchable conductors and wearable thin-film transistors. *Adv. Mater.* 28 (28): 5986–5996.

30 Coskun, S., Aksoy, B., and Unalan, H.E. (2011). Polyol synthesis of silver nanowires: an extensive parametric study. *Cryst. Growth Des.* 11 (11): 4963–4969.

31 Venkata Krishna, R.R., Venkata Abhinav, K., Karthik, P.S., and Singh, S.P. (2015). Conductive silver inks and their applications in printed and flexible electronics. *RSC Adv.* 5 (95): 77760–77790.

32 Langley, D., Giusti, G., Mayousse, C. et al. (2013). Flexible transparent conductive materials based on silver nanowire networks: a review. *Nanotechnology* 24 (45): 452001.

33 Lee, H., Dellatore, S.M., Miller, W.M., and Messersmith, P.B. (2007). Mussel-inspired surface chemistry for multifunctional coatings. *Science* 318 (5849): 426–430.

34 Cai, L., Song, A.Y., Wu, P. et al. (2017). Warming up human body by nanoporous metallized polyethylene textile. *Nat. Commun.* 8 (1): 496.

35 Hsu, P.C., Kong, D.S., Wang, S. et al. (2014). Electrolessly deposited electrospun metal nanowire transparent electrodes. *J. Am. Chem. Soc.* 136 (30): 10593–10596.

36 Sun, Y.G. and Xia, Y.N. (2002). Large-scale synthesis of uniform silver nanowires through a soft, self-seeding, polyol process. *Adv. Mater.* 14 (11): 833–837.

37 Li, B., Ye, S.R., Stewart, I.E. et al. (2015). Synthesis and purification of silver nanowires to make conducting films with a transmittance of 99%. *Nano Lett.* 15 (10): 6722–6726.

38 Zhu, B., Gong, S., and Cheng, W. (2019). Softening gold for elastronics. *Chem. Soc. Rev.* 48 (6): 1668–1711.

39 Huo, Z.Y., Tsung, C.K., Huang, W.Y. et al. (2008). Sub-two nanometer single crystal au nanowires. *Nano Lett.* 8 (7): 2041–2044.

40 Gong, S., Schwalb, W., Wang, Y. et al. (2014). A wearable and highly sensitive pressure sensor with ultrathin gold nanowires. *Nat. Commun.* 5: 3132.

41 Feng, H.J., Yang, Y.M., You, Y.M. et al. (2009). Simple and rapid synthesis of ultrathin gold nanowires, their self-assembly and application in surface-enhanced Raman scattering. *Chem. Commun.* (15): 1984–1986.

42 Maurer, J.H., Gonzalez-Garcia, L., Reiser, B. et al. (2016). Templated self-assembly of ultrathin gold nanowires by nanoimprinting for transparent flexible electronics. *Nano Lett.* 16 (5): 2921–2925.

43 Calvaresi, M., Quintana, M., Rudolf, P. et al. (2013). Rolling up a graphene sheet. *ChemPhysChem* 14 (15): 3447–3453.

44 Sun, D.M., Liu, C., Ren, W.C., and Cheng, H.M. (2013). A review of carbon nanotube- and graphene-based flexible thin-film transistors. *Small* 9 (8): 1188–1205.

45 Jin, L.H., Chortos, A., Lian, F.F. et al. (2018). Microstructural origin of resistance-strain hysteresis in carbon nanotube thin film conductors. *Proc. Natl. Acad. Sci. U.S.A.* 115 (9): 1986–1991.

46 Hong, S. and Myung, S. (2007). Nanotube electronics: a flexible approach to mobility. *Nat. Nanotechnol.* 2 (4): 207–208.

47 Cho, W.D., Schulz, M., and Shanot, V. (2014). Growth and characterization of vertically aligned centimeter long CNT arrays. *Carbon* 72: 264–273.

48 Hinds, B.J., Chopra, N., Rantell, T. et al. (2004). Aligned multiwalled carbon nanotube membranes. *Science* 303 (5654): 62–65.

49 Zhang, R.F., Zhang, Y.Y., and Wei, F. (2017). Controlled synthesis of ultralong carbon nanotubes with perfect structures and extraordinary properties. *Acc. Chem. Res.* 50 (2): 179–189.

50 Zhang, R.F., Zhang, Y.Y., Zhang, Q. et al. (2013). Growth of half-meter long carbon nanotubes based on Schulz-Flory distribution. *ACS Nano* 7 (7): 6156–6161.

51 Arvinte, A., Valentini, F., Radoi, A. et al. (2007). The NADH electrochemical detection performed at carbon nanofibers modified glassy carbon electrode. *Electroanalysis* 19 (14): 1455–1459.

52 Akinwande, D., Petrone, N., and Hone, J. (2014). Two-dimensional flexible nanoelectronics. *Nat. Commun.* 5: 5678.

53 Chen, S., Jiang, K., Lou, Z. et al. (2018). Recent developments in graphene-based tactile sensors and e-skins. *Adv. Mater. Technol.* 3 (2): 1700248.

54 Stankovich, S., Dikin, D.A., Piner, R.D. et al. (2007). Synthesis of graphene-based nanosheets via chemical reduction of exfoliated graphite oxide. *Carbon* 45 (7): 1558–1565.

55 Voiry, D., Yang, J., Kupferberg, J. et al. (2016). High-quality graphene via microwave reduction of solution-exfoliated graphene oxide. *Science* 353 (6306): 1413–1416.

56 Balint, R., Cassidy, N.J., and Cartmell, S.H. (2014). Conductive polymers: towards a smart biomaterial for tissue engineering. *Acta Biomater.* 10 (6): 2341–2353.

57 Kim, S., Sanyoto, B., Park, W.T. et al. (2016). Purification of PEDOT: PSS by ultrafiltration for highly conductive transparent electrode of all-printed organic devices. *Adv. Mater.* 28 (46): 10149–10154.

58 Lund, A., van der Velden, N.M., Persson, N.-K. et al. (2018). Electrically conducting fibres for e-textiles: an open playground for conjugated polymers and carbon nanomaterials. *Mater. Sci. Eng., R* 126: 1–29.

59 Stoppa, M. and Chiolerio, A. (2014). Wearable electronics and smart textiles: a critical review. *Sensors (Basel)* 14 (7): 11957–11992.

60 Gong, S. and Cheng, W. (2017). One-dimensional nanomaterials for soft electronics. *Adv. Electron. Mater.* 3 (3): 1600314.

61 Koo, J.H., Kim, D.C., Shim, H.J. et al. (2018). Flexible and stretchable smart display: materials, fabrication, device design, and system integration. *Adv. Funct. Mater.* 28 (35): 1801834.

62 Agcayazi, T., Chatterjee, K., Bozkurt, A., and Ghosh, T.K. (2018). Flexible interconnects for electronic textiles. *Adv. Mater. Technol.* 3 (10): 1700277.

63 Chen, X. (2017). Making electrodes stretchable. *Small Methods* 1 (4): 1600029.

64 Yao, S.S. and Zhu, Y. (2015). Nanomaterial-enabled stretchable conductors: strategies, materials and devices. *Adv. Mater.* 27 (9): 1480–1511.

65 Oh, K.W., Park, H.J., and Kim, S.H. (2003). Stretchable conductive fabric for electrotherapy. *J. Appl. Polym. Sci.* 88 (5): 1225–1229.

66 Liu, Z.F., Fang, S., Moura, F.A. et al. (2015). STRETCHY ELECTRONICS. Hierarchically buckled sheath-core fibers for superelastic electronics, sensors, and muscles. *Science* 349 (6246): 400–404.

67 Lund, A., Darabi, S., Hultmark, S. et al. (2018). Roll-to-roll dyed conducting silk yarns: a versatile material for e-textile devices. *Adv. Mater. Technol.* 3 (12): 1800251.

68 Wu, Y., Mechael, S.S., Chen, Y., and Carmichael, T.B. (2018). Solution deposition of conformal gold coatings on knitted fabric for e-textiles and electroluminescent clothing. *Adv. Mater. Technol.* 3 (3): 1700292.

69 Pani, D., Achilli, A., and Bonfiglio, A. (2018). Survey on textile electrode technologies for electrocardiographic (ECG) monitoring, from metal wires to polymers. *Adv. Mater. Technol.* 3 (10): 1800008.

70 Jin, H., Matsuhisa, N., Lee, S. et al. (2017). Enhancing the performance of stretchable conductors for e-textiles by controlled ink permeation. *Adv. Mater.* 29 (21): 1605848.

71 Shahariar, H., Kim, I., Soewardiman, H., and Jur, J.S. (2019). Inkjet printing of reactive silver ink on textiles. *ACS Appl. Mater. Interfaces* 11 (6): 6208–6216.

72 Kim, I., Shahariar, H., Ingram, W.F. et al. (2019). Inkjet process for conductive patterning on textiles: maintaining inherent stretchability and breathability in knit structures. *Adv. Funct. Mater.* 29 (7): 1807573.
73 La, T.G., Qiu, S., Scott, D.K. et al. (2018). Two-layered and stretchable e-textile patches for wearable healthcare electronics. *Adv. Healthc. Mater.* 7 (22): e1801033.
74 Yang, T., Li, X., Jiang, X. et al. (2016). Structural engineering of gold thin films with channel cracks for ultrasensitive strain sensing. *Mater. Horiz.* 3 (3): 248–255.
75 Tolvanen, J., Hannu, J., and Jantunen, H. (2018). Stretchable and washable strain sensor based on cracking structure for human motion monitoring. *Sci. Rep.* 8 (1): 13241.
76 De Rossi, D., Della Santa, A., and Mazzoldi, A. (1999). Dressware: wearable hardware. *Mater. Sci. Eng., C* 7: 31–35.
77 Li, Y., Cheng, X.Y., Leung, M.Y. et al. (2005). A flexible strain sensor from polypyrrole-coated fabrics. *Synth. Met.* 155 (1): 89–94.
78 Gullapalli, H., Vemuru, V.S., Kumar, A. et al. (2010). Flexible piezoelectric Zno-paper nanocomposite strain sensor. *Small* 6 (15): 1641–1646.
79 Li, X., Zhang, R., Yu, W. et al. (2012). Stretchable and highly sensitive graphene-on-polymer strain sensors. *Sci. Rep.* 2: 870.
80 Lee, J., Kwon, H., Seo, J. et al. (2015). Conductive fiber-based ultrasensitive textile pressure sensor for wearable electronics. *Adv. Mater.* 27 (15): 2433–2439.
81 Li, Y., Li, Y., Su, M. et al. (2017). Electronic textile by dyeing method for multiresolution physical kineses monitoring. *Adv. Electron. Mater.* 3 (10): 1700253.
82 Song, Y., Huang, W., Mu, C. et al. (2019). Carbon nanotube-modified fabric for wearable smart electronic-skin with exclusive normal-tangential force sensing ability. *Adv. Mater. Technol.* 4: 1800680. https://doi.org/10.1002/admt.201800680.
83 Wang, Y.L., Hao, J., Huang, Z.Q. et al. (2018). Flexible electrically resistive-type strain sensors based on reduced graphene oxide-decorated electrospun polymer fibrous mats for human motion monitoring. *Carbon* 126: 360–371.
84 Li, Y.M., Zhang, Z.T., Li, X.Y. et al. (2017). A smart, stretchable resistive heater textile. *J. Mater. Chem. C* 5 (1): 41–46.
85 Hong, S., Lee, H., Lee, J. et al. (2015). Highly stretchable and transparent metal nanowire heater for wearable electronics applications. *Adv. Mater.* 27 (32): 4744–4751.
86 Wang, Q.-W., Zhang, H.-B., Liu, J. et al. (2019). Multifunctional and water-resistant MXene-decorated polyester textiles with outstanding electromagnetic interference shielding and joule heating performances. *Adv. Funct. Mater.* 29 (7): 1806819.
87 Zhang, M., Wang, C., Liang, X. et al. (2017). Weft-knitted fabric for a highly stretchable and low-voltage wearable heater. *Adv. Electron. Mater.* 3 (9): 1700193.

88 Zeng, P., Tian, B., Tian, Q. et al. (2019). Screen-printed, low-cost, and patterned flexible heater based on Ag fractal dendrites for human wearable application. *Adv. Mater. Technol.* 4 (3): 1800453.
89 Zhao, N., Schofield, N., and Niu, W.Q. (2016). Energy storage system for a port crane hybrid power-train. *IEEE Trans. Transp Electr.* 2 (4): 480–492.
90 Hu, L., Pasta, M., Mantia, F.L. et al. (2010). Stretchable, porous, and conductive energy textiles. *Nano Lett.* 10 (2): 708–714.
91 Zhang, Z., Deng, J., Li, X. et al. (2015). Superelastic supercapacitors with high performances during stretching. *Adv. Mater.* 27 (2): 356–362.
92 Shi, H.H., Khalili, N., Morrison, T., and Naguib, H.E. (2018). Self-assembled nanorod structures on nanofibers for textile electrochemical capacitor electrodes with intrinsic tactile sensing capabilities. *ACS Appl. Mater. Interfaces* 10 (22): 19037–19046.
93 Wang, Z.L. and Song, J.H. (2006). Piezoelectric nanogenerators based on zinc oxide nanowire arrays. *Science* 312 (5771): 242–246.
94 Proto, A., Penhaker, M., Conforto, S., and Schmid, M. (2017). Nanogenerators for human body energy harvesting. *Trends Biotechnol.* 35 (7): 610–624.
95 Paosangthong, W., Torah, R., and Beeby, S. (2019). Recent progress on textile-based triboelectric nanogenerators. *Nano Energy* 55: 401–423.
96 Lund, A., Rundqvist, K., Nilsson, E. et al. (2018). Energy harvesting textiles for a rainy day: woven piezoelectrics based on melt-spun PVDF microfibres with a conducting core. *npj Flexible Electron.* 2 (1): 9.
97 Guo, H., Pu, X., Chen, J. et al. (2018). A highly sensitive, self-powered triboelectric auditory sensor for social robotics and hearing aids. *Sci. Robot.* 3 (20): eaat2516.
98 Venkatasubramanian, R., Siivola, E., Colpitts, T., and O'Quinn, B. (2001). Thin film thermoelectric devices with high room temperature figures of merit. *Nature* 413: 597–602.
99 Kim, S.J., We, J.H., and Cho, B.J. (2014). A wearable thermoelectric generator fabricated on a glass fabric. *Energy Environ. Sci.* 7 (6): 1959–1965.
100 Du, Y., Cai, K., Chen, S. et al. (2015). Thermoelectric fabrics: toward power generating clothing. *Sci. Rep.* 5: 6411.
101 Ryan, J.D., Lund, A., Hofmann, A.I. et al. (2018). All-organic textile thermoelectrics with carbon-nanotube-coated N-type yarns. *ACS Appl. Energy Mater* 1 (6): 2934–2941.
102 Wang, Z.L. (2018). Nanogenerators, self-powered systems, blue energy, piezotronics and piezo-phototronics - a recall on the original thoughts for coining these fields. *Nano Energy* 54: 477–483.
103 Wang, Z.L. (2017). On Maxwell's displacement current for energy and sensors: the origin of nanogenerators. *Mater. Today* 20 (2): 74–82.
104 Jeong, C.K., Lee, J., Han, S. et al. (2015). A hyper-stretchable elastic-composite energy harvester. *Adv. Mater.* 27 (18): 2866–2875.
105 He, S., Dong, W., Guo, Y. et al. (2019). Piezoelectric thin film on glass fiber fabric with structural hierarchy: an approach to high-performance, superflexible, cost-effective, and large-scale nanogenerators. *Nano Energy* 59: 745–753.

106 Fan, F.R., Tian, Z.Q., and Wang, Z.L. (2012). Flexible triboelectric generator. *Nano Energy* 1 (2): 328–334.

107 Kim, K.N., Chun, J., Kim, J.W. et al. (2015). Highly stretchable 2d fabrics for wearable triboelectric nanogenerator under harsh environments. *ACS Nano* 9 (6): 6394–6400.

108 Cao, R., Pu, X., Du, X. et al. (2018). Screen-printed washable electronic textiles as self-powered touch/gesture tribo-sensors for intelligent human-machine interaction. *ACS Nano* 12: 5190–5196.

109 Zhang, M., Zhao, M., Jian, M. et al. (2019). Printable smart pattern for multifunctional energy-management e-textile. *Matter* 1: 168–179. https://doi.org/10.1016/j.matt.2019.02.003.

110 Jo, J.W., Kim, K.H., Kim, J. et al. (2018). High-mobility and hysteresis-free flexible oxide thin-film transistors and circuits by using bilayer sol–gel gate dielectrics. *ACS Appl. Mater. Interfaces* 10 (3): 2679–2687.

111 Heo, J.S., Eom, J., Kim, Y.H., and Park, S.K. (2018). Recent progress of textile-based wearable electronics: a comprehensive review of materials, devices, and applications. *Small* 14 (3): 1703034.

112 Facchetti, A. (2011). π-Conjugated polymers for organic electronics and photovoltaic cell applications. *Chem. Mater.* 23 (3): 733–758.

113 Marrocchi, A., Facchetti, A., Lanari, D. et al. (2016). Current methodologies for a sustainable approach to π-conjugated organic semiconductors. *Energy Environ. Sci.* 9 (3): 763–786.

114 Zhou, N. and Facchetti, A. (2018). Naphthalenediimide (NDI) polymers for all-polymer photovoltaics. *Mater. Today* 21 (4): 377–390.

115 Rivnay, J., Inal, S., Collins, B.A. et al. (2016). Structural control of mixed ionic and electronic transport in conducting polymers. *Nat. Commun.* 7: 11287.

# Index



*Flexible and Wearable Electronics for Smart Clothing*, First Edition.
Edited by Gang Wang, Chengyi Hou, and Hongzhi Wang.

### T

## *U*